Extrachromosomal DNA in Cancer

This book provides a comprehensive review of the emerging role of extrachromosomal DNA (ecDNA) in cancer, highlighting its significance in tumor progression, oncogene amplification, drug resistance, and clinical applications. *Extrachromosomal DNA in Cancer: Mechanisms, Biomarkers, and Therapeutic Implications* explores the structural and functional characteristics of ecDNA, emphasizing its role in shaping the genetic and epigenetic landscapes of cancer cells. The book discusses the impact of ecDNA on tumor heterogeneity, metastasis, and immune evasion, providing insights into how ecDNA contributes to cancer evolution and therapeutic resistance. The text also examines cutting-edge imaging techniques, computational approaches, and non-invasive diagnostic applications of ecDNA, highlighting its potential as a biomarker in precision oncology. Additionally, it presents the novel therapeutic strategies targeting ecDNA, addressing challenges and opportunities in integrating these insights into cancer treatment.

FEATURES

- Explores the structural formation and oncogenic role of ecDNA in tumor development and progression
- Discusses the influence of ecDNA on gene expression, epigenetics, and immune modulation in cancer
- Examines the role of ecDNA in drug resistance, metastasis, and disease recurrence
- Highlights advanced imaging, computational tools, and non-invasive diagnostic applications of ecDNA
- Reviews emerging therapeutic strategies targeting ecDNA for improved cancer treatment and personalized medicine

This book is intended for researchers and students of cancer biology, molecular biology, and biotechnology.

Extrachromosomal DNA in Cancer

Mechanisms, Biomarkers, and Therapeutic Implications

Edited by
Sumel Ashique, Ayush Madan,
Biplab Debnath, and Gokhan Zengin

CRC Press
Taylor & Francis Group
Boca Raton London New York

CRC Press is an imprint of the
Taylor & Francis Group, an **informa** business

Designed cover image: Shutterstock.

First edition published 2026
by CRC Press
2385 NW Executive Center Drive, Suite 320, Boca Raton FL 33431

and by CRC Press
4 Park Square, Milton Park, Abingdon, Oxon, OX14 4RN

CRC Press is an imprint of Taylor & Francis Group, LLC

ISBN: 978-1-041-10483-4 (hbk)
ISBN: 978-1-041-10484-1 (pbk)
ISBN: 978-1-003-65529-9 (ebk)

DOI: 10.1201/9781003655299

Typeset in Times
by Apex CoVantage, LLC

Contents

Preface

The emergence of extrachromosomal DNA (ecDNA) as a pivotal player in cancer biology has transformed our understanding of tumorigenesis, genomic instability, and therapeutic resistance. Unlike chromosomal DNA, ecDNA exists outside the standard chromosomal framework, often carrying amplified oncogenes that drive aggressive cancer phenotypes. *Extrachromosomal DNA in Cancer: Mechanisms, Biomarkers, and Therapeutic Implications* provides a comprehensive overview of the origin, structure, and function of ecDNA in various cancer types. It explores the molecular mechanisms underlying ecDNA formation and its dynamic role in tumor evolution, heterogeneity, and adaptability.

Recent advances in next-generation sequencing and single-cell technologies have made it possible to detect ecDNA with greater precision, positioning it as a promising biomarker for cancer diagnosis, prognosis, and treatment monitoring. Moreover, targeting ecDNA-associated pathways offers novel therapeutic strategies, particularly in combating drug resistance and recurrence.

This book is intended for researchers, oncologists, molecular biologists, and graduate students seeking to understand the emerging significance of ecDNA in cancer. Through detailed chapters and updated evidence, we aim to bridge basic research with clinical translation, fostering innovation in cancer diagnostics and precision medicine. We hope this text inspires further exploration into the untapped potential of ecDNA as a transformative element in cancer therapy.

The Editors

Sumel Ashique is an assistant professor in the Department of Pharmaceutical Technology, Bharat Technology, West Bengal, India. He has 4 years of teaching experience and has authored 80 publications in internationally and nationally accredited, reputed journals. He is very knowledgeable in drug delivery, nanotechnology, and targeted treatment strategy and is a reviewer of many well-known publishers (Elsevier, Springer, Dove Press, and Bentham Sc.). He has 5 granted patents from IP and Australia and 20 published book chapters in Bentham, Springer, Elsevier, Bentham, and Taylor & Francis. Currently, Prof. Ashique is editing 14 books for CRC Press, Elsevier, Taylor & Francis, Wiley, Nova Science, and Springer. He has published more than 80 research papers, review papers, column articles, Scopus editorials, and WoS indexed. He is a guest editor for *JoVE* journal (IF 1.2, Scopus, and WOS indexed), and an editor for *Probiotics* (Bentham Science) and *Indian Science Journal*. He is a reviewer for Elsevier, Springer, Taylor & Francis, and Wiley journals for his expertise in the field of targeted drug delivery, nanotechnology, gut-microbiome, and postbiotics. Currently, he is pursuing his PhD from Lovely Professional University, Punjab, India (NAAC A++), in the Pharmaceutical Science Department.

Ayush Madan is an assistant professor in the Department of Biotechnology at People's University and a graduate research assistant at Universiti Malaysia Terengganu. He holds an MTech in biotechnology from Graphic Era University and has extensive expertise in microbial technology and environmental bioremediation. He has also completed his PhD work in the field of nanobiotechnology, where he has studied the cytotoxic effects of silver and gold nanoparticles on human cells with special reference to chromosomal aberrations. Prof. Madan has published numerous research and review articles in national and international journals and has presented his work at various conferences. He has also served as an editor of 14 books in prestigious publishing houses, including Taylor & Francis, CABI, Cambridge Scholars, Wiley Scrivener, and more. Previously, he was an assistant professor at Shobhit Institute of Engineering and Technology and completed a six-month internship at Biocon Research Limited. His current research involves integrated microalgae-shrimp cultivation for wastewater treatment and shrimp growth enhancement.

Dr. Biplab Debnath is a distinguished principal and professor at Bharat Technology, Uluberia, Howrah, West Bengal. His research expertise lies in silico synthesis and biological activities of novel heterocyclic molecules. He has made significant contributions to the field, publishing more than 60 research and review papers in esteemed journals, 2 granted patents, 5 book chapters, and 8 books with various publishing houses. Dr. Debnath is a dedicated mentor, guiding 18 MPharm scholars and 8 PhD students under his supervision. He serves on numerous advisory boards for scientific journals and is a member of several scientific associations. Dr. Debnath has also chaired sessions and presented at prestigious international and national conferences. Recent awards include the Recognition of Excellence in Pharmaceutical Education and Research Award 2024 from the Bioequivalence Study Centre, Jadavpur University, and the ABS Clinical Solution Best Young Principal Award in the Global Leadership Awards 2024, Dubai from the Pharmaceutical Royal International Society (PRISAL). Dr. Debnath has made innumerable achievements and contributions to pharmaceutical education and research.

Gokhan Zengin is a full professor of biology at Selcuk University. He received his biology diploma from the University of Selcuk (2007) and his PhD (2015) from Selcuk University. His research activity is documented by more than 900 scientific peer-reviewed papers and conferences and communications to scientific and industrial meetings, four books, and 10 book chapters. His main research interests have focused on chemical characterization, biological activities, innovative and sustainable extraction, and processing techniques of natural products.

1 Overview of ecDNA
Discovery and Emerging Significance in Cancer

Debajit Dewan,[1] *Lopamudra Chakraborty,*[2]
Kabirul Islam Mollah,[1] *and Soma Jana*[1]

1 Department of Pharmaceutical Technology
Bharat Technology
Uluberia, India
2 Department of Pharmacology
Calcutta Institute of Pharmaceutical Technology
and Allied Health Sciences
Uluberia, Howrah, India

1.1 INTRODUCTION TO ECDNA

1.1.1 Biogenesis

1.1.1.1 Breakage–Fusion–Bridge

Barbara McClintock first explained Breakage–fusion–bridge (BFB), one of the classical mechanisms of extrachromosomal DNA (ecDNA) formation. It starts with DNA breakage, subsequently followed by shortening of the telomere and joining of chromosomes at the ends to form an anaphase bridge with two centromeres. Then the bridge undergoes breakage and recombination, leading to structural alteration of the chromosome. Daughter with an aberrant centromere number continues the cycle, and after several rounds of BFB cycle the ecDNA forms [1].

1.1.1.2 Chromothripsis

In chromothripsis locally clustered DNA starts breaking within one or a few chromosomes, resulting in rearrangement of fragments and loss of DNA sequence. Formation of smaller circular ecDNA develops during this rearrangement process.

1.1.1.3 Episome

ecDNA can result from the excision of circular DNA and be enlarged by either replication or recombination [2].

1.1.2 Characteristic Features

In 1965 when the ecDNA was first documented, it was named double minutes or double bodies, but only 30% of ecDNAs occur as classic double minute pairs. So, the term "ecDNA" is now used to describe both singlet ecDNA and double minutes. ecDNA has a few unique characteristics that have turned the focus of cancer research recently. It's a circular, extrachromosomal, self-replicating DNA fragment originating from the chromosomal genome and can contain specific gene sequences and regulatory regions. During mitosis, ecDNA is inherited in a non-Mendelian manner because it lacks a centromere. ecDNA is characterized by its circular structure, ability to amplify oncogenes, high

DOI: 10.1201/9781003655299-1

accessibility of chromatin, and uneven segregation during cell division, leading to intra-tumoral heterogeneity [3, 4].

1.2 ROLE OF ECDNA IN CANCER PROGRESSION

1.2.1 Oncogene Amplification

ecDNA are oncogene-free centromereless DNA loops of approximately 1–3Mb diameter that often occur as doublets (double minutes) but also as individual elements. Oncogene presence on ecDNA was reported in different cancers. For example, in glioblastoma. Glioblastoma cells contain various oncogenes of ecDNA, including MYC, MYCN, EGFR, PDGFRA, MET, MECOM-PIK3CA-SOX2, and the CDK4-MDM2 gene. Allele-specific RNA-sequencing analysis showed that ecDNA can act as a template to regulate gene transcription. Oncogene encoded by ecDNA is typically highly expressed, which is a sign of the key role of ecDNA in oncogenic signal amplification in cancer. Compared with chromosomal DNA amplification, ecDNA copy number is typically higher, even sometimes higher than the BFB amplicons. ecDNA may enhance copy number by focal amplification, thereby upregulating oncogene expression [5].

1.2.2 Effect on Tumor Evolution

Since there is no presence of the structure of the centromere, the ecDNA inheritance is non-Mendelian, which results in an unequal distribution of ecDNA in the cells of the offspring. It was observed that increased copy numbers could be easily obtained when the proto-oncogenes were placed on ecDNAs. It is anticipated that ecDNA diversity will rise more quickly when it is contained in the tumor tissue compared to when it resides in HSRs, suggesting that ecDNAs can quickly instill genomic anarchy in tumors. Random ecDNA segregation drives rapid tumor evolution by at least two mechanisms.

1.2.3 Heterogeneity

After mitosis, unequal segregation leads to a rough Gaussian per-cell DNA content by inducing a discrepancy in the number of ecDNA units that are divided among daughter cells. Cell-to-cell distribution of ecDNAs among clinical samples agrees with these data. Besides the mechanism of amplification of the gene occurring on linear chromosomes, this random distribution leads to a subpopulation of cells with more than one copy of the gene, producing tumor heterogeneity.

1.3 SIGNALING PATHWAYS AND ECDNA INVOLVEMENT IN CANCER

In May 2019, an article first uncovered the structure of tumor-associated ecDNA and its characteristics in oncogene amplification [6]. In human cancer, extrachromosomal DNA plays a crucial role in amplified oncogenesis by modification of gene regulatory systems [7]. In another study, Hung and his team observed that approximately 10–100 folds of ecDNA form a cluster of proto-oncogenes, leading to cancer cells or primary tumors [8]. The mechanism of ecDNA in development of cancer is the key topic in this field. Different signaling pathway plays a crucial role in the regulation and development of cancer [9]. Previous investigations have indicated that the formation and degradation of ecDNA may be regulated by particular signaling pathways [10]. This suggests that ecDNA may be produced via chromosomal mechanisms and controlled by different signaling pathways. In this section, we will describe the possible signaling pathways previously demonstrated to be essential in the regulation of the ecDNA formation.

1.3.1 PI3K/Akt Signal Pathway

Previously, research has highlighted that the damage of double-stranded DNA is one of the causes of the formation of ecDNA. Rosen et al. reported that the BRCA1 gene has a major function in the

DNA damage repair process. The absence of BRCA1 leads to the inability in DNA repair [11]. Another study found that Sei1 plays a vital role in oncogene expression in ovarian cancer. It is also described that Sei1 accelerated the expression of ecDNA by activating the PI3K/Akt pathway and suppressed BRCA1 [9, 12]. However, the specific mechanism linking PI3K/Akt signaling to ecDNA formation is still not clearly understood.

1.3.2 MAPK Signal Pathway

Mitogen-activated protein kinase (MAPK) signaling pathway, or MAPK family, is a group of protein kinases that amplifies cancer progression by lowering genomic stability. Previous studies have shown that high gene amplification originates from ecDNA and the homogeneously staining region (HSR). Sun and her team investigated a microarray analysis on HSR and ecDNA by using a cancer cell line (UACC-1598) and observed that the generation of the MAPK pathway was significantly different between the two cell lines. The study also highlights that inhibiting the phosphorylation of ERK1/2 in MAPK pathway reduced the amplification of ecDNA in cancer cells. Another study performed with COLO320DM and UACC1598 cell lines of ecDNA genes containing MYC and MYCN, respectively, and reported that ecDNA selectively increased when subjected to a MAPK inhibitor. The amplified ecDNA is common in cancer cells, due to resistance to MAPKi (mitogen-activated protein kinase inhibitors) and activation of MAPK signaling pathway [13, 14]. The phosphorylation of ERK1/2 is a key factor in the formation of regulation of ecDNA carries gene (MYC, MYCN, EGFR, etc.). Hence, regulation of MAPK can be treated as a major pathway to inhibit ecDNA formation in cancer cells [15].

1.4 MECHANISM OF ECDNA IN CARCINOGENESIS

Extrachromosomal DNA or ecDNA is generated from chromosomes by various mechanisms. There are many ways that ecDNA helps cancer grow and change, such as by increasing the number of copies of oncogenes or the connection between chromosomes, and by serving as a storage space for DNA recombination by reintegrating into and being cut out of chromosomes [1].

1.4.1 Gene Amplification and Overexpression

Gene amplification is a prevalent characteristic of genomic instability, occurring in both ecDNA and HSRs. Extensive literature survey showed that oncogenes carried by ecDNA amplified more effectively than those in chromosomes, indicating the high mobility of ecDNA and its potential for heterogeneity. To understand the relationship between ecDNA and oncogenes, W. Ji et al. 2013, performed a study using the Affymetrix SNP Array 6.0 and noted that inhibiting the expression of double-minute genes reduces the presence of extrachromosomal DNA as it integrates into the cytoplasm. Previous research demonstrated that MYC, RAS, EGFR (epidermal growth factor receptor), MDM2 (murine double minute 2), Rel, MET (mesenchymal-epithelial transition factor), the MECOM-PIK3CA-SOX2 gene cluster, and the CDK4-MDM2 gene cluster were found on extrachromosomal DNA. It was found that 22 of the 25 ecDNAs contained an oncogene in several tumor samples from individual patients. Different studies have supported that ecDNA can lead to increased oncogene expression by focal amplification. Application of many advanced technologies suggested that the cancer cell lines and clinical samples revealed a high copy number of ecDNA, which correlates with higher expression of oncogenes [16].

1.4.2 ecDNA and Its Role in Tumor Adaptability

Previous studies indicated that tumor adaptability can be significantly enhanced by ecDNA across various cancers, due to unequal mitotic segregation. Numerous malignancies, including breast, pancreatic, and medulloblastoma, have been shown to have notable variations in ecDNA between their primary and metastatic sites. The longitudinal analysis of primary and recurrent tumor samples using fluorescent in situ hybridization also demonstrated the widespread presence and persistence of oncogenic ecDNA throughout the development of the tumor. Researchers observed that daughter cancer cells can change continuously with the changing environment. The previous discussion indicates that the ecDNA can regulate the heterogeneity and adaptability of tumor [16, 17].

1.5 ECDNA AND CHEMO DRUG RESISTANCES

ecDNA contributes substantially to cancer drug resistance, tumor evolution, and metastasis [18]. Due to rapid gene amplification, ecDNA provides increased promoter/transcriptional activity owing to a more open chromatin structure, and ecDNA viability and adaptability allows for rapid evolution under therapeutic pressure. In this manner, ecDNA facilitates intratumoral heterogeneity, escape from therapy, and the emergence of resistant subclones, all of which also affect the immune response and tumor microenvironment [6, 19]. For these reasons, ecDNA is an intriguing yet complicated therapeutic target in precision oncology [20].

1.5.1 Role of ecDNA in Adaptive Resistance to Drug Therapy

Another important aspect describing the formation of ecDNA contributing to adaptive resistance in cancer treatment involves accelerated and flexible genomic alterations. Such features enable rapid gene amplification, increased genetic variability, and the swift evolution and dominance of drug-resistant populations [21]. These discoveries underscore single-cell-multi-omic embattled therapeutic policies focused on fruition and ecDNA-mediated malleability instead of single-gene balanced targeting [22]. The unequal distribution during cell division exacerbates tumor heterogeneity and promotes rapid drug-resistant subclones [3]. In addition, the high levels of transcriptional activity and flexibility are backed by the open chromatin state of ecDNA, allowing tumor cells to adapt to the stress of treatments [23]. These intriguing findings suggest that medication withdrawal may reverse ecDNA-mediated resistance, allowing tumor re-sensitization [24]. Random segregation of ecDNA because it lacks a centromere creates high heterogeneity within tumors and facilitates fast non-Mendelian inheritance of amplified oncogenes like EGFR, MYC, and DHF. This diversity supports survival and regrowth after treatment is halted by allowing quick and reversible shifts in gene copy numbers under drug pressure [1]. Within precision oncology, ecDNAs rank among the top priorities, with hopes that methods designed to recognize, disrupt, or otherwise exploit their weaknesses (such as through CHK1- or RNR-inhibitor treatments) can reverse drug resistance plaguing aggressive cancer types [25].

1.5.2 The Influence of ecDNA on Chemotherapy and Targeted Drug Resistance

ecDNA is a key factor for drug resistance in cancer, particularly in chemotherapy and targeted therapy. Unlike chromosomal DNA, ecDNA can repeat genes without the structure and regulation of the chromosome, allowing quick adaptability and therapeutic pressure resistance [26]. One of the main mechanisms of ecDNA includes amplification of oncogenes like EGFR, MYC, and MDM2, which leads to their overexpression and cancer cell survival under drug stress [27]. This type of amplification is especially virulent because of uncontrolled, random segregation of ecDNA during mitosis, resulting in intertumoral heterogeneity. ecDNA is able to increase copy number during treatment and decrease copy number when pressure is removed, thus allowing transient and reversible drug tolerance [28]. In targeted therapy, ecDNA drives resistances by amplifying and overexpressing mutated oncogenic drivers such as CORD or CDK4. In leukemia, this overexpression disrupts cell-cycle phase arrest, allowing cancer cell to evade therapy. Furthermore, ecDNA's very structure lacks chromatin, allowing for higher transcriptional activity [29]. Inhibiting ecDNA formation, replication, or segregation could unlock new therapeutic approaches, especially if combined with drugs that target primary oncogenic drivers [30]. In conclusion, ecDNA is a potent driver of resistance in cancer treatment, and further research into its detection and blockade may result in more effective, long-lasting treatment plans and improved patient outcomes [31]. It can promote overexpression of detoxifying enzymes and DNA repair genes, facilitating the survival of cancer cells against DNA-damaging agents such as alkylating agents, taxanes, and platinum drugs [32, 33]. Therapeutic targeting of ecDNA is a promising new frontier in enhancing cancer therapy outcomes [34].

1.5.3 Interaction between ecDNA and Immune Response

The nexus between ecDNA and immune response stands central to the pathophysiology of the appearance of multiple sclerosis (MS) in the investigation "Extracellular DNA and Deoxyribonuclease Activity as Potential Biomarkers of Inflammation in Multiple Sclerosis." It presents major roles of the immune activity that are ecDNA-associated in MS [35]. Once released into the extracellular area, ecDNA acts as a damage-associated molecular pattern (DAMP) that binds with pattern recognition receptors (PRRs), with the most studied being Toll-like receptor 9 (TLR-9) on immune cells. This interaction triggers a cascade resulting in the generation of pro-inflammatory cytokines, thus paving the way for neuroinflammation on sustained levels [36]. Neutrophils, which are shown to be instrumental in the workings of MS, release ecDNA as part of their NETs to trap pathogens [1]. In physiological conditions, ecDNA is eliminated by deoxyribonucleases (DNases) to sustain immune homeostasis [37]. Interleukins like IL-1 β and IL-18 were—on the other hand—the impaired clearance of ecDNA by low DNase activity is expected to allow for the accumulation of ecDNA, thus fueling inflammation and possibly driving disease progression [38].

The study observed increased concentrations of ecDNA and decreased DNase activity in plasma of MS patients compared to healthy controls [39]. ecDNA correlated positively with inflammatory cytokines such as IL-8, MCP-1, and IL-18, oxidative stress markers as TBARS, and clinical indicators of disease severity as EDSS. High concentrations of ecDNA and mtDNA were detected in MS patients with active inflammatory lesions on Gadolinium-enhancing MRI [40]. These findings carry implications with respect to the study of MS pathophysiology. In this setting, ecDNA accumulation would probably amplify immune responses via TLR-9 signaling and, hence, serve as a biomarker of active inflammation and correlate to the severity of illnesses [41].

1.5.4 Effect of ecDNA Alter Chromatin Accessibility

ecDNA is a circular and acentric DNA form prevalent in cancer, markedly altering chromatin architecture to promote tumor progression (24). Unlike chromosomal DNA, ecDNA lacks higher-order chromatin compaction, resulting in a relaxed, highly accessible structure enriched with active histone modifications such as H3K27ac and H3K4me3 [42]. Techniques like ATAC-Seq and Circle hunter leverage tools elevate chromatin accessibility to detect ecDNA. The transcriptional permissiveness of ecDNA not only drives excessive oncogene expression but also leads to transcription-replication conflicts, single-stranded DNA accumulation, and DNA damage [43]. Collectively, these properties make ecDNA a potent driver of tumor aggressiveness and heterogeneity while also presenting exploitable vulnerabilities for therapeutic intervention [44].

1.5.5 Role in Metastatic Potential

ecDNA increases cancer metastasis potential through several interrelated processes. Studies show that ecDNA enables aggressive phenotypes in various cancers, including glioblastoma and neuroblastoma, where it carries potent oncogenes such as MYC and MYCN that drive invasion and distant spread [45]. Certain ecDNAs like chr8 ecDNA 64950741–114379093 appear to progressively increase with tumor growth in colorectal cancer, and poorer outcomes indicate that these specific ecDNAs are implicated in metastasis. ecDNA has been shown to possess higher amounts of super-enhancers and regulatory elements, which allow enhanced transcription of oncogenic programs as well as restructuring of DNA that enables cell migration and survival [46]. The presence of ecDNA in a tumor is clinically associated with lower relapse and recovery times, resistance to treatment, and overall, advanced stages of the tumor [47].

1.6 CLINICAL APPLICATIONS OF ECDNA IN CANCER

ecDNA is an emerging biomarker and therapeutic target in cancer. It can be detectable using a variety of diagnostic platforms, including whole genome sequencing (WGS), whole exome sequencing (WES), fluorescence in situ hybridization (FISH), and novel single-cell and enrichment sequencing methods in a variety of clinically relevant contexts. ecDNA has the additional benefit that it is often found in blood, which would provide a non-invasive biomarker. ecDNA acting as an essential role in therapeutic resistance by rapid adaptation in the cancer cells through oncogene amplification. Targeting ecDNA may be a new therapeutic intervention via CHK1 inhibition, by manipulating the biogenesis or clustering of ecDNA.

1.6.1 Diagnostic Methods and Clinical Applications of ecDNA in Cancer

ecDNA is a key player in cancer biology, whereby it drives oncogene amplification, tumor evolution, drug resistance, and heterogeneity [3]. Advances in detection technologies and bioinformatic tools have expedited habitat as an oncological avenue of exploration—applying ecDNA to diagnostics and therapeutics [48] has shown in Table 1.1.

1.6.2 Biomarker

ecDNA, a type of extrachromosomal circular DNA (eccDNA), is an emerging class of novel and highly dynamic biomarker in oncology. ecDNA can be detected non-invasively through liquid biopsies and in real time to evaluate tumor burden, minimal residual disease, and therapeutic response. Moreover, ecDNA profiling will allow precision therapy to treat patients for amplified oncogenes and regulatory elements. Detecting ecDNA has become easier due to technological advances, particularly next-generation sequencing, such as FISH and digital PCR. Nevertheless, with accumulated data continuing to improve our understanding of its molecular biology, ecDNA will lead the way for cancer diagnostics and targeted treatment, which are discussed in Table 1.2 [51].

TABLE 1.1
Diagnostic and Therapeutic Approach of ecDNA

Diagnostic Methods	Details	Ref.
1. Cytogenetics (FISH/DAPI)	Fluorescence in situ hybridization (FISH) and DAPI staining identify ecDNA by locating gene amplifications.	[1, 49]
2. Whole Genome Sequencing (WGS)	AmpliconArchitect and Amplicon Classifier reconstruct ecDNA structures and detect amplifications.	[35]
3. Whole Exome Sequencing (WES)	Computational models (e.g., GCAP) extract ecDNA patterns from exome data in clinical cohorts.	[49]
4. Circle-Seq	Enrichment-based sequencing method that isolates and sequences circular DNA with high specificity.	[49]
5. Single-Cell Sequencing (scGTP-seq)	Captures ecDNA and structural variants at single-cell resolution, providing insights into tumor heterogeneity.	[50]
6. ecTag	A live-cell imaging method using CRISPR-Cas9 system to fluorescently tag ecDNA sequences.	[49]

TABLE 1.2
Clinical Applications of ecDNA in Cancer—Uses as Biomarker

Application Area	Details	Ref
Cancer Diagnosis (Liquid Biopsy)	ecDNA is detectable in serum/plasma; serves as a non-invasive biomarker for early detection.	[52]
Prognosis Indicator	High ecDNA levels correlate with poor clinical outcomes and aggressive tumor phenotypes.	
Monitoring Treatment Response	Changes in circulating ecDNA reflect tumor burden, resistance, or relapse in real time.	
Predicting Therapeutic Resistance	ecDNA contributes to drug resistance via oncogene amplification and tumor heterogeneity.	[53]
Therapeutic Targeting	Potential to develop ecDNA-specific inhibitors targeting replication or transcription machinery.	
Stratifying Tumor Subtypes	ecDNA profiles help identify specific oncogene amplifications, aiding personalized therapy decisions.	[52]
Prognostic Stratification	Different ecDNA content across cancers assists in clinical stratification for outcome prediction.	[54]

1.6.3 Drug Resistance

Extrachromosomal DNA is a major contributor to drug resistance in cancer through numerous biological mechanisms [54]. In divergence to chromosomal DNA, ecDNA is not uniformly inherited with each cell division event, giving rise to substantial intratumoral heterogeneity and providing the tumor with opportunities for rapid evolutionary adaptation to chemotherapeutic and targeted interventions [53]. Studies showing ecDNA mediating resistance to HER2-targeted therapies in esophagogastric adenocarcinoma would be highly relevant here as well, especially given the evidence that analogue ecDNA carrying ERBB2 has given rise to higher copy number and worse prognosis [55]. These studies identify ecDNA as a contributor to drug resistance and a potential therapeutic target to overcome treatment failure in cancer [56], which are summarized in Table 1.3.

1.6.4 Therapeutic Target

ecDNA's unique structure and replication-transcription profile make it vulnerable to targeted therapies. CHK1 inhibitors like BBI-2779 represent one of the most promising therapeutic approaches under preclinical investigation, and therapies aimed at transcriptional regulation, DNA replication stress, or cluster disruption are being developed to exploit the weaknesses of ecDNA-positive cancer cells [46], which are summarized in Table 1.4.

1.7 BIOINFORMATIC TOOLS FOR ANALYZING ECDNA

Extrachromosomal DNA is a key factor for oncogene amplification and tumor evolution. A range of bioinformatic tools has been developed to analyze ecDNA from sequencing data, especially WGS. Different genomic and bioinformatic techniques play an important role in DNA sequencing-based analysis to understand the biology of ecDNA. In this section, we highlight different tools that are used for imaging data. AmpliconArchitect (AA) is a tool designed for reconstructing ecDNA amplicons through WGS data. It was further modified to the Amplicon Suite and Amplicon Classifer, along with CNVkit, to detect shorter circular DNA. A research study reported that this tool shows a sensitivity of 83% in detecting circular ecDNA. Long-read sequencing (LRS) with optical mapping, along with analysis using Amplicon BNReconstructor, is more precise for the characterization of

TABLE 1.3
Different Types of Cancer-Related ecDNA Associated with Drug Resistances

Cancer Type	ecDNA-Related Resistance Features	Key Gene(s)	References
Ovarian Cancer (HGSOC)	ecDNA carrying *HIF1A* enhances resistance to cisplatin by regulating EMT and DNA repair pathways.	HIF1A	[57]
Glioblastoma	ecDNA harboring *EGFRvIII* leads to resistance to EGFR inhibitors; dynamic copy number changes observed.	EGFRvIII	
Esophagogastric Adenocarcinoma	ERBB2+ecDNA correlated with differential response to HER2-targeted therapies.	ERBB2	[58]

TABLE 1.4
Therapeutic Approaches to Target ecDNA

Therapeutic Strategy	Mechanism	Clinical Implication	Source
CHK1 Inhibition	Targets *replication stress and transcription-replication conflict* prevalent in ecDNA-positive tumors.	Selectively kills ecDNA+ tumor cells; e.g., CHK1 inhibitor *BBI-2779* causes tumor regression in gastric cancer.	[18]
Synthetic Lethality Approaches	Exploits ecDNA-specific vulnerabilities such as *open chromatin*, *replication timing*, and *DNA damage load.*	Precision targeting without affecting normal cells; promising for *ecDNA-dependent oncogene addiction.*	
Targeting Transcriptional Activity	ecDNAs exhibit *hypertranscription* of oncogenes due to open chromatin.	Inhibition of transcription factors (e.g., *BRD4*) or *RNA Pol II* may suppress oncogene expression.	[1]
Disruption of ecDNA Hubs/ Clusters	Interferes with *ecDNA clustering*, which amplifies oncogenic signaling.	Inhibiting interactions among ecDNA hubs may reduce gene overexpression.	[52]
Blocking ecDNA Biogenesis Pathways	Targets mechanisms like *chromothripsis*, *NHEJ*, and *DNA repair errors*, leading to ecDNA formation.	Prevents formation of new ecDNA, limiting tumor evolution and drug resistance.	[46]
Liquid Biopsy for Therapy Monitoring	Uses circulating ecDNA as a *biomarker* to evaluate treatment response in real time.	Enables *non-invasive monitoring* and early detection of therapeutic failure or recurrence.	[59]

ecDNA. This tool is more efficient for spanning breakpoint, which provide detailed data on structural variants to detect tandem repeats like HSRs. LRS has been combined with in vitro techniques for analyzing the heterogeneity of ecDNA at ever-higher resolution. However, the aforementioned tools have some limitations for multi-copy oncogene analysis. To address this limitation, researchers have utilized single nucleotide polymorphism (SNP) analysis technique. Some researchers analyzed exonic SNPs in ecDNA-amplified oncogenes from heterozygous cell lines to compare allele ratios in RNA-seq. and WGS data, revealing similar EGFR transcriptional efficiency from ecDNA and chromosomal loci [60].

1.8 THERAPEUTIC MODALITIES TO TARGET ECDNA

Since ecDNA contributes significantly to cancer progression and drug resistance, it is becoming a promising target for novel cancer therapies. The therapeutic modalities to target ecDNA have been described in Table 1.5.

TABLE 1.5
Summary of Therapeutic Modalities to Target ecDNA [1]

Clinical Modalities	Potential Application
Diagnostic Biomarker	• ecDNA could serve as diagnostic biomarkers in plasma samples for lung cancer.
Prognostic Biomarker	• Some ecDNAs were associated with compromised survival in various cancers, e.g., glioblastoma and HGSOC. • Some ecDNA were associated with better prognosis in gastric cardia adenocarcinoma.
Drug Resistance	• Oncogene amplicons carried by ecDNA can accelerate the development of intra-tumoral heterogeneity. • ecDNA can serve as a latent reservoir for genomic alterations.
Therapeutic Target	• Hydroxyurea restored drug sensitivity. • PARP inhibitors may target the reintegration of ecDNA. • BET inhibitors may target ecDNA hubs. • Treatments could be developed to induce antigen presentation pathways.

1.9 CONCLUSION

Extrachromosomal DNA (ecDNA) has emerged as a pivotal player in cancer biology, offering critical insights into tumor evolution, heterogeneity, and therapeutic resistance. Its unique biogenesis and structural features distinguish it from chromosomal DNA, allowing for oncogene amplification and enhanced adaptability of tumor cells. The involvement of ecDNA in key signaling pathways and its role in promoting chemoresistance, immune evasion, and metastasis underscore its significance in cancer progression. Clinically, ecDNA holds promise as a diagnostic biomarker, a predictor of drug response, and a therapeutic target. With advancements in bioinformatics and imaging technologies, the detection and analysis of ecDNA have become increasingly precise. Therapeutic strategies aimed at targeting ecDNA—ranging from small molecules to personalized immunotherapies—are opening new avenues for overcoming resistance and improving cancer treatment outcomes. As research continues to unravel the complexities of ecDNA, its integration into clinical oncology could revolutionize personalized cancer care.

ABBREVIATIONS

BET: bromodomain and extra-terminal domain

CHK1: checkpoint kinase 1
DAPI: 4’,6-diamidino-2-phenylindole
DHF: 7,8-dihydroxyflavone
DM: double minutes
EGFR: epidermal growth factor receptor
ERBB2: Erb-B2 receptor tyrosine kinase 2
FISH: fluorescence in situ hybridization

HSR: homogeneously staining region
LRS: long-read sequencing
MDM2: murine double minute
MYC: myelocytomatosis
PARP: poly-ADP ribose polymerase
WES: whole exome sequencing
WGS: whole genome sequencing

REFERENCES

1. Dong, Y., He, Q., Chen, X., Yang, F., He, L., & Zheng, Y. (2023). Extrachromosomal DNA (ecDNA) in cancer: Mechanisms, functions, and clinical implications. *Frontiers in Oncology*, *13*, 1194405.
2. Wu, S., Bafna, V., Chang, H. Y., & Mischel, P. S. (2022). Extrachromosomal DNA: An emerging hallmark in human cancer. *Annual Review of Pathology: Mechanisms of Disease*, *17*(1), 367–386.
3. Pecorino, L. T., Verhaak, R. G., Henssen, A., & Mischel, P. S. (2022). Extrachromosomal DNA (ecDNA): An origin of tumor heterogeneity, genomic remodeling, and drug resistance. *Biochemical Society Transactions*, *50*(6), 1911–1920.
4. Gu, X., Yu, J., Chai, P., Ge, S., & Fan, X. (2020). Novel insights into extrachromosomal DNA: Redefining the onco-drivers of tumor progression. *Journal of Experimental & Clinical Cancer Research*, *39*, 1–10.
5. Yi, E., Chamorro González, R., Henssen, A. G., & Verhaak, R. G. (2022). Extrachromosomal DNA amplifications in cancer. *Nature Reviews Genetics*, *23*(12), 760–771.
6. Wu, S., Tao, T., Zhang, L., Zhu, X., & Zhou, X. (2023). Extrachromosomal DNA (ecDNA): Unveiling its role in cancer progression and implications for early detection. *Heliyon*, *9*(11).
7. Wu, S., Turner, K. M., Nguyen, N., Raviram, R., Erb, M., Santini, J., . . . & Mischel, P. S. (2019). Circular ecDNA promotes accessible chromatin and high oncogene expression. *Nature*, *575*(7784), 699–703.
8. Hung, K. L., Yost, K. E., Xie, L., Shi, Q., Helmsauer, K., Luebeck, J., . . . & Chang, H. Y. (2021). ecDNA hubs drive cooperative intermolecular oncogene expression. *Nature*, *600*(7890), 731–736.
9. Tian, X., Liu, C., Wang, X., Wang, F., Wang, L., Xu, L., . . . & Zhang, C. (2018). Sei-1 promotes double minute chromosomes formation through activation of the PI3K/Akt/BRCA1-Abraxas pathway and induces double-strand breaks in NIH-3T3 fibroblasts. *Cell Death & Disease*, *9*(3), 341.
10. Sun, W., Quan, C., Huang, Y., Ji, W., Yu, L., Li, X., . . . & Fu, S. (2015). Constitutive ERK1/2 activation contributes to production of double minute chromosomes in tumour cells. *The Journal of Pathology*, *235*(1), 14–24.
11. Stewart, G. S., Wang, B., Bignell, C. R., Taylor, A. M. R., & Elledge, S. J. (2003). MDC1 is a mediator of the mammalian DNA damage checkpoint. *Nature*, *421*(6926), 961–966.
12. Xiang, T., Ohashi, A., Huang, Y., Pandita, T. K., Ludwig, T., Powell, S. N., & Yang, Q. (2008). Negative regulation of AKT activation by BRCA1. *Cancer Research*, *68*(24), 10040–10044.
13. Dharanipragada, P., Zhang, X., Liu, S., Lomeli, S. H., Hong, A., Wang, Y., . . . & Lo, R. S. (2023). Blocking genomic instability prevents acquired resistance to MAPK inhibitor therapy in melanoma. *Cancer Discovery*, *13*(4), 880–909.
14. Decarvalho, A. C., Kim, H., Poisson, L. M., Winn, M. E., Mueller, C., Cherba, D., . . . & Verhaak, R. G. (2018). Discordant inheritance of chromosomal and extrachromosomal DNA elements contributes to dynamic disease evolution in glioblastoma. *Nature Genetics*, *50*(5), 708–717.
15. Wu, H., Liu, S., Wu, D., Zhou, H., & Wu, G. (2024). Tumor extrachromosomal DNA: Biogenesis and recent advances in the field. *Biomedicine & Pharmacotherapy*, *174*, 116588.
16. Zhao, L., Jiang, Y., Lei, X., & Yang, X. (2023). Amazing roles of extrachromosomal DNA in cancer progression. *Biochimica et Biophysica Acta (BBA)-Reviews on Cancer*, *1878*(1), 188843.
17. Huang, Q., Zhang, S., Wang, G., & Han, J. (2024). Insight on ecDNA-mediated tumorigenesis and drug resistance. *Heliyon*.
18. Tang, J., Weiser, N. E., Wang, G., Chowdhry, S., Curtis, E. J., Zhao, Y., . . . & Chang, H. Y. (2024). Enhancing transcription-replication conflict targets ecDNA-positive cancers. *Nature*, *635*(8037), 210–218.
19. Wu, S., Tao, T., Zhang, L., Zhu, X., & Zhou, X. (2023). Extrachromosomal DNA (ecDNA): Unveiling its role in cancer progression and implications for early detection. *Heliyon*, *9*(11).
20. Yan, X., Mischel, P., & Chang, H. (2024). Extrachromosomal DNA in cancer. *Nature Reviews Cancer*, *24*(4), 261–273.
21. Yang, M., Zhang, S., Jiang, R., Chen, S., & Huang, M. (2023). Circlehunter: A tool to identify extrachromosomal circular DNA from ATAC-Seq data. *Oncogenesis*, *12*(1), 28.
22. Zeng, X., Wan, M., & Wu, J. (2020). ecDNA within tumors: A new mechanism that drives tumor heterogeneity and drug resistance. *Signal Transduction and Targeted Therapy*, *5*(1), 277.
23. Meng, X. N., Ma, J. F., Liu, Y. H., Li, S. Q., Wang, X., Zhu, J., . . . & Fu, S. B. (2024). Dynamic genomic changes in methotrexate-resistant human cancer cell lines beyond DHFR amplification suggest potential new targets for preventing drug resistance. *British Journal of Cancer*, *130*(11), 1819–1827.
24. Lange, J. T., Rose, J. C., Chen, C. Y., Pichugin, Y., Xie, L., Tang, J., . . . & Mischel, P. S. (2022). The evolutionary dynamics of extrachromosomal DNA in human cancers. *Nature Genetics*, *54*(10), 1527–1533.
25. Chung, O. W., Yao, S., Yang, F., Wang, L., Cerda-Smith, C., Hutchinson, H. M., . . . & Zhang, Z. Z. (2025). BRCA1-A and LIG4 complexes mediate ecDNA biogenesis and cancer drug resistance. *bioRxiv*.

26. Bukowski, K., Kciuk, M., & Kontek, R. (2020). Mechanisms of multidrug resistance in cancer chemotherapy. *International Journal of Molecular Sciences*, *21*(9), 3233.
27. Zuo, S., Yi, Y., Wang, C., Li, X., Zhou, M., Peng, Q., . . . & He, Q. (2022). Extrachromosomal circular DNA (eccDNA): From chaos to function. *Frontiers in Cell and Developmental Biology*, *9*, 792555.
28. Germano, S., & O'driscoll, L. (2009). Breast cancer: Understanding sensitivity and resistance to chemotherapy and targeted therapies to aid in personalised medicine. *Current Cancer Drug Targets*, *9*(3), 398–418.
29. Wu, T., Wu, C., Zhao, X., Wang, G., Ning, W., Tao, Z., . . . & Liu, X. S. (2022). Extrachromosomal DNA formation enables tumor immune escape potentially through regulating antigen presentation gene expression. *Scientific Reports*, *12*(1), 3590.
30. Turner, K. M., Deshpande, V., Beyter, D., Koga, T., Rusert, J., Lee, C., . . . & Mischel, P. S. (2017). Extrachromosomal oncogene amplification drives tumour evolution and genetic heterogeneity. *Nature*, *543*(7643), 122–125.
31. Wei, J., Wu, C., Meng, H., Li, M., Niu, W., Zhan, Y., . . . & Zhou, M. (2020). The biogenesis and roles of extrachromosomal oncogene involved in carcinogenesis and evolution. *American Journal of Cancer Research*, *10*(11), 3532.
32. Li, Z., & Qian, D. (2024). Extrachromosomal circular DNA (eccDNA): From carcinogenesis to drug resistance. *Clinical and Experimental Medicine*, *24*(1), 83.
33. Weiser, N. E., Watkins, T. B., Chang, H. Y., & Mischel, P. S. (2025). A guide to extrachromosomal DNA: Cancer's dynamic circular genome. *Cancer Discovery*, *15*(6), 1105–1114.
34. Qiu, H., Shao, Z. Y., Wen, X., & Zhang, L. Z. (2020). New insights of extrachromosomal DNA in tumorigenesis and therapeutic resistance of cancer. *American Journal of Cancer Research*, *10*(12), 4056.
35. Bailey, C., Pich, O., Thol, K., Watkins, T. B., Luebeck, J., Rowan, A., . . . & Swanton, C. (2024). Origins and impact of extrachromosomal DNA. *Nature*, *635*(8037), 193–200.
36. Ludwig, M. P., Wilson, J. R., Galbraith, M. D., Bhandari, N., Dunn, L. N., Black, J. C., & Sullivan, K. D. (2025). NF-κB signaling directs a program of transient amplifications at innate immune response genes. *bioRxiv*, 2025–03.
37. Stöber, M. C., González, R. C., Brückner, L., Conrad, T., Wittstruck, N., Szymansky, A., . . . & Haase, K. (2024). Intercellular extrachromosomal DNA copy-number heterogeneity drives neuroblastoma cell state diversity. *Cell Reports*, *43*(9).
38. Kim, H., Nguyen, N. P., Turner, K., Wu, S., Gujar, A. D., Luebeck, J., . . . & Verhaak, R. G. (2020). Extrachromosomal DNA is associated with oncogene amplification and poor outcome across multiple cancers. *Nature Genetics*, *52*(9), 891–897.
39. Wang, F., Zhang, H., Wu, Z., Hu, D., Zhou, Z., Girault, J. B., . . . & Li, G. (2023). Fine-grained functional parcellation maps of the infant cerebral cortex. *elife*, *12*, e75401.
40. Marček, P., Kadlic, P., Adamová, L. M., Tóthova, Ľ., Pastorek, M., Kovalčíkova, A. G., . . . & Slezáková, D. (2025). Extracellular DNA and deoxyribonuclease activity as potential biomarkers of inflammation in multiple sclerosis. *Molecular Neurobiology*, 1–10.
41. Wu, S., Bafna, V., Chang, H. Y., & Mischel, P. S. (2022). Extrachromosomal DNA: An emerging hallmark in human cancer. *Annual Review of Pathology: Mechanisms of Disease*, *17*(1), 367–386.
42. Hung, K. L., Mischel, P. S., & Chang, H. Y. (2022). Gene regulation on extrachromosomal DNA. *Nature Structural & Molecular Biology*, *29*(8), 736–744.
43. Nakagawa, T., Luebeck, J., Zhu, K., Lange, J. T., Sasik, R., Phillips, C., . . . & Califano, J. A. (2025). Inhibition of human-HPV hybrid ecDNA enhancers reduces oncogene expression and tumor growth in oropharyngeal cancer. *Nature Communications*, *16*(1), 2964.
44. Yi, E., Gujar, A. D., Guthrie, M., Kim, H., Zhao, D., Johnson, K. C., . . . & Verhaak, R. G. (2022). Live-cell imaging shows uneven segregation of extrachromosomal DNA elements and transcriptionally active extrachromosomal DNA hubs in cancer. *Cancer Discovery*, *12*(2), 468–483.
45. Fang, J., Ying, L., Ma, Z., Yang, Y., Zhu, R., & Su, D. (2024). The distribution of the extrachromosomal DNA molecules in early lung cancer. *Science Progress*, *107*(3), 00368504241276771.
46. Rehman, S. U. (2025). The multifaceted role of extrachromosomal DNA in cancer: Functions and clinical implications. *Bangladesh Journal of Medicine*, *36*(2), 82–91.
47. Nguyen, D. D., Hooper, W. F., Liu, W., Chu, T. R., Geiger, H., Shelton, J. M., . . . & Faltas, B. M. (2024). The interplay of mutagenesis and ecDNA shapes urothelial cancer evolution. *Nature*, *635*(8037), 219–228.
48. Li, R., Wang, Y., Li, J., & Zhou, X. (2022). Extrachromosomal circular DNA (eccDNA): An emerging star in cancer. *Biomarker Research*, *10*(1), 53.
49. Wang, S., Wu, C. Y., He, M. M., Yong, J. X., Chen, Y. X., Qian, L. M., . . . & Zhao, Q. (2024). Machine learning-based extrachromosomal DNA identification in large-scale cohorts reveals its clinical implications in cancer. *Nature Communications*, *15*(1), 1515.

50. Chang, L., Deng, E., Wang, J., Zhou, W., Ao, J., Liu, R., . . . & Fan, X. (2023). Single-cell third-generation sequencing-based multi-omics uncovers gene expression changes governed by ecDNA and structural variants in cancer cells. *Clinical and Translational Medicine*, *13*(8), e1351.
51. Arshadi, A., Tolomeo, D., Venuto, S., & Storlazzi, C. T. (2023). Advancements in focal amplification detection in tumor/liquid biopsies and emerging clinical applications. *Genes*, *14*(6), 1304.
52. Deng, E., & Fan, X. (2024). Categorizing extrachromosomal circular DNA as biomarkers in serum of cancer. *Biomolecules*, *14*(4), 488.
53. Li, J., Luo, P., Li, Z., Wang, Q., Huang, X., Wang, K., . . . & Chen, R. (2025). Extrachromosomal circular DNA in cancer: Mechanisms and clinical applications. *Cell Proliferation*, e70040.
54. Purshouse, K., Pollard, S. M., & Bickmore, W. A. (2024). Imaging extrachromosomal DNA (ecDNA) in cancer. *Histochemistry and Cell Biology*, *162*(1), 53–64.
55. Li, Z., Wang, B., Liang, H., & Han, L. (2022). Pioneering insights of extrachromosomal DNA (ecDNA) generation, action and its implications for cancer therapy. *International Journal of Biological Sciences*, *18*(10), 4006–4025.
56. Luo, J., Li, Y., Zhang, T., Xv, T., Chen, C., Li, M., . . . & Wan, S. (2023). Extrachromosomal circular DNA in cancer drug resistance and its potential clinical implications. *Frontiers in Oncology*, *12*, 1092705.
57. Xing, J., Ning, Q., Tang, D., Mo, Z., Lei, X., & Tang, S. (2021). Progress on the role of extrachromosomal DNA in tumor pathogenesis and evolution. *Clinical Genetics*, *99*(4), 503–512.
58. Tsai, C., Suman, S., Chu, C., Chou, J.F., Krein, P.M., Capanu, M., Ilson, D.H., Ku, G.Y., Janjigian, Y.Y., Drilon, A.E., & Berger, M.F. Association of ERBB2 extrachromosomal DNA (ecDNA) with first-line HER2-targeted treatment outcomes in advanced esophagogastric adenocarcinoma (EGC).
59. Wang, S., Li, H., Zhang, Y., Yan, J., & Li, W. (2025). Extrachromosomal circular DNA: Enlightenment to widespread clinical application. *Holistic Integrative Oncology*, *4*(1), 8.
60. Purshouse, K., Pollard, S. M., & Bickmore, W. A. (2024). Imaging extrachromosomal DNA (ecDNA) in cancer. *Histochemistry and Cell Biology*, *162*(1), 53–64.

2 Formation and Structure of ecDNA in Cancer Cells

Sanjesh Kumar,[1] Mansi Singh,[2] M. V. N. L. Chaitanya,[3] Joy Das,[3] Tapan Kumar Shaw,[4] Mohini Mondal,[5] and Sumel Ashique[6]

1 Chandigarh University
Nawabganj, Unnao, India
2 Rakshpal Bahadur College of Pharmacy
Bareilly, India
3 School of Pharmaceutical Sciences
Lovely Professional University
Phagwara, Punjab, India
4 Department of Pharmaceutical Technology
JIS University, Agarpara Campus
Kolkata, India
5 Department of Pharmaceutical Technology
JIS University
Kolkata, India
6 Department of Pharmaceutical Technology
Bharat Technology
Uluberia, Howrah, India

2.1 INTRODUCTION

2.1.1 Background of ecDNA in Cancer

The nucleus of eukaryotic cells contains circular, acentric DNA elements that are not related to the chromosomal genome; this is called extrachromosomal DNA (ecDNA). ecDNAs, in contrast to chromosomal DNA, are devoid of centromeres, which enables them to circumvent conventional mitotic segregation and results in an asymmetrical distribution among daughter cells. At first, witnessed in human neuroblastoma and colorectal cancer cells using electron microscopy, ecDNA components were first referred to in the 1960s as "double minutes" (DMs). These structures appeared as small, paired chromatin bodies and were later classified with homogeneously staining regions (HSRs) based on cytogenetic patterns—both hallmarks of gene amplification in malignancies [1].

The recent renaissance of ecDNA research has been propelled by next-generation sequencing and single-cell genomic technologies, highlighting ecDNA as a crucial factor in tumor evolution. ecDNA often has oncogenes that have been increased, like MYC, EGFR, MDM2, and CDK4. It is also more common in high-grade glioblastomas, neuroblastomas, small-cell lung carcinomas, and esophageal squamous cell carcinomas. Importantly, their acentric nature results in non-Mendelian inheritance, enhancing intratumoral genetic diversity. A study on glioblastoma multiforme (GBM) reveals that EGFRvIII-carrying ecDNA confers significant proliferative advantages, with its variable expression resulting in differing resistance and susceptibility to treatment [2].

Transcriptional regulation and chromatin accessibility are mechanistically varied by the availability of ecDNA. ecDNA constitutes an accessible and hyperacetylated chromatin state, which

DOI: 10.1201/9781003655299-2

supports super-enhancer-driven transcription, according to ATAC-seq and Hi-C experiments performed in U87 and HF3016 GBM cells. Furthermore, their ability to cluster into transcriptional "hubs" within the nucleus facilitates interactions between multiple ecDNA molecules and the transcriptional machinery, thereby amplifying oncogene expression beyond the levels achieved by chromosomal amplification. Therefore, ecDNA constitutes a crucial element in the adaptive landscape of tumors. By fostering genetic heterogeneity, enhancing oncogene dosage, and circumventing clonal constraints, ecDNA promotes therapy resistance and tumor evolution in a wide range of epithelial and mesenchymal cancers, making it a central focus in contemporary cancer biology [3].

2.1.2 Significance of Studying ecDNA Structure and Formation

It is essential to comprehend the structure and genesis of ecDNA, as it plays a central role in the development of malignant phenotypes. ecDNA's topological architecture, which is circular, non-centromeric, and transcriptionally permissive, enables the dynamic regulation of gene expression. Circle-seq, nanopore sequencing, and FISH structural mapping have demonstrated that ecDNA can contain both oncogenes and enhancer elements nearby, resulting in the formation of "enhancer hijacking" complexes that subvert normal epigenetic control. The amplification of MYC and FGFR1 by ecDNA is associated with aggressive development and poor clinical outcomes in patients with medulloblastoma and triple-negative breast cancer [4].

The mechanisms underlying ecDNA formation are closely linked to genome instability events such as chromothripsis, breakage–fusion–bridge (BFB) cycles, religation of DNA, fork stalling, and replication stress (Figure 2.1). Osteosarcoma and GBM cells exhibit chromothripsis, which is characterized by catastrophic chromosome fragmentation followed by aberrant re-ligation, resulting in ecDNA with altered oncogenic loci. BFB cycles—prominent in epithelial tumors such as esophageal carcinoma—generate palindromic sequences and inverted repeats that facilitate circularization. Amplification of oncogenes, notably CCNE1, in ovarian cancer generates replication stress, resulting in micronuclei that serve as sites for ecDNA synthesis via nuclear envelope rupture and defective DNA repair processes. ecDNA's contribution to epigenetic plasticity enables cancer cells to react differently to treatments such as chemotherapy. Its extrachromosomal status facilitates the rapid amplification or deletion in response to environmental components, thereby enabling its "dynamic gene amplification" characteristics. The targeted reduction of ecDNA copies of EGFRvIII in EGFR-amplified GBM following treatment with tyrosine kinase inhibitors suggests their reversible and responsive characteristics. ecDNA possesses potential as a biomarker for early identification, prognosis, and treatment stratification in clinical settings. Circulating ecDNA in plasma has been detected in pancreatic and lung cancers using digital PCR and cell-free DNA sequencing. Moreover, the structural configuration of ecDNA provides novel therapeutic targets—such as disrupting ecDNA maintenance via ATR or PARP inhibitors in synthetic lethality approaches [5].

2.2 CLASSIFICATION AND FEATURES OF ECDNA

2.2.1 Types of ecDNA in Cancer Cells

Circular ecDNA and double minutes (DMs) are the two primary forms of extrachromosomal DNA (ecDNA) observed in cancer. Each of these entities possesses unique functional and structural characteristics. Double minutes in neuroblastoma and glioblastoma multiforme (GBM) are initially observed cytogenetically as small, circular, acentric DNA pieces that do not have centromeres or telomeres [7]. This results in their asymmetrical segregation during mitosis. These elements are often enriched with oncogenes such as MYC, EGFR, and MDM2, contributing to high-level gene amplification independent of chromosomal loci. Their presence in malignancies such as colorectal and small-cell lung cancer has been related to enhanced tumor aggressiveness and drug resistance. The majority of oncogenic ecDNAs exhibit a circular topology, typically ranging from a few kilobases to several megabases, and are generated through complex mechanisms such as

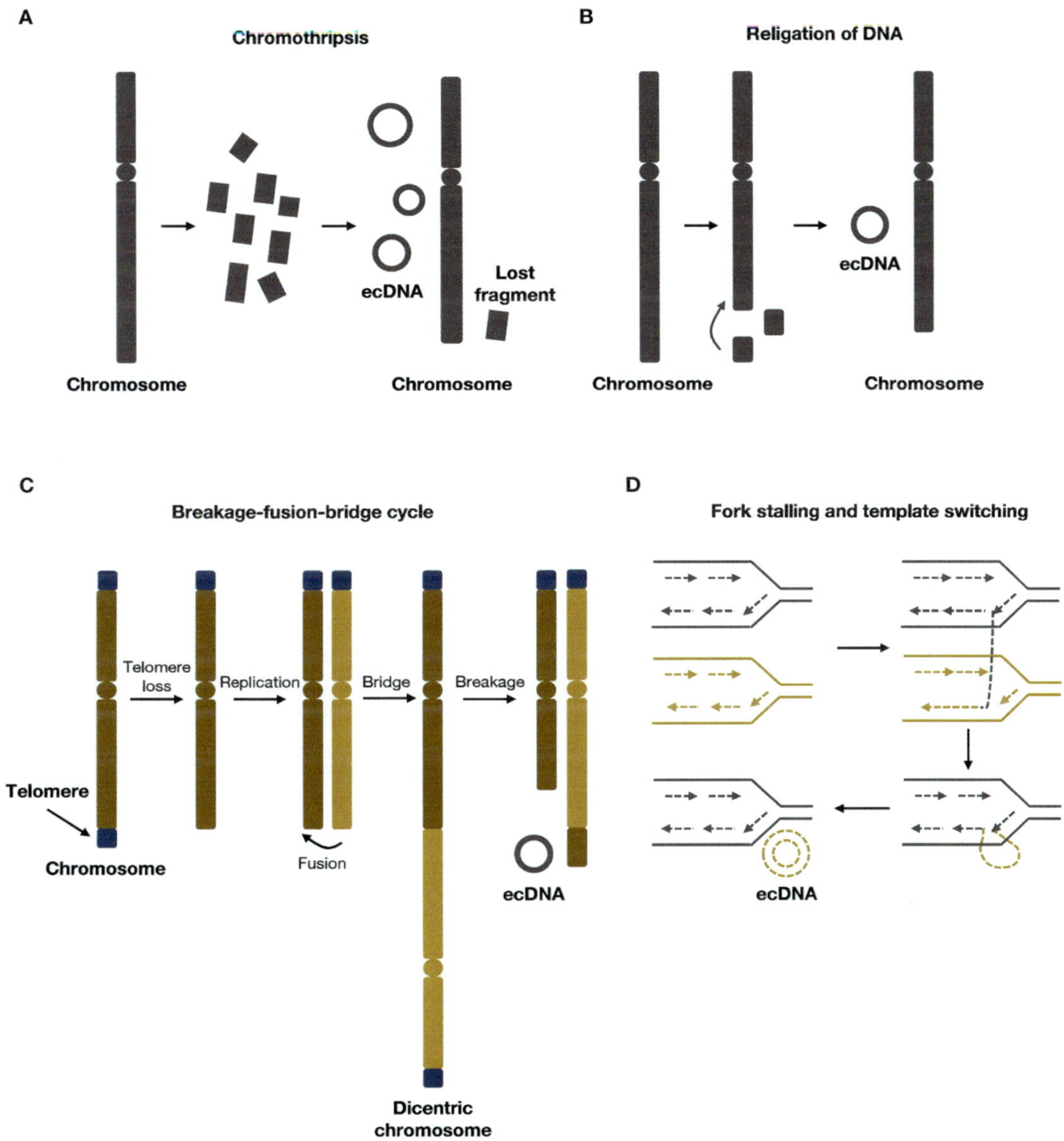

FIGURE 2.1 Mechanisms behind the formation of ecDNA. (A) In chromothripsis, chromosomes are fragmented and rejoined to create ecDNA. (B) Localized rejoining of DNA can lead to ecDNA formation. (C) Breakage–fusion–bridge (BFB) cycles produce altered chromosomes and ecDNA. (D) Fork stalling and template switching can result in the formation of ecDNA [6]. @CCBY.

chromothripsis, breakage–fusion–bridge cycles, and replication fork collapse. Extensive profiling of circular ecDNAs in several cancers, including glioblastoma, triple-negative breast cancer, and esophageal carcinoma, has shown they promote the creation of transcriptionally active hubs using enhancer hijacking and promoter-enhancer reconfiguration. These circular structures lack centromeric sequences, which results in their mitotic instability; however, they possess the ability for exponential clonal expansion through unequal segregation.

On the other hand, although they are rarely described, linear extrachromosomal fragments are acknowledged as the consequence of telomeric crisis and DNA damage-induced chromosomal fragmentation. These forms are more likely to be temporary and can be broken down by exonucleolysis or reintegrated into chromosomal DNA through non-homologous end joining (NHEJ) [8]. Typically smaller in size (varying from 100 bp to a few kilobases), eccDNA (extrachromosomal circular DNA) is a separate but related substance seen in both normal and cancerous cells. In healthy tissues such as human skeletal muscle and liver, eccDNAs are thought to arise from apoptotic DNA fragmentation or replication errors. However, eccDNAs are substantially concentrated in cancer cells and usually contain repetitive elements like LINE-1 and ALU or incomplete sequences of oncogenes. This implies that these eccDNAs contribute to adaptive evolution and genomic flexibility. Interestingly, concatemerization and rolling circle amplification in tumor cells have been linked to the development of larger ecDNA structures by eccDNAs, suggesting a possible evolutionary continuum between normal and cancer-associated extrachromosomal components [9].

2.2.2 General Properties

ecDNA molecules vary in size, ranging from a few hundred base pairs to more than 10 megabases. The majority of cancer-related ecDNAs range in size from one to three megabases. Their design allows them to transport whole oncogenes or several gene cassettes, which are usually bordered by enhancer-rich regions. Extrachromosomal DNAs (ecDNAs) are recognized as extensively rearranged structures featuring intricate chimeric junctions, circularized genomic fragments, and repetitive elements crucial for their independent preservation and proliferation, as demonstrated by structural analyses utilizing optical mapping and long-read sequencing technologies (Nanopore, PacBio).

Unlike chromosomal DNA, ecDNAs replicate independently of the host cell cycle using both origin-dependent and recombination-mediated mechanisms. ecDNA replication can begin with the recruitment of origin recognition complexes (ORCs) and minichromosome maintenance (MCM) helicase complexes, or with break-induced replication (BIR), particularly under replication stress. It has been investigated that ecDNA replication in cancer cells, including A549 lung carcinoma and U87 glioblastoma, happens asynchronously and is responsive to ATR inhibition, highlighting its reliance on the DNA damage response (DDR) system [10]. Due to the absence of centromeres, non-Mendelian segregation occurs during mitosis in ecDNA, rendering the inheritance process highly unpredictable. This results in significant intratumoral heterogeneity, distinguished by subclones possessing diverse ecDNA copy levels. The significant variability in ecDNA copy numbers within clusters of cells in patient-derived GBM xenografts was related to differences in gene expression and phenotypic plasticity. Given that ATAC-seq and ChIP-seq analyses showed a notable enrichment of H3K27ac and H3K4me3 histone marks, the open chromatin orientation of ecDNA is one of its most distinctive features. This accessibility facilitates the binding of transcriptional co-activators such as BRD4, MED1, and RNA polymerase II, leading to elevated transcriptional activity of oncogenes compared to their chromosomal counterparts. Moreover, Hi-C data from ecDNA-rich tumors show that ecDNAs can create distinct three-dimensional hubs, spatially grouping with other ecDNA molecules and nuclear substructures, hence increasing gene expression using cooperative enhancer-promoter interactions. The integration of ecDNA into such transcriptional microenvironments enhances the adaptability of cancer cells to therapeutic pressure and environmental stress, contributing to aggressive tumor behavior and poor clinical outcomes [11].

2.3 MECHANISMS OF ECDNA FORMATION

2.3.1 Chromosomal Breakage and Repair Pathways

The production of ecDNA in cancer cells is closely associated with chromosomal instability, which is specifically caused by aberrant DNA repair pathways and large-scale genomic fragmentation events. One of the most extensively characterized mechanisms is the breakage–fusion–bridge (BFB) cycle, which was first recognized in early cytogenetic investigations. BFB cycles are initiated by telomere attrition or deprotection and promote dicentric chromosome formation. This chromosome undergoes recurrent breakage during anaphase as a result of mitotic spindle tension. This process has been documented in osteosarcoma, glioblastoma, and esophageal squamous cell carcinoma, where disrupted chromosomal arms containing oncogenes like MYC and EGFR undergo repeated breakage and circularization, resulting in the emergence of ecDNA harboring amplified gene segments.

ecDNA biogenesis is also significantly influenced by DNA double-strand break (DSB) repair mechanisms, particularly microhomology-mediated end joining (MMEJ) and non-homologous end joining (NHEJ). Although NHEJ is essential for the preservation of genome integrity, it is error-prone and operates independently of sequence homology. Often leading to deletions or insertions at the junction points, it enables the direct ligation of broken DNA ends. This has been interlinked with the production of ecDNA in lung cancer and melanoma when NHEJ-induced ligation of distant genomic loci results in circularized extrachromosomal fragments. In contrast, MMEJ utilizes 5–25 bp regions of microhomology to align and anneal DNA ends, often leading to deletions and rearrangements. In pancreatic ductal adenocarcinoma and triple-negative breast cancer, MMEJ-driven ecDNA formation has been characterized, often at chromatin loop anchors and replication fork-stalling regions. These repair pathways are often activated under replication stress or in cells lacking proficient homologous recombination (HR), such as BRCA1/2-deficient tumors [12].

A particularly catastrophic mechanism contributing to ecDNA generation is chromothripsis, wherein chromosomes shatter into dozens to hundreds of fragments and are haphazardly reassembled. Chromothripsis frequently arises from mitotic errors, the rupture of micronuclei, or exposure to ionizing radiation, and has a direct association with the formation of extrachromosomal DNA in medulloblastoma, glioblastoma, and prostate cancer. The comprehensive analysis of these tumors through whole-genome sequencing uncovers fluctuating copy number patterns and grouped breakpoints, frequently situated within the confines of ecDNA. These findings suggest that post-shattering DNA fragments can circularize independently, especially when shielded from canonical DNA damage checkpoints. Collectively, these mechanisms underscore the complex interplay of repair failure and chromosomal chaos in generating ecDNA and promoting oncogenic amplification [13].

2.3.2 Replication Stress and DNA Damage Response

Replication stress is a critical mechanism that promotes the generation of ecDNA in cancer cells by causing the delay and eventual collapse of replication forks under oncogenic pressure. Overexpression of oncogenes such as MYC, CCNE1, and ERBB2 accelerates DNA replication, leading to fork asymmetry and DNA breaks, particularly when nucleotide reserves or origin licensing are insufficient. Replication stress is associated with increased levels of γH2AX foci at fragile sites and early replicating fragile regions (ERFS) in HeLa, U2OS, and BT-474 breast cancer cells. These regions are also enriched for ecDNA hotspots. The ATR kinase serves as a crucial mediator, activated by the buildup of single-stranded DNA (ssDNA) at halted replication forks. ATR phosphorylates downstream effectors, such as CHK1 and RAD17, to stabilize replication machinery and prevent fork collapse. Suppression of ATR in U251 glioma cells intensifies replication-induced double-strand breaks and promotes the formation of acentric, circular DNA structures. The p53 pathway is similarly employed by the ATM kinase to regulate homologous recombination and cell cycle arrest upon activation by double-strand breaks (DSBs). Chromosomal fragmentation and circularization are significantly elevated in lymphomas that lack ATM, suggesting that ATM loss facilitates the emergence of ecDNA [14].

The tumor suppressor p53 acts as a genomic gatekeeper in response to genotoxic stress by modulating senescence, apoptosis, and DNA repair. Its inactivation, which is a specific characteristic of more than 50% of malignancies, establishes a favorable environment for the genesis of ecDNA by disabling checkpoint controls. In p53-null murine embryonic fibroblasts (MEFs) that are exposed to replication stressors such as aphidicolin or hydroxyurea, extrachromosomal DNA circles have been observed. These circles are frequently derived from common vulnerable sites such as FRA3B and FRA16D. Such circles are a significant accumulation. Moreover, structure-specific nucleases, including MUS81 and SLX1, frequently process unresolved replication intermediates, which in turn promote DSBs and recombination events. When these enzymes are dysregulated, they contribute to junctional instability and the release of DNA fragments into the cytoplasm or nuclear periphery, where they can circularize through end-joining. In ovarian cancer cells with high replication stress,

MUS81 knockdown reduces ecDNA frequency, confirming its mechanistic role. This has been associated with the production of ecDNA in lung cancer and melanoma when NHEJ-induced ligation of distant genomic loci results in circularized extrachromosomal fragments [15].

2.3.3 Role of Mobile Genetic Elements and Telomere Dysfunction

Furthermore, the production of extrachromosomal DNA (ecDNA) in cancer cells is influenced by the activity of mobile genetic elements (MGEs), particularly retrotransposons, which induce genomic instability and rearrangements that promote circular DNA origination. LINE-1 (Long Interspersed Nuclear Element-1), a major class of autonomous retrotransposons, plays a pivotal role in promoting insertional mutagenesis and DNA double-strand breaks (DSBs). LINE-1 elements are transcriptionally reactivated in a variety of malignancies, such as lung adenocarcinoma, hepatocellular carcinoma, and triple-negative breast cancer, frequently as a result of hypomethylation in their promoter regions. When active, LINE-1 endonuclease and reverse transcriptase often guide DNA insertion into unsuspected genomic sites, resulting in chromosomal fragmentation and subsequent elimination of circular DNA components using target site duplications, deletions, or rearrangements [16]. In HeLa and HCT116 colorectal cancer cells, enhanced LINE-1 activity has been interlinked with the generation of circular DNA intermediates harboring oncogenic loci such as KRAS, CCND1, and TERT. These fragments are frequently flanked by short homologous sequences, exposing microhomology-mediated circularization facilitated by LINE-1-driven breaks. The sequencing of ecDNA in glioblastoma revealed that numerous breakpoints coincided with regions abundant in retrotransposons, thereby reinforcing their potential role in disrupting chromatin architecture and facilitating biogenesis [17]. Telomere collapse is an additional critical mechanism for the production of ecDNA, in addition to transposon activity. Telomeres, which are composed of repetitive TAG sequences and affiliated shelterin proteins (e.g., TRF1, TRF2, POT1), protect chromosome ends from degradation and end-to-end fusion. Telomere shortening, which is frequently the result of replicative stress and insufficient telomerase activity, is the primary cause of chromosome uncapping and fusion. This often sets off breakage–fusion–bridge cycles. These cycles, discussed in Section 3.1, inherently predispose the genome to ecDNA formation. Extrachromosomal telomeric circles (t-circles) and other ecDNA elements are highly abundant in ALT (Alternative Lengthening of Telomeres)-positive cancer cells, including the U2OS and SAOS-2 osteosarcoma lines. This strongly suggests a connection between defective telomere maintenance and the formation of circular DNA. Additionally, in the absence of adequate shelterin function or ATM/ATR surveillance, dicentric chromosomes are susceptible to mechanical rupture during mitosis as a consequence of telomere-telomere fusions. The acentric chromosomal fragments that result are frequently excised and ligated into extrachromosomal circles. Telomere crisis has been documented to initiate chromothripsis-like events in pancreatic cancer and prostate cancer, where ecDNA-containing oncogenes such as AR and MYCN emerge following telomere-driven genomic rearrangements. In summary, the synergistic degradation of genomic integrity by telomere erosion and retrotransposon reactivation leads to fragmented DNA regions with substantial circularization potential. These routes highlight the dynamic and flexible character of cancer genomes in reaction to genotoxic stress and increase the diversity of ecDNA genesis mechanisms [18].

2.4 STRUCTURAL ORGANIZATION OF ECDNA

The circular topology of extrachromosomal DNA (ecDNA) in cancer cells distinguishes it from other DNA structures and has a significant impact on its function. ecDNA is a supercoiled, closed-loop molecule that does not contain centromeres or telomeres, allowing it to freely rotate. This reduces torsional stress during transcription and replication, in contrast to linear chromosomes. These DNA circles vary in size from 1 to 2 kb eccDNA fragments to megabase-scale ecDNAs harboring full oncogenes like MYC, EGFR, CDK4, and MDM2, as seen in glioblastoma multiforme (GBM), neuroblastoma, and NSCLC. The structural heterogeneity and topology of ecDNA have been verified by high-resolution instruments such as electron microscopy, Circle-Seq, and Nano-C.

The arbitrary nuclear localization of ecDNA, which is frequently located near transcriptionally active zones, has been demonstrated by optical mapping and FISH. This circular structure allows for rolling-circle replication and tandem gene amplification, rendering ecDNA resistant to chromosomally targeted tools like CRISPR-Cas9. ecDNAs act as dynamic genomic amplifiers, increasing intratumoral heterogeneity and resistance to treatment [19]. The chromatin architecture of ecDNA is also unique. The strong transcriptional potential is indicated epigenetically by the highly accessible chromatin in ecDNA regions that have active histone marks, particularly H3K27ac and H3K4me3. ecDNA locations are easier to reach than chromosomal loci, which facilitate the binding of transcription factors, including MYC, NF-κB, and AP-1, according to ATAC-seq studies on cancer lines like COLO320 and SNU16. Methods of chromosome conformation capture (Hi-C, 4C) have shown more enhancer-promoter rewiring, hence enabling super-enhancers to interact with oncogene promoters in both cis and trans orientations; MET-driven ecDNA in HCC827 cells offers a typical illustration. Moreover, multiple ecDNAs can aggregate into transcriptional "hubs" or condensates enriched in RNA polymerase II and coactivators, promoting transcriptional bursts, as shown in glioblastoma stem-like cells (Bafna & Mischel, 2022). These hubs, sensitive to BET inhibitors, represent focal points of oncogene regulation. The permissive chromatin state of ecDNA is also influenced by global hypomethylation, CTCF border loss, and disturbed nucleosome density. ecDNA functions independently of centromere-based processes in terms of replication and segregation. In cell lines such as U87MG and IMR-32, ecDNA replicates early in S-phase and may undergo multiple replication rounds per cycle. Repli-Seq and BrdU labeling reveals dynamic timing and rolling-circle mechanisms that amplify gene dosage. Nonetheless, the absence of centromeres results in the random segregation of ecDNA during mitosis, as demonstrated through live-cell imaging in PC3 and HeLa cells, which in turn leads to non-Mendelian inheritance and clonal diversity. Sullivan et al. (Nature, 2023) reported that ecDNA preferentially localizes to nuclear peripheries and micronuclei, perhaps promoting chromothripsis and ecDNA regeneration. Defects in the p53 or ATM/ATR pathways allow these segregation mistakes to persist, worsening genomic instability. Furthermore, ecDNA clusters recruit replication machineries such as PCNA and MCM complexes, indicating partial mitotic memory. Collectively, the circular architecture, open chromatin landscape, and erratic replication-segregation dynamics of ecDNA underpin its role as a key driver of tumor evolution, transcriptional plasticity, therapeutic resistance, and intratumoral heterogeneity—offering both mechanistic insights and novel therapeutic targets in oncology [20].

2.5 TECHNIQUES FOR DETECTION AND CHARACTERIZATION OF ECDNA

2.5.1 Cytogenetic Techniques

Cytogenetic approaches have historically provided the foundation for the initial discovery and visualization of extrachromosomal DNA (ecDNA), particularly in the form of double minutes (DMs). Fluorescence in situ hybridization (FISH) is among the most commonly employed techniques, allowing for the direct identification of specific ecDNA molecules within metaphase chromosome spreads or interphase nuclei. Metaphase FISH in various cancer cell lines, including COLO320 (colorectal carcinoma) and U87MG (glioblastoma), has demonstrated punctate, acentric signals indicative of DMs situated distally from chromosomal arms, hence proving their extrachromosomal characteristics. Coupling DAPI (4′,6-diamidino-2-phenylindole) staining with metaphase analysis allows for the identification of circular ecDNA lacking centromeres and kinetochores [21].

The co-localization of oncogene loci such as MYC or EGFR with ecDNA is enabled by metaphase spread analysis, which is frequently performed in conjunction with immuno-FISH. However, cytogenetic techniques are inherently limited by their resolution and are often unable to distinguish small ecDNA species (<100 kb), leading to underestimation of their abundance in tumor samples. In addition, sample preparation artifacts—such as DNA fragmentation, poor metaphase quality, or hybridization failure can obscure or distort ecDNA detection. DMs may also be misclassified due to nuclear blebbing or missegregation artifacts, particularly in rapidly dividing tumor cells like HeLa or PC3 [22].

Cytogenetic analyses appear to be valuable for determining the presence of clinical ecDNA, particularly in drug-resistant tumors where DM accumulation has been associated with chemoresistance, as demonstrated in ovarian carcinoma (SKOV3) and neuroblastoma (SK-N-BE(2)), despite these limitations. Recent modifications that employ 3D-FISH and RNA-FISH to enhance spatial resolution have facilitated the investigation of transcriptional activity on ecDNA in situ [23].

2.5.2 Molecular and Imaging-Based Techniques

The structural characterization and high-throughput detection of ecDNA have been transformed by molecular techniques. Circle-Seq is a certain method for selectively enriching circular DNA that uses exonuclease to cut linear DNA and rolling circle amplification to profile ecDNA across cancer genomes. Applied in cancers such as glioblastoma, lung adenocarcinoma, and triple-negative breast cancer (MDA-MB-231), Circle-Seq has identified oncogene-containing ecDNAs, including EGFR, MYC, and CCND1. However, it may overlook large circular amplicons due to size bias in amplification.

By mapping focal amplifications and circular architecture, AmpliconArchitect, a computational pipeline, reconstructs ecDNA structures using short-read whole-genome sequencing. It has proven especially powerful in characterizing ecDNA in neuroblastoma and esophageal carcinoma, identifying rearrangement junctions and structural variants. For greater resolution, long-read sequencing technologies like PacBio SMRT and Oxford Nanopore provide complete ecDNA structure mapping, capturing complex head-to-tail repeats and inversion junctions missed by short reads.

To gain a better understanding of the structure of high-molecular-weight ecDNA in primary tumor samples, optical mapping (e.g., with the use of Bionano Genomics platforms) uses fluorescent labeling and nanochannel stretching to visualize big ecDNA molecules. Hi-C (chromosome conformation capture) analysis enhances this by mapping ecDNA-chromatin interactions. In glioblastoma, Hi-C showed that ecDNA localizes to transcriptionally active nuclear compartments and creates enhancer-promoter clusters with strong transcriptional output [24]. On the imaging front, super-resolution microscopy, including STORM (Stochastic Optical Reconstruction Microscopy) and SIM (Structured Illumination Microscopy), allows nanoscale visualization of ecDNA organization and clustering within the nucleus. Live-cell imaging using CRISPR-based tagging and fluorescent ecDNA probes has uncovered ecDNA mobility and mitotic inheritance dynamics in real time, especially in models like HeLa and A549 cells [25]. Table 2.1 describes overview of approaches for ecDNA identification and profiling [26–33].

TABLE 2.1
Summary of Techniques for ecDNA Detection and Characterization

Technique	Target Feature	Resolution/Advantage	Notable Use Cases	References
FISH	Localization of ecDNA	Single-cell, locus-specific	MYC, EGFR ecDNA in GBM, colon CA	[26]
Circle-Seq	Circular DNA enrichment	Genome-wide, circular-specific	Lung, breast, glioblastoma	[27]
AmpliconArchitect	Structure reconstruction	Detects rearranged ecDNA amplicons	Neuroblastoma, esophageal CA	[28]
Long-Read Sequencing	Full ecDNA architecture	Maps large structural variants	GBM, prostate CA	[29]
Optical Mapping	Physical structure mapping	High-resolution large fragment detection	Breast and prostate tumors	[30]
Hi-C	Chromatin interaction	Maps 3D organization and enhancer hubs	GBM, medulloblastoma	[31]
Super-Resolution Microscopy	Spatial ecDNA clustering	Visualizes nanoscale localization	HeLa, PC3 cells	[32]
Live-Cell Imaging	Dynamics of ecDNA	Real-time tracking of segregation	A549, U2OS cells	[33]

2.5.3 Imaging Approaches

It has become essential to study the spatial and temporal dynamics of extrachromosomal DNA (ecDNA) within cancer cells using advanced imaging techniques. The latest advances in super-resolution imaging and live-cell microscopy have significantly heightened our capacity to observe and study the architecture and behavior of ecDNA in real time, overcoming the limitations of traditional microscopy methods in terms of resolution and specificity.

Super-resolution microscopy, which comprises Stochastic Optical Reconstruction Microscopy (STORM) and Structured Illumination Microscopy (SIM), has enabled the imaging of ecDNA at the nanoscale. These methods surpass the diffraction limit of conventional fluorescence microscopy, providing resolutions as low as approximately 20 nm.

STORM has identified the presence of ecDNA clusters, also known as ecDNA hubs, in glioblastoma stem-like cells (GSCs) and lung adenocarcinoma lines (A549, H1975). RNA polymerase II, BRD4, and H3K27ac are among the active transcriptional indicators that co-localize with these clusters. These hubs are thought to improve oncogene expression from ecDNA by facilitating enhancer-promoter interactions in trans [34].

Developments in CRISPR/dCas9-based fluorescent tagging systems have made live-cell imaging possible, allowing for tracking of ecDNA dynamics across the cell cycle. During mitosis in HeLa, U2OS, and PC3 cells, researchers have observed ecDNA migrations and segregation using dCas9-EGFP fusion proteins that target EGFR or MYC areas. These studies have demonstrated asymmetric ecDNA segregation, contributing to intratumoral heterogeneity and therapy resistance. ecDNAs are extremely mobile in the interphase nucleus and tend to accumulate in transcriptionally permissive locations such as nuclear speckles and euchromatic compartments, according to live imaging.

Furthermore, 3D confocal microscopy combined with FISH has enabled volumetric quantification of ecDNA, showing that ecDNA often exists as spatially distinct clusters within nuclei, separated from chromosomal DNA. These clusters exhibit increased accessibility to transcription factors and chromatin remodelers, such as p300, CBP, and SWI/SNF complexes. Research on triple-negative breast cancer (TNBC) and medulloblastoma has demonstrated that increased transcriptional output, particularly of amplified oncogenes, is correlated with the proximity of ecDNA hubs to active chromatin areas [35].

2.6 FUNCTIONAL IMPLICATIONS OF ECDNA STRUCTURE IN CANCER BIOLOGY

2.6.1 Oncogene Amplification and Expression

A prominent functional implication of ecDNA in cancer biology is its ability to amplify and hyperactivate oncogenes. Numerous prominent oncogenes, such as MYC, EGFR, MDM2, and CDK4, have been consistently detected on ecDNAs in various cancers. For example, in glioblastoma multiforme (GBM), EGFR is frequently found on circular ecDNA molecules, correlating with aggressive tumor behavior and resistance to EGFR inhibitors. Similarly, in neuroblastoma, medulloblastoma, and colorectal cancer, MYC-carrying ecDNAs increase proliferation and genomic instability [36].

A major distinction between ecDNA and chromosomal DNA is the level of transcriptional activity. A study by Wu et al. (2019) using ATAC-seq and RNA-seq in COLO320DM and U87 glioma cells revealed that oncogenes located on ecDNAs exhibit up to 40-fold higher expression levels compared to the same genes on linear chromosomal loci. This elevated transcription is partly due to the open chromatin architecture of ecDNA and its ability to form transcriptional hubs by spatial clustering with enhancers, as shown by Hi-C and super-resolution microscopy in HCT116 colon cancer cells [37].

Additionally, enhancer hijacking is facilitated by ecDNA, which occurs when distal super-enhancers are repositioned near oncogenes on circular DNA. The CDK4 ecDNA identified in liposarcoma possesses active enhancers absent in the native genomic context, which results in significant enhancement of gene expression. Further, the absence of centromeres and telomeres in

TABLE 2.2
Key Oncogenes and Their ecDNA Associations across Various Cancer Types

Oncogene	Cancer Type	ecDNA Role	Reference Model	References
MYC	Neuroblastoma, CRC	High-level amplification	COLO320DM, IMR-32	[38]
EGFR	Glioblastoma	Increased expression, therapy resistance	U87, GBM PDX models	[39]
CDK4	Dedifferentiated Liposarcoma	Enhancer rewiring, aggressive growth	LPS141, DL221	[40]
MDM2	Sarcoma, Glioma	p53 inhibition via overexpression	SF763, SJSA1	[41]

ecDNAs facilitates their evasion of mitotic checkpoints, hence allowing the fast amplification of oncogene copies in reaction to growth stimuli or stressors. Table 2.2 describes oncogene amplification through ecDNA in cancers [38–41].

2.6.2 Intratumoral Heterogeneity and Clonal Evolution

The depletion of centromeric sequences in ecDNA results in random segregation during mitosis, which leads to non-Mendelian inheritance. Live-cell imaging studies in U2OS and HeLa cells using fluorescent dCas9-tagged probes proved that ecDNAs segregate unevenly between daughter cells, forming micronuclei or persisting in the cytoplasm randomly, contributing to clonal variability. Such unequal distribution promotes cancer plasticity, wherein ecDNA-enriched subclones can dominate under selective pressures such as chemotherapy or targeted therapy. In EGFR-amplified gliomas, ecDNA-positive subclones exhibit enhanced proliferation and treatment resistance, as shown by ex vivo cultures and single-cell RNA-seq analyses. Notably, an investigation by Nathanson et al. (2014) found that anti-EGFR therapy reduced ecDNA-positive clones, but their re-emergence post-treatment indicated reversible adaptive resistance.

This plasticity advances clonal evolution, a Darwinian process allowing cancer cells to dynamically modulate oncogene levels based on microenvironmental cues. Hypoxia amplifies MYC-containing ecDNAs in medulloblastoma, enabling tumor development and adaptability. ecDNA-driven transcriptional heterogeneity provides phenotypic variety and allows a few resistant clones to survive cytotoxic therapy, contributing to relapse and metastasis [42]. Single-cell sequencing approaches like scDNA-seq and scATAC-seq have revealed significant variation in ecDNA content across individual tumor cells in triple-negative breast cancer (TNBC), indicating that ecDNA distribution is not only uneven but also dynamically regulated. These insights establish ecDNA as a potent contributor to tumor evolution, therapeutic failure, and relapse [43].

2.6.3 Immune Evasion and Stress Responses

In addition to its role in transcriptional regulation, ecDNA plays a significant part in immune modulation and the adaptation of tumor cells to stress. In instances of fragmentation of ecDNA, its localization to the cytoplasm can occur, subsequently leading to the activation of innate immune pathways. The cGAS-STING pathway identifies cytosolic double-stranded DNA (dsDNA) as a danger-associated molecular pattern (DAMP). In lung cancer (A549) and melanoma (B16-F10) models, the build-up of cytosolic ecDNA activates cGAS, resulting in STING-dependent interferon responses. However, many tumors actively suppress this response. Mutations in STING or loss of cGAS expression—observed in pancreatic cancer, glioma, and colorectal carcinoma—blunt immune activation, enabling immune evasion. Additionally, studies on KRAS-mutant NSCLC show that ecDNA-induced replication stress triggers ATR/Chk1, which helps cells respond to DNA damage and simultaneously increases the expression of immunological checkpoints such as PD-L1, which helps cells evade the immune system. Accumulation of ecDNA enhances the effectiveness of pharmacological STING agonists in boosting immune responses in immune-competent mouse models, such as the MC38 colon cancer model, highlighting a potential therapeutic strategy. Recent research reveals that ecDNA-containing micronuclei break during mitotic stress, release DNA into the cytoplasm, and activate inflammasome pathways like AIM2, creating inflammatory but tumor-promoting microenvironments. Thus, ecDNA amplifies genes and affects innate immunity, replication stress, and immunogenic cell death. Combinatorial cancer immunotherapy targets it for its immune recognition and evasion functions [44].

2.7 CLINICAL AND THERAPEUTIC PERSPECTIVES

2.7.1 Diagnostic and Prognostic Value of ecDNA

The detection and characterization of extrachromosomal circular DNA (ecDNA) have profound clinical implications in cancer diagnostics and prognostics. A promising non-invasive biomarker that can be obtained through liquid biopsy is circulating ecDNA (cDNA), which is released into the bloodstream by apoptosis, necrosis, or active secretion. In contrast to linear circulating tumor DNA (ctDNA), CeDNA maintains a covalently closed circular structure that enhances stability in circulation and resists exonucleolytic degradation.

According to recent research that employed plasma DNA extraction techniques and Circle-Seq, CeDNA is abundant in patients with highly aggressive malignancies, including glioblastoma, non-small cell lung cancer (NSCLC), and triple-negative breast cancer (TNBC). For example, Kim et al. (2020) found that the plasma of GBM patients contained EGFR- and MYC-amplified ecDNA, which was associated with tumor burden, disease progression, and poor overall survival (OS). Furthermore, metastatic prostate cancer research has demonstrated a robust correlation between androgen-independent proliferation, castration resistance, and elevated levels of AR-driven ecDNA in cDNA. The predictive usefulness of ecDNA goes beyond detection. Quantitative evaluations using nanopore sequencing and digital droplet PCR (ddPCR) linked high ecDNA copy counts to early metastases, drug resistance, and clonal fitness. Trastuzumab resistance in breast cancer is revealed by HER2 amplification in cDNA, which assists clinicians in the development of treatment plans.

Furthermore, ATAC-seq and ChIP-seq have identified active histone marks (e.g., H3K27ac, H3K4me1) linked with enhancer hijacking, which serve as indicators of transcriptionally active ecDNA hubs. These epigenetic patterns are now being explored for disease stratification and early relapse prediction. CeDNA collectively provides a dynamic, tumor-specific representation of oncogenic activity, tumor heterogeneity, and treatment response. The incorporation of this as a real-time diagnostic inside clinical workflows could transform early detection, monitoring, and personalized therapy planning for ecDNA-driven malignancies [45].

2.7.2 Targeting ecDNA-Driven Cancers

The objective of therapeutic strategies that target ecDNA-driven malignancies is to capitalize on the distinctive structural, transcriptional, and replication characteristics of ecDNA. ecDNA undergoes rapid, unregulated replication, often stimulated from replication origins with high chromatin accessibility, leading to replication stress and activation of the ATR-Chk1 pathway. ATR inhibitors (e.g., VE-821, AZD6738) decrease ecDNA replication in preclinical models, including U87MG glioma and SK-MEL-28 melanoma cells, hence boosting cell death and preferentially decreasing oncogene amplification. Additionally, the epigenetic landscape of ecDNA provides therapeutic vulnerabilities. ecDNA lacks canonical chromosomal boundaries, enabling enhancer rewiring and transcriptional hyperactivation. In ecDNA-rich neuroblastoma and colorectal cancer xenografts, histone deacetylase inhibitors (HDACi) and bromodomain inhibitors (BETi) degrade this structure, resulting in reduced MYC expression and tumor growth. In COLO320DM cells, BET inhibitor JQ1 down-regulates ecDNA-associated MYC but not chromosomal MYC. Emerging therapies include CRISPR/Cas9-based ecDNA disruption. Guide RNAs (gRNAs) can be engineered to specifically target ecDNA without disrupting the native genome, as ecDNA molecules frequently have unique junction sequences. In patient-derived xenografts (PDXs) of glioblastoma, CRISPR-mediated cleavage of EGFR-ecDNA significantly impairs proliferation. Furthermore, recent work has demonstrated targeted degradation of ecDNA using exonucleases, such as TREX1, introduced via liposomal nanoparticles, leading to DNA damage responses and apoptosis in cancer cells. Another potential approach is to disrupt the segregation of ecDNA. ecDNA maintenance can be influenced by the modification of chromatin tethering proteins (such as CTCF and BRD4), as ecDNAs segregate asymmetrically as a result of their lack of centromeres. Small molecule cohesin complex inhibitors prevent oncogene spread by reducing ecDNA clustering and propagation. These preliminary methods demonstrate the translational potential of ecDNA targeting, which may transform a genetic abnormality that is often disregarded into a precision oncology focus for a range of aggressive and resistant malignancies [46].

2.8 FUTURE DIRECTIONS AND CHALLENGES

Despite the substantial progress made in the study of extrachromosomal DNA (ecDNA), there are still significant concerns regarding its biosynthesis, control, and tissue-specific emergence. The selective exaggeration of oncogenes such as MYC, EGFR, and CDK4 on ecDNA suggests that few genomic loci exhibit intrinsic vulnerability to excision and circularization, likely affected by chromatin accessibility, timing of replication, and genomic fragile sites like FRA3B and FRA16D [47]. However, the precise molecular determinants governing this selectivity remain undefined. In contrast to chromosomal DNA, ecDNA replicates independently and asynchronously, with Repli-seq and DNA fiber assays indicating the involvement of host replication components, including ORC1, CDC45, and the MCM complex. The integration of ecDNAs into conventional replication factories or their interference with chromosomal DNA synthesis raises mechanistic problems with therapeutic implications, especially in the context of exploiting synthetic lethality. The tissue-specific prevalence of ecDNA, as seen in glioblastoma, lung adenocarcinoma, and neuroblastoma, is further underscored by TCGA and PCAWG data indicating high ecDNA abundance in brain and adrenal tumors, potentially due to intrinsic properties like neural progenitor plasticity or extrinsic stressors such as hypoxia [48]. Concerns have also been raised about the changing nature of ecDNA, such as reintegration events through chromothripsis, and how they might cause mutations and contribute to the development of tumors. However, real-time tracking remains limited [24]. Although super-resolution microscopy (e.g., STORM, SIM) and CRISPR-dCas9 imaging have enabled visualization, their resolution, phototoxicity, and multiplexing limitations hinder full elucidation of ecDNA dynamics during mitosis and clustering. Further obstacles are presented by sequencing technologies: short-read methods, such as Illumina, are unable to address junctional complexity, while long-read platforms (Oxford Nanopore, PacBio HiFi) are hindered by scalability and error rates. AmpliconArchitect and Circle-Map are computational tools that provide insights; however, they currently lack standardization across cohorts. Integrating multi-omics data (ATAC-seq, Hi-C, ChIP-seq, RNA-seq) is essential for elucidating the regulatory landscape of ecDNA; nevertheless, disparities in depth and resolution hinder interpretability. Moreover, current diagnostic methods like DAPI and FISH lack sensitivity and specificity for small or low-abundance ecDNA species, and therapeutic manipulation via nucleases or replication inhibitors is hampered by off-target effects. The paradigm of ecDNA research is being transformed by new technologies to address these limitations. AI-driven strategies, such as DeepCircle and ecNet, improve the accuracy of ecDNA identification from NGS data. The fabrication of synthetic ecDNAs in HEK293, U2OS, and NIH3T3 cells, enabled by CRISPR/Cas9 and recombinase systems, allows for a more controlled analysis of enhancer usurpation and replication stress. Dynamic spatiotemporal monitoring is made possible by live-cell CRISPR imaging, and high-resolution systems such as Bionano Genomics and DNA curtains enable single-molecule mapping of ecDNA topologies. Collectively, these advances are crucial for decoding ecDNA biology, resolving its oncogenic mechanisms, and leveraging its unique features for precision diagnostics and targeted therapeutics [21].

2.9 CONCLUSIONS AND TRANSLATIONAL SIGNIFICANCE

A key component of cancer biology, ecDNA has redefined our knowledge of tumor heterogeneity, adaptability, and progression. ecDNA, in contrast to chromosomal DNA, is found outside of the linear chromosomal framework and frequently takes the form of circular, acentromeric, and telomeric components that range in size from 1 kb to several Mb [49]. These structures predominantly originate from double-strand breaks (DSBs) at genomic fragile sites and are processed through mechanisms, including microhomology-mediated end joining (MMEJ) and breakage–fusion–bridge (BFB) cycles, as evidenced by genomic reconstructions from AmpliconArchitect and Circle-Map analyses. ecDNA formation is further promoted by replication stress, aberrant mitosis, and chromothripsis—a catastrophic chromosomal shattering event, commonly observed in high-grade gliomas and triple-negative breast cancers. ecDNA structurally contains amplified oncogenes, including EGFR, MYC, CDK4, and PDGFRA, as well as active enhancers and replication origins, allowing it to operate as an independent transcriptional and replicative entity [12]. Its open chromatin state—marked by H3K27ac, H3K4me1, and BRD4 occupancy—facilitates the formation of neo-TADs (topologically associating domains) and enhancer hijacking, dramatically upregulating oncogene expression independent of the host chromosomal architecture [50]. These properties contribute to tumor aggressiveness, therapy resistance, and poor prognosis, as confirmed in multi-cohort studies across glioblastoma (GBM), esophageal carcinoma, and non-small cell lung cancer (NSCLC). By facilitating intra-tumoral heterogeneity via clonal development, stress adaptability, and uneven segregation during mitosis, ecDNA functions as an oncogenic driver and genomic destabilizer. However, this very feature also renders ecDNA a therapeutic vulnerability [51]. The absence of centromeres leads to mitotic missegregation, allowing exploitation through agents targeting replication stress (e.g., ATR inhibitors) and epigenetic readers (e.g., BET inhibitors) that preferentially affect ecDNA-bearing cells. Furthermore, CRISPR-Cas systems that focus on ecDNA-specific junctions, along with exonuclease-mediated degradation approaches, demonstrate significant effectiveness in the targeted removal of ecDNA, while preserving chromosomal DNA [22].

In light of these multifaceted roles, ecDNA represents not just a biomarker but a cornerstone in precision oncology. Its detection through liquid biopsy (e.g., cfDNA profiling and FISH-based methods), its transcriptional activity via single-cell RNA-seq, and its spatial organization using 3D genome mapping (Hi-C) are progressively being integrated into clinical diagnostics. The integration of ecDNA profiling into therapeutic decision-making will facilitate the real-time monitoring of tumor evolution, the identification of resistance mechanisms, and the rational selection of targeted therapies as technological advancements and mechanistic discoveries converge. Finally, ecDNA is both an oncogenesis engine and a possible Achilles' heel. A new paradigm for cancer diagnosis, stratification, and individualized treatment is revealed. Precision oncology requires full integration of ecDNA biology into translational research and clinical procedures.

REFERENCES

1. Curtis EJ, Rose JC, Mischel PS, et al. (2024) Extrachromosomal DNA: Biogenesis and functions in cancer. Annu Rev Cancer Biol 8:135–153.
2. Chen Y, Qiu Q, She J, et al. (2023) Extrachromosomal circular DNA in colorectal cancer: Biogenesis, function and potential as a therapeutic target. Oncogene 42:941–951.
3. Dhar R, Kumar A, Karmakar S (2024) Extrachromosomal DNA: New players in oncology. Asian J Med Sci 15:1–2.

4. Pecorino LT, Verhaak RGW, Henssen A, et al. (2022) Extrachromosomal DNA (ecDNA): An origin of tumor heterogeneity, genomic remodeling, and drug resistance. Biochem Soc Trans 50:1911–1920.
5. Pecorino LT, Verhaak RGW, Henssen A, et al. (2022) Extrachromosomal DNA (ecDNA): An origin of tumor heterogeneity, genomic remodeling, and drug resistance. Biochem Soc Trans 50:1911–1920.
6. Dong Y, He Q, Chen X, et al. (2023) Extrachromosomal DNA (ecDNA) in cancer: Mechanisms, functions, and clinical implications. Front Oncol. https://doi.org/10.3389/fonc.2023.1194405
7. Yang Q-L, Xie Y, Qiao K, et al. (2025) Modern biology of extrachromosomal DNA: A decade-long voyage of discovery. Cell Res 35:11–22.
8. Hung KL, Mischel PS, Chang HY (2022) Gene regulation on extrachromosomal DNA. Nat Struct Mol Biol 29:736–744.
9. Kumar V, Stewart JH (2023) Immunometabolic reprogramming, another cancer hallmark. Front Immunol. https://doi.org/10.3389/fimmu.2023.1125874
10. Kim H, Nguyen N-P, Turner K, et al. (2020) Extrachromosomal DNA is associated with oncogene amplification and poor outcome across multiple cancers. Nat Genet 52:891–897.
11. Li Z, Wang B, Liang H, Han L (2022) Pioneering insights of extrachromosomal DNA (ecDNA) generation, action and its implications for cancer therapy. Int J Biol Sci 18:4006–4025.
12. Wu S, Bafna V, Mischel PS (2021) Extrachromosomal DNA (ecDNA) in cancer pathogenesis. Curr Opin Genet Dev 66:78–82.
13. Lange JT, Rose JC, Chen CY, et al. (2022) The evolutionary dynamics of extrachromosomal DNA in human cancers. Nat Genet 54:1527–1533.
14. Wu S, Bafna V, Chang HY, et al. (2022) Extrachromosomal DNA: An emerging hallmark in human cancer. Annual Review of Pathology: Mechanisms of Disease 17:367–386.
15. Saxena S, Zou L (2022) Hallmarks of DNA replication stress. Mol Cell 82:2298–2314.
16. Ragu S, Matos-Rodrigues G, Lopez BS (2020) Replication stress, DNA damage, inflammatory cytokines and innate immune response. Genes (Basel) 11:409.
17. Tanaka H, Abe S, Huda N, et al. (2012) Telomere fusions in early human breast carcinoma. Proceedings of the National Academy of Sciences 109:14098–14103.
18. Yoon I, Kim U, Jung KO, et al. (2024) 3C methods in cancer research: Recent advances and prospects. Exp Mol Med 56:788–798.
19. Cybulla E, Vindigni A (2023) Leveraging the replication stress response to optimize cancer therapy. Nat Rev Cancer 23:6–24.
20. Bafna V, Mischel PS (2022) Extrachromosomal DNA in cancer. Annu Rev Genomics Hum Genet 23:29–52.
21. Song Y, Guan C, Zhang Y, et al. (2025) A novel CRISPR-Cas9 nickase-mediated rolling circle amplification (CRIRCA) technique for gene identification and quantitative analysis of extrachromosomal DNA. J Adv Res. https://doi.org/10.1016/j.jare.2025.04.031
22. Noorani I, Mischel PS, Swanton C (2022) Leveraging extrachromosomal DNA to fine-tune trials of targeted therapy for glioblastoma: Opportunities and challenges. Nat Rev Clin Oncol 19:733–743.
23. Irdianto SA, Fadhillah F, Lestari R, et al. (2025) Extrachromosomal DNA in breast cancer cell lines: Detection and characterization. Microsc Res Tech 88:1239–1247.
24. Yoon HY, Koo H, Kim K, et al. (2017) Molecular imaging based on metabolic glycoengineering and bioorthogonal click chemistry. Biomaterials 132:28–36.
25. Rao A, Barkley D, França GS, et al. (2021) Exploring tissue architecture using spatial transcriptomics. Nature 596:211–220.
26. Hu D, Cui Y, Markillie LM, et al. (2021) Counting mRNA copies in intact bacterial cells by fluctuation localization imaging-based fluorescence. In Situ Hybridization (fliFISH). pp 237–247.
27. Yoon HY, Koo H, Kim K, et al. (2017) Molecular imaging based on metabolic glycoengineering and bioorthogonal click chemistry. Biomaterials 132:28–36.
28. Tian M, He X, Jin C, et al. (2021) Transpathology: Molecular imaging-based pathology. Eur J Nucl Med Mol Imaging 48:2338–2350.
29. Petukhov V, Xu RJ, Soldatov RA, et al. (2022) Cell segmentation in imaging-based spatial transcriptomics. Nat Biotechnol 40:345–354.
30. Kesim C, Bektas SN, Kulali Z, et al. (2022) HENLE FIBER LAYER MAPPING WITH DIRECTIONAL OPTICAL COHERENCE TOMOGRAPHY. Retina 42:1780–1787.
31. Marcolungo L, Vincenzi L, Ballottari M, et al. (2023) Structural refinement by direct mapping reveals assembly inconsistencies near Hi-C junctions. Plants 12:320.
32. Yang Z, Samanta S, Yan W, et al. (2021) Super-Resolution Microscopy for Biological Imaging. pp 23–43.
33. Shroff H, Testa I, Jug F, et al. (2024) Live-cell imaging powered by computation. Nat Rev Mol Cell Biol 25:443–463.

34. Liu S, Hoess P, Ries J (2022) Super-resolution microscopy for structural cell biology. Annu Rev Biophys 51:301–326.
35. Collings DA (2015) Optimisation approaches for concurrent transmitted light imaging during confocal microscopy. Plant Methods 11:40.
36. Press MF (1990) Oncogene amplification and expression: Importance of methodologic considerations. Am J Clin Pathol 94:240–241.
37. Hung KL, Yost KE, Xie L, et al. (2021) ecDNA hubs drive cooperative intermolecular oncogene expression. Nature 600:731–736.
38. Dhanasekaran R, Deutzmann A, Mahauad-Fernandez WD, et al. (2022) The MYC oncogene—the grand orchestrator of cancer growth and immune evasion. Nat Rev Clin Oncol 19:23–36.
39. Guo G, Gong K, Beckley N, et al. (2022) EGFR ligand shifts the role of EGFR from oncogene to tumor suppressor in EGFR-amplified glioblastoma by suppressing invasion through BIN3 upregulation. Nat Cell Biol 24:1291–1305.
40. Ruscetti M, Morris JP, Mezzadra R, et al. (2020) Senescence-induced vascular remodeling creates therapeutic vulnerabilities in pancreas cancer. Cell 181:424–441.e21.
41. Haines DS (1997) The mdm2 Proto-Oncogene. Leuk Lymphoma 26:227–238.
42. Losic B, Craig AJ, Villacorta-Martin C, et al. (2020) Intratumoral heterogeneity and clonal evolution in liver cancer. Nat Commun 11:291.
43. McGranahan N, Swanton C (2017) Clonal heterogeneity and tumor evolution: Past, present, and the future. Cell 168:613–628.
44. Sun Y-F, Wu L, Liu S-P, et al. (2021) Dissecting spatial heterogeneity and the immune-evasion mechanism of CTCs by single-cell RNA-seq in hepatocellular carcinoma. Nat Commun 12:4091.
45. Lavik KO, Veseth M, Frøysa H, et al. (2023) This is what I need a clinical feedback system to do for me: A qualitative inquiry into perspectives of adolescents and their therapists. J Clin Psychol 79:729–747.
46. Pan H, Ho SE, Xue C, et al. (2024) Atherosclerosis is a smooth muscle cell-driven tumor-like disease. Circulation 149:1885–1898.
47. Liao Z, Jiang W, Ye L, et al. (2020) Classification of extrachromosomal circular DNA with a focus on the role of extrachromosomal DNA (ecDNA) in tumor heterogeneity and progression. Biochimica et Biophysica Acta (BBA)—Reviews on Cancer 1874:188392.
48. Liao Z, Jiang W, Ye L, et al. (2020) Classification of extrachromosomal circular DNA with a focus on the role of extrachromosomal DNA (ecDNA) in tumor heterogeneity and progression. Biochimica et Biophysica Acta (BBA)—Reviews on Cancer 1874:188392.
49. Wu T, Wu C, Zhao X, et al. (2022) Extrachromosomal DNA formation enables tumor immune escape potentially through regulating antigen presentation gene expression. Sci Rep 12:3590.
50. Irdianto SA, Fadhillah F, Lestari R, et al. (2025) Colcemid duration effects on extrachromosomal DNA analysis in breast cancer cell line. Tissue Cell 96:102952.
51. Irdianto SA, Fadhillah F, Lestari R, et al. (2025) Colcemid duration effects on extrachromosomal DNA analysis in breast cancer cell line. Tissue Cell 96:102952.

3 ecDNA in Oncogene Amplification and Tumor Progression

Soukat Chandra,[1] *Souvik Biswas,*[1] *Nikunj Solanki,*[2] *Sonal Solanki,*[3] *and S. Mohana Lakshmi*[4]

1 Department of Pharmaceutical Technology
Bharat Technology
Uluberia, India
2 Department of Pharmaceutics
HSBPVT's, Group of Institutions, Faculty of Pharmacy
Kashti, Ta-Shrigonda, India
3 Department of Pharmaceutics
HSBPVT's, Group of Institutions, Faculty of Pharmacy
Kashti, Ta-Shrigonda, India
4 Amity Institute of Pharmacy, Amity University Bengaluru
Devanahalli Taluk
Bengaluru, Karnataka

3.1 INTRODUCTION

3.1.1 Overview on Extrachromosomal DNA (ecDNA)

Extrachromosomal DNA (ecDNA) refers to circular DNA fragments that exist independently of chromosomes within a cell, often found in cancer cells. These fragments can contain genes and regulatory regions, contributing to tumor growth, drug resistance, and overall cancer evolution. ecDNA's ability to amplify oncogenes, drive intratumoral heterogeneity, and evade therapeutic interventions makes it a significant area of research in cancer biology (1). Basically, ecDNA is a circular fragment of genetic material that breaks away from chromosomes and floats freely in the cell nucleus (2). It is created by DNA damage, chromosomal rearrangements, or errors during cell replication. Commonly present in cancer cells. It is detected in 17% of tumor samples, with higher prevalence in liposarcomas, brain tumors, and breast cancers (1,2). In prokaryotes, nonviral extrachromosomal DNA is primarily found in plasmids, whereas in eukaryotes extrachromosomal DNA is primarily found in organelles. Mitochondrial DNA is a main source of this extrachromosomal DNA in eukaryotes. The fact that this organelle contains its own DNA supports the hypothesis that mitochondria originated as bacterial cells engulfed by ancestral eukaryotic cells. Extrachromosomal DNA is often used in research into replication because it is easy to identify and isolate (3). ecDNA is a distinct entity, and it is identified in the nuclei of cancer cells and has been shown to carry many copies of driver oncogenes. ecDNA is considered to be a primary mechanism of gene amplification, resulting in many copies of driver oncogenes and very aggressive cancers (4). ecDNA in the cytoplasm has been found to be structurally different from nuclear DNA. Cytoplasmic DNA is less methylated than DNA found within the nucleus. It is also confirmed that the sequences of cytoplasmic DNA were different from nuclear DNA in the same organism, showing that cytoplasmic DNAs

DOI: 10.1201/9781003655299-3

are not simply fragments of nuclear DNA. In cancer cells, ecDNA has been shown to be primarily isolated to the nucleus (3,4). In cancer biology, extrachromosomal DNA (ecDNA) has become a ground-breaking field that is challenging fundamental genetic theories and providing fresh perspectives on drug resistance and cancer progression (6). Since ecDNA is more open than chromosomal DNA, it has been found to be a significant carrier of amplified oncogenes, leading to enhanced transcriptional activity. Some tumor cells possess large copy numbers of oncogenes that can significantly affect tumor heterogeneity and medication resistance because ecDNA lacks a centromere and is inherited in a non-Mendelian manner during mitosis. Recent research employing cytogenetics and whole genome sequencing (WGS) techniques revealed that ecDNA is rarely detected in normal tissues, but it is present in a variety of cancer types. Additionally, research has shown that ecDNA may be linked to tumors that are more aggressive and have a poorer clinical prognosis (5).

3.1.2 ecDNA Involved in Cancer Progression

Extrachromosomal DNA (ecDNA), circular DNA found outside chromosomes, plays a significant role in cancer progression, particularly in driving treatment resistance and tumor heterogeneity. ecDNA is a common origin for amplified oncogenes in many human cancers and can be found in various malignant tumors. Its presence is associated with increased tumor heterogeneity, drug resistance, and a poorer prognosis (7). Oncogenes are genes that promote uncontrolled proliferation of cells and the development of cancer. ecDNA has the ability to amplify and overexpress these genes. Tumorigenesis can be amplified by high-level oncogenic transcription caused by the circular structure of ecDNA (6). Increased genetic heterogeneity within a tumor is caused by non-chromosomal inheritance of ecDNA and random segregation during cell division. Because of this heterogeneity, tumors can develop more resistance to treatment and are better able to adapt to various environments. For instance, in glioblastoma, ecDNA encourages treatment resistance. Studies have shown that ecDNAs are more common in metastatic cancers (5,6). As seen in Barrett's esophagus, ecDNA can exist even before cancer develops, indicating that it may be the cause of early tumorigenic processes. As cancers advance from their early to late stages, their ecDNA volume increases. It is highly probable that ecDNA plays a role in the development and ongoing spread of esophageal cancer, given its presence in precancerous tissue. ecDNA provides tumors with competitive advantages during cancer progression and metastasis (6,7).

3.2 GENETIC HETEROGENEITY, ECDNA, AND TUMOR PROGRESSION

When various genetic processes result in the same or similar phenotypes (observable characteristics), this is referred to as genetic heterogeneity. This basically means that multiple mutations in the same gene, or even different mutations in different genes, could result in the same disease or condition (8). One of the main causes of genetic heterogeneity in cancer is extrachromosomal DNA (ecDNA), especially when it contains oncogenes. This leads to tumor progression and resistance to treatment. This is because during cell division, ecDNAs can undergo unequal segregation, which can result in quick changes in the number of oncogene copies and gene expression. These changes may subsequently affect the behavior and phenotype of cells. Significant genetic heterogeneity within the tumor results from the unequal segregation of ecDNAs during cell division, which causes each daughter cell to contain a different number of copies of the ecDNA (8,9).

Extrachromosomal DNA (ecDNA) plays a critical role in driving intratumoral heterogeneity through several key mechanisms. One major factor is its uneven segregation during cell division, as ecDNA lacks centromeres and is randomly distributed to daughter cells. This results in varying oncogene copy numbers across different cells, giving rise to genetically distinct subpopulations. Moreover, ecDNA often harbors high copy numbers of oncogenes, leading to their overexpression. Cells with elevated ecDNA content may gain a proliferative advantage, further contributing to clonal diversity. Another key element is rapid genetic evolution, as ecDNA enables swift amplification or deletion of genes, promoting the emergence of therapy-resistant or aggressive tumor clones. In addition, genes located on ecDNA are frequently in a more open chromatin state, facilitating epigenetic modifications and greater transcriptional accessibility, which allows for dynamic regulation of gene expression. Under environmental or therapeutic stress, stress-induced ecDNA rearrangements can occur, including restructuring or fusion events that create new oncogenic combinations and foster the formation of novel, genetically distinct subclones. Intriguingly, horizontal gene transfer-like behavior has also been observed, with studies suggesting that ecDNA may occasionally transfer between cells or integrate into chromosomes, thereby spreading oncogenic traits and enhancing tumor heterogeneity (8,9,10). Together, these properties make ecDNA a powerful driver of cancer evolution and a challenging target in oncology.

3.3 ECDNA AND ONCOGENE EXPRESSION

Oncogene expression refers to the process by which these altered genes are transcribed and translated into proteins, leading to abnormal cellular behavior, such as uncontrolled cell proliferation and tumor formation (10).

3.3.1 High Copy Number Amplification of Oncogenes

Oncogene amplification on ecDNA has been found in a number of cancer types. For instance, a number of oncogenes, such as *MYC*, *MYCN*, *EGFR*, *PDGFRA*, *MET*, the *MECOM-PIK3CA-SOX2* gene cluster, and the *CDK4-MDM2* gene cluster, have been identified on ecDNA in a study of glioblastoma cells. In addition, 22 of the 25 ecDNAs detected were found to carry an oncogene in multiple tumor samples from different patients (10,11). ecDNA can enhance oncogene expression by increasing the number of copies through focal amplification. There is proof that ecDNA helped oncogenes be amplified focally in order to increase the number of copies. Oncogenes carried by ecDNA may account for the focused amplification in almost half of the WGS samples from tumor tissues and patient cell lines, according to a prior study. In a neuroblastoma cell line, ecDNA-related oncogenes are consistently associated with the majority of focal amplifications (85.7%) detected by whole genome sequencing (WGS). Haplotype phasing provided compelling evidence for significance of ecDNA, which is a source of focused amplification, by demonstrating that it is exclusively derived from the amplified *MYCN* allele. Moreover, higher oncogene expression has been associated with a higher copy number on ecDNA. For instance, analysis of allele-specific expression (ASE) in the same cell line confirmed that the increased gene expression was derived from ecDNA. Similarly, examination of clinical samples and cancer cell lines revealed that a high copy of ecDNA has been identified in cancer and that higher oncogene expression is associated with a higher copy number of ecDNA (11). Furthermore, there is evidence that ecDNA can enhance chromatin accessibility in order to increase the expression of oncogenes. When ecDNA gets released from chromatin, it exhibits significantly enhanced chromatin accessibility and does not have the higher-order compaction characteristic of chromosomes. A study that investigated the chromatin accessibility of ecDNA using *ATAC-seq* and *ATAC-see* found that there are noticeably more *ATAC-seq* peaks in all cell cycle phases than chromosomal DNA. According to a combined analysis of the RNA-seq results, the oncogene expression was significantly influenced by the enhanced accessibility. Oncogenes on ecDNA can express themselves more than oncogenes on linear chromosomes, even after controlling for copy quantity. This may be because chromatin is more accessible on ecDNA (12).

3.3.2 Key Oncogenes Associated with ecDNA

It has been shown that by local or distal enhancer hijacking, ecDNA can form ultra-long chromatin relations that can promote oncogene expression. According to the local enhancer hijacking model, oncogenes and local upstream enhancers have the ability to circularize and create ecDNA, which promotes enhancer-oncogene interaction to promote expression. For example, in glioblastoma samples from *TCGA*, *EGFR* and two active enhancers are co-amplified and preserved on circular ecDNA. According to *ChIP-Seq*, ecDNA synthesis may be able to bring together non-adjoining segments and facilitate communication between *EGFR* promoters and upstream enhancers that are initially separated by insulators. Subsequent analysis revealed that distal enhancer co-amplification is present in 30 out of 43 amplified oncogenes in different tumor types (13).

According to the distal enhancer hijacking model, oncogenes and distal enhancers on the same or different chromosomes circularize to produce ecDNA. *MYCN* and distal enhancer e4 are co-amplified in neuroblastoma cell lines, and chromatin capture analysis indicated that e4 interacted with the *MYCN* promoter to increase *MYCN* expression (12,13). Furthermore, as shown in the cases of *EGFR* in glioblastoma and *MYCN* in neuroblastoma, hijacking enhancers by creating ecDNA may alter cell survival by controlling oncogene expression (12). An essential cell cycle protein, CDK4, is frequently found amplified on extrachromosomal DNA (ecDNA) in various kinds of malignancies. This correlation indicates that ecDNA triggers CDK4 carcinogenesis and treatment resistance. Increased immune cell infiltration and tumor growth have been associated with CDK4 amplification, especially when it is associated with other oncogenes like *ERBB2* (13).

3.3.3 Transcriptional Regulation and Overexpression

In order to control gene expression, ecDNA may either produce an ecDNA hub or interact with chromosomal DNA. By interacting with chromosomal DNA, ecDNA acts as a mobile trans-acting component that enhances transcription. Two of the three cell lines obtained from glioblastoma patients tested positive for ecDNA, according to Hi-C analysis, and ecDNA interacted with chromosomal non-coding sequences, which are abundant in H3K27ac signals. Furthermore, transcriptional activity is associated with the interaction between ecDNA and chromatin, and chromosomal oncogene expression is elevated in ecDNA-negative cells by synthetic circular DNA that mimics ecDNA. Thus, the previously discussed information demonstrated that ecDNA interacts with chromosomal DNA to function as a strong trans-acting regulator of oncogene expression (14). Additionally, by interacting to form ecDNA hubs, ecDNAs may control the production of oncogenes. The bromodomain and extra-terminal domain (BET) protein has been found to be crucial for preserving the structure of ecDNA hubs, according to a recent study that proposed that approximately 10 to 100 ecDNAs could create ecDNA hubs. The absence of enhancers in *MYC* ecDNA may be due to the regulation of enhancers in ecDNA hubs, indicating that the development of ecDNA hubs facilitated oncogene-enhancer interaction to increase oncogene expression (13,14).

3.3.4 ecDNA Serves as a Reservoir for DNA Recombination

The capability of ecDNA to serve as a reservoir for DNA recombination is a further essential function. According to one study, ecDNA can be produced from extrachromosomal circular elements, and excision or ligation of DNA fragments may also produce ecDNA from another ecDNA. Furthermore, intra- and inter-chromosomal rearrangements may result from ecDNA's ability to reintegrate into linear chromosomes. Through reintegration into nuclear chromosomes, ecDNA has been shown to cause oncogenic genome remodeling in neuroblastoma cells; this mechanism has been linked to unfavorable clinical outcomes. The *EGFRvIII* mutant allele is present in high copy number on ecDNA with a consistent circular shape of 1.29 Mb in a glioblastoma multiforme (GBM) cell line. A subclone containing *EGFRvIII* exclusively on chromosomal homogeneous staining regions (HSRs) has been identified to have tandem duplications, which are multiple copies of *EGFRvIII* that result from the reintegration of the *EGFRvIII*-containing ecDNA elements into the genome. According to the structural study, the fine structure of the *EGFRvIII* amplicon has been found to be conserved in both the parental and subcloned cells. This suggests that ecDNA dynamically relocates into chromosomal HSRs while maintaining important structural characteristics (14,15). According to several assessments of genomic rearrangements in cancer cells, genome

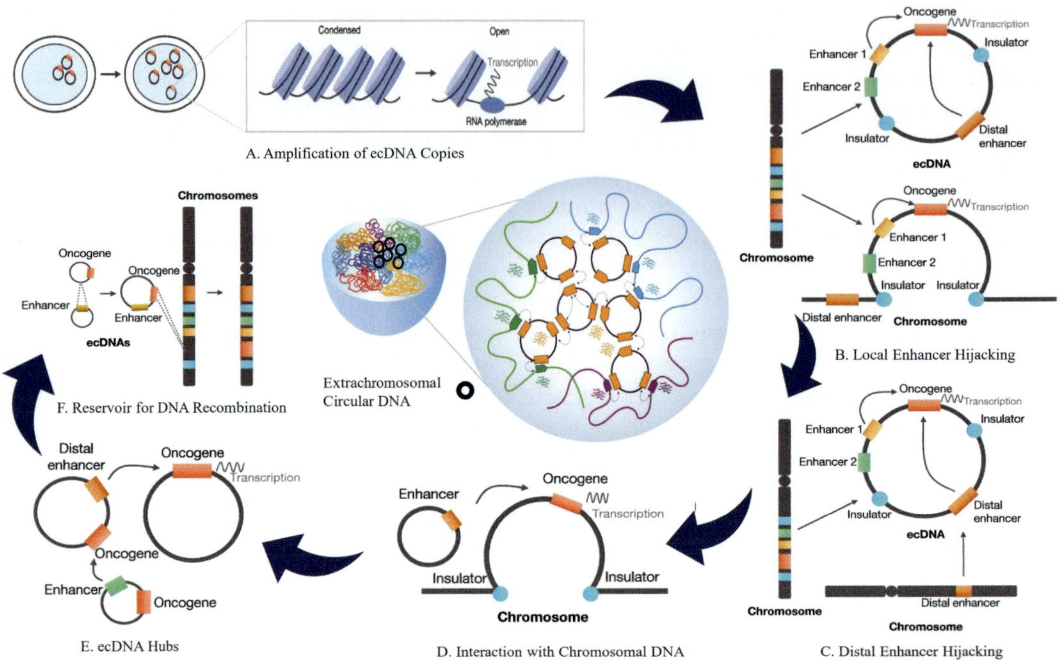

FIGURE 3.1 ecDNA and oncogene expression and the pathway of ecDNA in tumor progression.

remodeling is mediated by ecDNA reintegration. First, ecDNA-mediated genome remodeling was supported by an analysis of WGS data that revealed the majority of intra- and inter-chromosomal rearrangements in neuroblastoma genomes correlated with extrachromosomal circularization areas. Furthermore, examining Circos plots of multiple neuroblastoma cell lines visually revealed that inter-chromosomal rearrangements at circularization loci often exhibited a tree-like pattern, indicating clusters of inter-chromosomal rearrangements that start at the same loci and spread to other distant regions of the genome. Therefore, it is supported that ecDNA can act as a substrate for genomic rearrangements by the correlation between circularization loci and tree-shaped patterns in Circos plots. Furthermore, the majority of rearrangement recipient sites (83.3%) have been determined as circular integrations by using allele-specific polymerase chain reaction (PCR) to analyze phased heterozygous single-nucleotide polymorphisms (SNPs), which are close to integration breakpoints. This further reinforces the concept that ecDNA is a significant source of inter-chromosomal rearrangement by integration (16) (Figure 3.1).

3.4 ROLE OF ECDNA IN TUMOR PROGRESSION

3.4.1 Rapid Tumor Growth and Evolution

Circular DNA fragments known as extrachromosomal DNA (ecDNA) are found outside of chromosomes and often include amplified oncogenes. Tumor cells have a major growth advantage due to their capacity for independent replication and asymmetric distribution during cell division. Since ecDNA is not bound by the restrictions of mitotic segregation like chromosomal DNA is, it can exhibit dynamic and non-Mendelian inheritance patterns (17). By facilitating the high-level production of oncogenes such as MYC, EGFR, CDK4, and MDM2, ecDNA accelerates the growth of tumors (18). These oncogenes promote cancer by altering metabolic pathways, preventing apoptosis, and promoting cellular proliferation. The transcriptional hyperactivity of ecDNA is a result of its large copy number and epigenetically active chromatin state (19). Furthermore, ecDNA-containing tumor cells can proliferate clonally under selection pressure, speeding up tumor progression, according to research employing single-cell sequencing and imaging (20). This quick evolution gives it a survival edge in the face of immunological and therapeutic obstacles, which frequently leads to tumor progression and recurrence.

3.4.2 Intratumoral Heterogeneity and Adaptation

One of the main challenges in the treatment of cancer and a major contributor to treatment resistance is intratumoral heterogeneity. By establishing variation in oncogene dose and gene expression among tumor cells, ecDNA plays a crucial role in this heterogeneity (19). During mitosis, ecDNA segregates unevenly due to its absence of centromeres, resulting in daughter cells with varying ecDNA contents and, consequently, varying carcinogenic potential. The tumor can quickly adjust to environmental changes such as food deprivation, hypoxia, and immune surveillance, thanks to this mosaicism in oncogene dose. Under pharmacological pressure, for example, subclones with elevated EGFR ecDNA can multiply, while other subclones might endure by using different survival strategies (21). To further enhance genetic plasticity, recent research has demonstrated that ecDNA can either reintegrate into the genome or go through cycles of circularization and decircularization (22). Tumors can more effectively modify gene expression and create resistance mechanisms because of their dynamic remodeling ability than if they were limited to chromosomal DNA.

3.4.3 Contribution to Metastasis

By boosting genes involved in invasion, migration, and colonization, ecDNA plays a crucial role in metastasis, which accounts for most cancer-related deaths. Genes including MYC, AKT, and PD-L1 that are overexpressed on ecDNA promote tumor cell plasticity and the epithelial-mesenchymal transition (EMT), which is necessary for the spread of cancer (17,18). Additionally, by modifying cytokine expression and encouraging angiogenesis, ecDNA can rewire the tumor microenvironment (TME). For example, enhanced blood vessel development has been associated with ecDNA-driven VEGFA amplification, which promotes tumor cell intravasation and metastasis (23). ecDNA-bearing cells have a higher chance of surviving in circulation and developing secondary tumors by promoting these pro-metastatic characteristics (Figure 3.1).

3.5 ECDNA AND THERAPY RESISTANCE

3.5.1 Impact on Drug Resistance Mechanisms

A major problem in oncology is treatment resistance, which frequently results from the overexpression or mutation of therapeutic targets. ecDNA aids in the proliferation of genes linked to resistance, which contributes to this resistance. For instance, resistance to kinase inhibitors employed in glioblastoma and non-small cell lung cancer (NSCLC) is conferred by MET and EGFR genes on ecDNA (15,17). Because of the frequent reversibility of this amplification, tumor cells can adjust to the presence or absence of the treatment. According to Turner et al. (2017), this flexibility has been shown in preclinical settings where resistant clones regain sensitivity after losing ecDNA due to drug withdrawal. The dynamic nature of ecDNA sets it apart from steady chromosomal amplifications. Furthermore, ecDNA may carry several gene amplifications at once, allowing for redundant survival pathways that make therapeutic targeting more difficult (19). According to Shoshani et al. (2020), this complex resistance profile increases the tumors' resistance to monotherapies and promotes tumor recurrence (22).

3.5.2 Role in Targeted Therapy Failure

The purpose of targeted therapy is to block particular molecular drivers, although this strategy is compromised by the existence of ecDNA. For example, despite pharmacological suppression, tumors with ecDNA-driven EGFRvIII amplification maintain significant transcriptional output, demonstrating resistance to EGFR inhibitors (17). Furthermore, new oncogenic fusions and enhancer hijacking events might be produced by structural changes in ecDNA, which avoid identification by traditional sequencing methods (18). By activating alternative signaling cascades, these changes may make targeted inhibitors ineffective. In order to uncover resistance mechanisms, more thorough diagnostic techniques are required due to the hidden layer of genetic complexity that ecDNA represents.

3.5.3 Strategies to Overcome ecDNA-Driven Resistance

Addressing ecDNA-driven resistance involves targeting the unique features of ecDNA biology. Promising strategies include the following:

1. **Inhibition of ecDNA replication**: ecDNA replication requires DNA polymerase theta (Polθ), a crucial enzyme in alternative end-joining repair. It has been demonstrated that Polθ inhibitors can specifically target ecDNA maintenance (20).
2. **Induction of ecDNA degradation**: Cytosolic ecDNA can trigger innate immune responses that result in ecDNA degradation by activating the cGAS-STING pathway. Preclinical research is being done on STING agonists (24).
3. **Epigenetic targeting**: By focusing on super-enhancers and bromodomains, BET inhibitors like JQ1 can reduce the transcriptional activity of ecDNA-associated oncogenes (19).
4. **Combination therapies**: Targeting ecDNA oncogenes and related bypass routes at the same time may stop resistant clones from emerging. Dual EGFR and PI3K inhibition has been used to validate this strategy in glioblastoma models (11).

3.6 CLINICAL IMPLICATIONS AND THERAPEUTIC STRATEGIES

3.6.1 ecDNA as a Biomarker for Cancer Diagnosis

ecDNA is a useful biomarker for cancer diagnosis and prognosis because of its unique properties. ecDNA differs from chromosomal DNA in that it is greatly amplified, circular, and centromere-free. By using methods like liquid biopsy, optical mapping, and Circle-Seq, these characteristics can be used for detection (16). Tumor ecDNA circulation in plasma provides a non-invasive way to identify and track the development of cancer. According to a number of studies, individuals with advanced malignancies had higher blood levels of ecDNA, which is associated with a higher tumor burden and a worse prognosis (11,20). Additionally, ecDNA profiles can be used as indicators of how well a treatment will work. High ecDNA concentration tumors are typically more aggressive and resistant to conventional treatments, necessitating more intense or focused therapeutic approaches.

3.6.2 Potential Drug Targets to Suppress ecDNA

Therapeutic suppression of ecDNA involves targeting its formation, maintenance, and transcriptional activity through several key mechanisms:

3.6.2.1 DNA Repair and Replication Proteins

DNA damage response (DDR) pathways, especially those involving error-prone repair processes like alternative end-joining (alt-EJ), are intimately linked to the production of ecDNA. DNA polymerase theta (Polθ) inhibitors, such as novobiocin, can interfere with DNA double-strand break repair, which results in the production of ecDNA. Particularly in cells with homologous recombination deficit, PARP inhibitors, which prevent single-strand break repair, have also demonstrated promise in lowering the synthesis of ecDNA (24).

3.6.2.2 Chromatin Remodelers

High oncogene expression is facilitated by the open chromatin states that ecDNA frequently displays. The transcriptional activity of ecDNA can be suppressed by medications that alter histone modifications, such as histone methyltransferase (HMT) and histone deacetylase (HDAC) inhibitors. These substances may reduce oncogene overexpression and hinder ecDNA-driven tumor growth by changing chromatin accessibility.

3.6.2.3 Harnessing Immune Pathways

The innate immune system can recognize ecDNA in the cytoplasm as a warning indication. Immune responses are triggered by the cyclic GMP-AMP synthase (cGAS)-stimulator of interferon genes (STING) pathway, which recognizes cytosolic DNA. The cGAS-STING axis can be activated by therapeutic drugs to improve immune surveillance and encourage the death of tumor cells that contain ecDNA. For instance, STING agonists may be utilized in conjunction with checkpoint inhibitors and are presently undergoing clinical studies for solid tumors (18).

3.6.3 Future Directions in Cancer Treatment

The study of ecDNA biology is developing quickly, and its therapeutic application could revolutionize the way cancer is treated. Important avenues for the future include the following:

Emerging research on extrachromosomal DNA (ecDNA) is paving the way for significant advancements in cancer care. Advanced diagnostics, such as routine genetic profiling that includes ecDNA detection, could greatly enhance risk assessment and diagnostic accuracy. Targeted therapeutics are being developed, focusing on biologics and small molecules that selectively disrupt ecDNA maintenance pathways. Additionally, combination regimens that integrate ecDNA-targeting drugs with immunotherapy or chemotherapy are showing potential for improved outcomes. To address treatment resistance, longitudinal studies of ecDNA dynamics during therapy are essential, enabling adaptive therapeutic strategies. Incorporating ecDNA profiling into precision oncology further supports personalized medicine, allowing treatment plans to be tailored based on oncogene burden and ecDNA content. Furthermore, preclinical modeling that accurately simulates ecDNA behavior in complex in vitro and in vivo systems is advancing drug development efforts. Clinical trials are increasingly focused on integrating ecDNA-targeting agents with existing therapeutic frameworks. Lastly, improving the sensitivity and specificity of liquid biopsy platforms for ecDNA detection could revolutionize real-time monitoring of tumor progression and response to therapy.

3.7 CONCLUSION

Extrachromosomal DNA (ecDNA) has become a key factor in the growth of tumors and the amplification of oncogenes. Oncogenes like *MYC*, *EGFR*, and *MDM2* can be amplified quickly and at a high level because of ecDNA, which is distinct from chromosomal DNA and takes place outside the normal genomic architecture. The unique structure promotes dynamic gene expression and facilitates quick response to environmental pressures, which leads to tumor heterogeneity, therapeutic resistance, and aggressive cancer phenotypes. Intratumoral diversity and evolution are further stimulated by the non-Mendelian inheritance of ecDNA during cell division. Lastly, ecDNA is a central player in tumor progression, therapy resistance, and metastasis. Its unique properties provide both challenges and opportunities for innovative cancer therapies. Targeting ecDNA represents a promising frontier in the pursuit of more effective and durable cancer treatments.

ABBREVIATIONS

AKT: AK Strain Transforming, where "AK" refers to the AKR mouse strain and "T" indicates the transforming ability.
ASE: allele-specific expression
ATAC-seq: Assay for Transposase-Accessible Chromatin using sequencing
CDK4: cyclin-dependent kinase 4
ChIP-Seq: Chromatin Immunoprecipitation Sequencing

ecDNA: extrachromosomal deoxyribonucleic acid
EGFR: epidermal growth factor receptor
EMT: epithelial-mesenchymal transition
ERBB2: Erb-B2 receptor tyrosine kinase 2, also known as HER2 (human epidermal growth factor receptor 2).
H3K27ac: histone H3 lysine 27 acetylation
MDM2: murine double minute type 2
MECOM: MDS1 and EVI1 complex locus and is also known as ecotropicviral integrationsite 1 (EVI1) and myelodysplastic syndrome 1 (MDS1-EVI1)
MET: mesenchymal-epithelial transition factor
MYC: myelocytomatosis oncogene
MYCN: MYCN proto-oncogene
NSCLC: non-small cell lung cancer

PCR: polymerase chain reaction
PIK3CA: Phosphatidylinositol-4,5-bisphosphate 3-kinase catalytic subunit alpha
SNPs: single nucleotide polymorphisms
SOX2: SRY-box transcription factor 2, member of Sry-related HMG box (SOX) gene family
TCGA: The Cancer Genome Atlas
TME: tumor microenvironment
VEGFA: Vascular Endothelial Growth Factor A
WGS: whole genome sequencing

REFERENCES

1. Pecorino, L. T., Verhaak, R. G. W., Henssen, A., & Mischel, P. S. (2022). Extrachromosomal DNA (ecDNA): An origin of tumor heterogeneity, genomic remodeling, and drug resistance. Biochemical Society Transactions, 50(6), 1911. https://doi.org/10.1042/BST20221045
2. Dong, Y., He, Q., Chen, X., Yang, F., He, L., & Zheng, Y. (2023). Extrachromosomal DNA (ecDNA) in cancer: Mechanisms, functions, and clinical implications. Frontiers in Oncology, 13, 1194405. https://doi.org/10.3389/FONC.2023.1194405/BIBTEX
3. Wang, T., Zhang, H., Zhou, Y., & Shi, J. (2021). Extrachromosomal circular DNA: A new potential role in cancer progression. Journal of Translational Medicine, 19(1), 1–16. https://doi.org/10.1186/S12967-021-02927-X
4. Bailey, C., Pich, O., Thol, K., Watkins, T. B. K., Luebeck, J., Rowan, A., Stavrou, G., Weiser, N. E., Dameracharla, B., Bentham, R., Lu, W.-T., Kittel, J., Yang, S. Y. C., Howitt, B. E., Sharma, N., Litovchenko, M., Salgado, R., Hung, K. L., Cornish, A. J., . . . Swanton, C. (2024). Origins and impact of extrachromosomal DNA. Nature, 635(8037), 193–200. https://doi.org/10.1038/s41586-024-08107-3
5. Wu, S., Tao, T., Zhang, L., Zhu, X., & Zhou, X. (2023). Extrachromosomal DNA (ecDNA): Unveiling its role in cancer progression and implications for early detection. Heliyon, 9(11), e21327. https://doi.org/10.1016/J.HELIYON.2023.E21327
6. Wu, T., Wu, C., Zhao, X., Wang, G., Ning, W., Tao, Z., Chen, F., & Liu, X. S. (2022). Extrachromosomal DNA formation enables tumor immune escape potentially through regulating antigen presentation gene expression. Scientific Reports, 12(1), 1–9. https://doi.org/10.1038/S41598-022-07530-8

7. Chang, H. Y., Mischel Co-Corresponding, P. S., Chang, H. Y., Weiser, N. E., Watkins, T. B. K., & Mischel, P. S. (2025). A guide to extrachromosomal DNA: Cancer's dynamic circular genome. Cancer Discovery, 15(6), 1105–1114. https://doi.org/10.1158/2159-8290.CD-25-0230
8. Bailey, C., Shoura, M. J., Mischel, P. S., & Swanton, C. (2020). Extrachromosomal DNA—relieving heredity constraints, accelerating tumour evolution. Annals of Oncology, 31(7), 884–893. https://doi.org/10.1016/J.ANNONC.2020.03.303
9. Purshouse, K., Pollard, S. M., & Bickmore, W. A. (2024). Imaging extrachromosomal DNA (ecDNA) in cancer. Histochemistry and Cell Biology, 162(1–2), 53–64. https://doi.org/10.1007/S00418-024-02280-2/TABLES/1
10. Karami Fath, M., Karimfar, N., Fazlollahpour Naghibi, A., Shafa, S., Ghasemi Shiran, M., Ataei, M., Dehghanzadeh, H., Nabi Afjadi, M., Ghadiri, T., Payandeh, Z., & Tarhriz, V. (2022). Revisiting characteristics of oncogenic extrachromosomal DNA as mobile enhancers on neuroblastoma and glioma cancers. Cancer Cell International, 22(1), 1–14. https://doi.org/10.1186/S12935-022-02617-8
11. Hong, J., Zheng, S., & Jiang, D. (2021). The contributions of extrachromosomal DNA elements in neoplasm progression. American Journal of Cancer Research, 11(6), 2417. https://pmc.ncbi.nlm.nih.gov/articles/PMC8263671/
12. Zhao, L., Jiang, Y., Lei, X., & Yang, X. (2023). Amazing roles of extrachromosomal DNA in cancer progression. Biochimica et Biophysica Acta (BBA)—Reviews on Cancer, 1878(1), 188843. https://doi.org/10.1016/J.BBCAN.2022.188843
13. Ott, C. J. (2020). Circles with a point: New insights into oncogenic extrachromosomal DNA. Cancer Cell, 37(2), 145–146. https://doi.org/10.1016/J.CCELL.2020.01.008
14. Ur Rehman, S. (2025). The multifaceted role of extrachromosomal DNA in cancer: Functions and clinical implications. Bangladesh Journal of Medicine, 36(2), 82–91. https://doi.org/10.3329/bjm.v36i2.79673
15. Wu, H., Liu, S., Wu, D., Zhou, H., & Wu, G. (2024). Tumor extrachromosomal DNA: Biogenesis and recent advances in the field. Biomedicine and Pharmacotherapy, 174, 116588. https://doi.org/10.1016/j.biopha.2024.116588
16. Kim, H., Kim, S., Wade, T., Yeo, E., Lipsa, A., Golebiewska, A., Johnson, K. C., An, S., Ko, J., Nam, Y., Lee, H. Y., Kang, S., Chung, H., Niclou, S. P., Moon, H. E., Paek, S. H., Bafna, V., Luebeck, J., & Verhaak, R. G. W. (2024). Mapping extrachromosomal DNA amplifications during cancer progression. Nature Genetics, 56(11), 2447–2454. https://doi.org/10.1038/S41588-024-01949-7
17. Nathanson, D. A., Gini, B., Mottahedeh, J., Visnyei, K., Koga, T., Gomez, G., . . . & Mischel, P. S. (2014). Targeted therapy resistance mediated by dynamic regulation of extrachromosomal mutant EGFR DNA. Science, 343(6166), 72–76. https://doi.org/10.1126/science.1241328
18. Kim, H., Nguyen, N. P. D., Turner, K., Wu, S., Gujar, A. D., Luebeck, J., . . . & Mischel, P. S. (2020). Extrachromosomal DNA is associated with oncogene amplification and poor outcome across multiple cancers. Nature Genetics, 52(9), 891–897. https://doi.org/10.1038/s41588-020-0678-2
19. Turner, K. M., Deshpande, V., Beyter, D., Koga, T., Rusert, J., Lee, C., . . . & Mischel, P. S. (2017). Extrachromosomal oncogene amplification drives tumour evolution and genetic heterogeneity. Nature, 543(7643), 122–125. https://doi.org/10.1038/nature21356
20. Koche, R. P., Rodriguez-Fos, E., Helmsauer, K., Burkert, M., MacArthur, I. C., Maag, J., . . . & Bernstein, B. E. (2020). Extrachromosomal circular DNA drives oncogenic genome remodeling in neuroblastoma. Nature Genetics, 52(1), 29–34. https://doi.org/10.1038/s41588-019-0547-z
21. Verhaak, R. G., Bafna, V., & Mischel, P. S. (2019). Extrachromosomal oncogene amplification in tumour pathogenesis and evolution. Nature Reviews Cancer, 19(5), 283–288. https://doi.org/10.1038/s41568-019-0128-6
22. Shoshani, O., Brunner, S. F., Yaeger, R., Ly, P., Nechemia-Arbely, Y., Kim, D. H., . . . & Santaguida, S. (2020). Chromothripsis drives the evolution of gene amplification in cancer. Nature, 591(7848), 137–141. https://doi.org/10.1038/s41586-020-03064-z
23. Hung, K. L., Yost, K. E., Xie, L., Shi, Q., Helmsauer, K., Luebeck, J., . . . & Mischel, P. S. (2021). ecDNA hubs drive cooperative intermolecular oncogene expression. Nature, 600(7890), 731–736. https://doi.org/10.1038/s41586-021-04272-y
24. Wang, Y., Wang, M., Djekidel, M. N., & Chen, H. (2021). Circular ecDNAs promote accessible chromatin and high oncogene expression. Nature, 599(7883), 731–736. https://doi.org/10.1038/s41586-021-03992-4

4 Gene Expression and Epigenetic Landscapes of ecDNA in Cancer

Vijay Singh,[1] *Sapna Saini,*[2] *Sumit Jangra,*[3] *and Mohammad Muztaba*[4]

1 Department of Medical Laboratory Technology, School of Allied Health Sciences, Galgotias University, Uttar Pradesh, India
2 Department of Biochemistry, All India Institute of Medical Sciences, Gorakhpur, India
3 Department of Anatomy, Ajay Sanghal Institute of Medical Sciences and Research and Ayushmaan Hospital, Village Rajjak Nagar, Shamli Uttar Pradesh, India
4 Department of Pharmacology, Praduman Singh SPS Pharmacy College, Phutahiya Sansarpur, Basti Uttar Pradesh, India

4.1 INTRODUCTION

4.1.1 Overview of ecDNA in Cancer

ecDNA, or extrachromosomal DNA, is circular genetic material found outside of the main chromosomal DNA within eukaryotic cell nuclei. While mitochondrial DNA is also extrachromosomal, it is typically inherited from the mother. In contrast, ecDNAs in cancer are mainly non-mitochondrial and result from abnormal changes in the genome. These fragments can vary in size, ranging from a few thousand bases to millions of bases. One common characteristic of ecDNA is its tendency to contain oncogenes, regulatory elements, and genes that confer resistance to drugs (L. Yang et al., 2022).

ecDNA is a crucial factor in cancer, as it enables oncogenes to amplify and enhance gene expression without modifying the chromosomal DNA itself. This amplification contributes significantly to tumor heterogeneity and plasticity, ultimately aiding cancer cells in adapting to various treatments. However, due to the absence of centromeres, ecDNA does not distribute equally during mitosis, resulting in uneven distribution among daughter cells. This process further increases intratumoral heterogeneity, which is a characteristic feature of aggressive and difficult-to-treat cancers (S. Wu et al., 2022).

Through the use of advanced sequencing technologies and imaging tools, researchers have been able to identify and map the structure of ecDNA. This has dispelled the notion that ecDNAs are simply a byproduct of genomic instability, instead revealing them to be active and ever-changing components that play a significant role in tumor progression and evolution (Espinosa et al., 2024).

4.1.2 Historical Context and Discovery

The idea of additional genetic elements in human cancer was first discovered in the 1960s through the observation of tiny chromatin bodies known as "double minutes" (DMs) under light microscopy. Although initially considered abnormalities in cancer cells, further investigation revealed that DMs

DOI: 10.1201/9781003655299-4

could contain amplified oncogenes such as MYC, EGFR, and MDM2 and could self-replicate without the involvement of chromosomes (S. Wu et al., 2022).

Nevertheless, the limited availability of sophisticated molecular techniques hindered the thorough analysis of these formations. It wasn't until advanced methods such as high-throughput sequencing, chromatin conformation assays, and fluorescence in situ hybridization (FISH) were introduced that scientists were able to confirm the circular form of these DNA fragments and differentiate them from linear amplicons. As knowledge expanded on the significance of these elements in biological processes beyond their structural abnormalities, the term "ecDNA" gained wider usage (Rashid & Stingl, 2015).

Recent research, specifically from The Cancer Genome Atlas (TCGA) and other extensive genomics initiatives, has revealed the prevalence of ecDNA in a variety of cancer forms, such as glioblastoma, neuroblastoma, colorectal, and lung cancers. This has reignited interest in the potential diagnostic, prognostic, and therapeutic implications of ecDNA (Cooper et al., 2018).

4.1.3 Biological Importance and Relevance in Oncology

ecDNA has been acknowledged as a significant contributor to the intricate nature of cancer. This is due to its distinct features, one being the presence of numerous copies of oncogenes such as MYC, CDK4, and EGFR. The lack of regulatory mechanisms present in chromosomes allows these genes to be transcribed at excessively high levels (Wei et al., 2020). Research has revealed that ecDNA regions are more accessible to transcription factors, leading to hyperactive transcription. As a result, the open chromatin state and enhanced enhancer-promoter interactions contribute to the excessive expression of oncogenes, ultimately supporting uncontrolled cell proliferation (Q.-L. Yang et al., 2025).

As a result of its non-chromosomal segregation during cell division, ecDNA is dispersed randomly among progeny cells, promoting variation within the tumor population. This diversity leads to the identification of resistant clones when faced with treatment, ultimately causing relapse and unfavorable prognosis (Bailey et al., 2024).

Cancer cells with ecDNA are better equipped to handle environmental stress, including chemotherapy. This type of DNA can either be removed entirely or integrated into chromosomes, resulting in a constantly changing and highly adaptable genetic structure (Dong et al., 2023). The presence of ecDNA is often linked to aggressive tumor behavior, unfavorable clinical outcomes, and resistance to targeted treatments. Additionally, detecting ecDNA in circulating tumor DNA (ctDNA) has the potential for non-invasive cancer diagnosis and monitoring (Zhu et al., 2022).

With these implications in mind, ecDNA presents itself as a promising yet challenging target for cancer therapy. Gaining a deeper understanding of its biology not only sheds light on tumor progression but also offers new therapeutic approaches that aim to disrupt its formation, maintenance, or function (Q.-L. Yang et al., 2025).

4.2 STRUCTURAL FEATURES OF ECDNA

4.2.1 Circular Architecture

One of the defining characteristics of extrachromosomal DNA (ecDNA) is its circular configuration. Unlike linear chromosomes, ecDNA exists as closed-loop structures that are both structurally unique and functionally advantageous for cancer cells. This circular architecture allows for high levels of gene transcription, as it lacks the constraints imposed by chromatin compaction and often contains active enhancers and promoter regions. Additionally, the close physical proximity of regulatory elements to oncogene sequences enhances transcriptional efficiency and accessibility to the transcriptional machinery (Z. Li et al., 2022; Zhang et al., 2024). Furthermore, circular ecDNA can autonomously replicate using the host cell's DNA replication machinery, facilitating rapid amplification of oncogenic material (M. Wu & Rai, 2022). Recent advancements in long-read sequencing and circular DNA enrichment protocols have revealed that many ecDNAs consist of segments from multiple chromosomal regions that are connected through complex rearrangements into circular forms (Warburton & Sebra, 2023).

4.2.2 Lack of Centromeres and Telomeres

In contrast to chromosomes, ecDNAs do not possess centromeres and telomeres, making them acentric and atelomeric. Centromeres serve as crucial attachment sites for the mitotic spindle during mitosis, ensuring proper chromosome segregation. On the other hand, telomeres play a vital role in protecting linear chromosomes and maintaining genomic stability (Achrem et al., 2020). The absence of these structures in ecDNA has significant implications. Its random distribution between daughter cells during cell division, due to the lack of centromeres, contributes to intratumoral heterogeneity. This allows for the selection of subpopulations of cells with varying levels of oncogenes and adaptive advantages (Lange et al., 2022). Additionally, ecDNA does not have the same regulated lifespan as chromosomes due to the absence of telomeres. This grants it greater flexibility and durability in terms of replication and stability, at least in the short term. While this may seem disadvantageous, it enables rapid genomic remodeling in cancer, leading to drug resistance and disease progression (Jones-Weinert et al., 2025).

4.2.3 Copy Number Variability

An essential aspect of ecDNA in cancer is its ability to exist in varying numbers within and between tumor cells. This variation stems from the independent replication of ecDNA and its uneven distribution during mitosis due to the absence of centromeres. In contrast to chromosomal amplification, which typically results in stable regions, ecDNA-mediated amplification leads to fluctuating levels of oncogene expression (Hung et al., 2021). This adaptability allows cancer cells to regulate the expression of crucial survival genes in response to external pressures, such as chemotherapy or hypoxia. For instance, an abundance of ecDNA carrying EGFR or MYC genes can lead to excessive expression that drives aggressive tumor growth, while a decrease in copy number after treatment can help evade targeted therapies. Not only is the variability of ecDNA a contributor to tumor evolution, but it also serves as a potential prognostic biomarker (Stöber et al., 2024). Research has shown that tumors with high levels of ecDNA have a higher resistance to treatment and are linked with poorer clinical outcomes. Therefore, keeping track of ecDNA levels is a significant area for further investigation in cancer diagnostics and individualized treatment plans (Ashique et al., 2022).

4.3 MECHANISMS OF ECDNA FORMATION

4.3.1 Breakage–Fusion–Bridge Cycles

The breakage–fusion–bridge (BFB) cycle, an early recognized contributor to genomic instability, is believed to have a crucial role in the creation of ecDNA in cancer cells. This process begins when a chromosome experiences a double-strand break at its telomeric ends. If the repair process does not accurately restore the telomeres, the broken portions could fuse, forming a dicentric chromosome with two centromeres (Singh et al., 2024). These dicentric chromosomes are then pulled toward opposite poles during cell division, causing mechanical strain that leads to further breakage. As this cycle repeats itself, it results in significant changes to the genome, including rearrangements, amplifications, and deletions of DNA segments (MacKinnon & Campbell, 2011).

With time and constant BFB cycles, specific regions of chromosomes containing oncogenes or regulatory sequences may be removed. These excised fragments can form circular ecDNA as a result. Since BFB cycles are initiated by DNA damage and perpetuated by faulty DNA repair mechanisms, they are commonly observed in tumor cells where genomic instability is prevalent. The circular DNA produced by these cycles often contains high levels of oncogene amplification, which contribute to accelerated tumor growth and advancement (Tanaka & Watanabe, 2020).

4.3.2 Chromothripsis and Genomic Rearrangement

Chromothripsis, meaning "chromosome shattering" in Greek, describes a severe genetic incident where a small number of chromosomes are broken into numerous pieces during a single cellular crisis. Rather than undergoing organized repair, these fragments are joined together in random order using error-prone mechanisms like non-homologous end joining (NHEJ). This disordered reassembly leads to a patchwork of deletions, duplications, and inversions, creating a significantly altered genomic pattern (de Groot et al., 2023).

New findings indicate that chromothripsis may play a direct role in producing ecDNA (Cortés-Ciriano et al., 2020). In the aftermath of a chromothriptic event, specific DNA segments may be omitted from reattaching to the linear chromosome and instead form circular ecDNA molecules. These rounded fragments frequently contain oncogenes or other active components that promote tumor growth (Dong et al., 2023). Unlike the gradual build-up of mutations, chromothripsis causes a sudden spike in genetic alterations, resulting in oncogene-containing ecDNA being formed in one swift motion. This rapid evolutionary process is especially significant in aggressive cancers and is associated with unfavorable medical outcomes [23,24].

4.3.3 Replication-Based Models

Replication-based models offer an alternative perspective on ecDNA creation, highlighting errors that arise during DNA replication and repair. In these models, the common occurrence of replication stress in cancer cells can result in halted or damaged replication forks, leading to double-strand breaks. These breaks are then subjected to abnormal repair through mechanisms such as FoSTeS or MMBIR (Lovett, 2017), resulting in large segments of duplicated, rearranged, or displaced DNA. If these altered fragments are not reinserted into the chromosomal structure, they may form circular ecDNA forms. The presence of short homologous sequences or micro-homology at the junctions of ecDNA supports the role of replication-based processes in their development. Unlike BFB or chromothripsis, which often involve major chromosomal abnormalities, the formation of replication-based ecDNA is more localized and gradual (Malkova & Ira, 2013).

This is important because it helps explain the structural diversity of ecDNA, which often contains merged fragments from different chromosomal regions. Ultimately, these models highlight how interactions between replication dynamics and genomic instability contribute to shaping the cancer genome and promoting the emergence of ecDNA as a significant mechanism for oncogene amplification (Casas-Delucchi et al., 2022).

4.4 GENE EXPRESSION REGULATION BY ECDNA

4.4.1 Oncogene Amplification and Expression

One defining characteristic of extrachromosomal DNA (ecDNA) is its ability to greatly amplify oncogenes. While chromosomal amplification is typically limited by the structure and environment of chromosomes, ecDNA allows for unrestricted gene amplification due to its independent nature. This leads to significantly heightened levels of oncogene transcripts, directly contributing to the development of tumors, rapid cell growth, and malignant transformation (Hung et al., 2022; Q.-L. Yang et al., 2025). Notably, genes such as MYC, EGFR, CDK4, and MDM2 are commonly found on ecDNA and are expressed at much higher levels compared to their chromosomal counterparts. This overexpression is not solely a result of increased gene quantity but is also aided by the accessible chromatin structure of ecDNA that promotes transcription. This unique ability of ecDNA to enhance oncogene expression without altering the architecture of chromosomal DNA highlights its crucial role in cancer progression and resistance to treatment (Qiu et al., 2020).

4.4.2 Enhancer Hijacking and Transcriptional Activation

Enhancer hijacking is a crucial process utilized by ecDNA to regulate gene expression, often with a cancer-promoting impact. Rather than being randomly composed, ecDNA often contains enhancers—regulatory DNA sequences that enhance the transcription of target genes—from diverse genomic locations. When these enhancers are positioned near oncogenes within the ecDNA circle, they can cause abnormal and excessive transcription. This rearrangement allows ecDNA to bring potent enhancers close to genes they typically do not control, resulting in the "hijacking" of the enhancer's function (Claringbould & Zaugg, 2021).

This process not only boosts levels of oncogenes but also sidesteps typical regulatory protocols found in the linear genome. The 3D structure of ecDNA allows for the development of fresh enhancer-promoter connections, resulting in highly active regulatory centers that surpass their chromosomal equivalents in terms of transcriptional output. Thus, ecDNA serves as a movable base for rearranging gene regulation to promote tumor proliferation and longevity [6,31].

4.4.3 Nuclear Localization and ecDNA Clustering

Not only does ecDNA exhibit a distinct spatial organization within the nucleus, but recent imaging studies have also revealed its tendency to form clusters in specific nuclear domains, known as "ecDNA hubs." These hubs, which are enriched with active transcription regions, typically localize near the nuclear interior, a region associated with high levels of transcriptional activity (Yi et al., 2022). This clustering is not simply a physical gathering but rather serves a functional purpose in enhancing gene expression. By bringing ecDNAs into proximity, cancer cells may create microenvironments optimized for transcriptional amplification (D'haeseleer, 2005).

In addition to its role in gene expression, the nuclear positioning of ecDNA also affects its interaction with other components within the nucleus, such as RNA polymerases and chromatin remodelers. Due to its dynamic mobility, ecDNA can access regions that are more conducive to transcriptional activity, avoiding repressive chromatin zones found at the nuclear periphery. Therefore, it can be said that both nuclear positioning and clustering add another layer of regulation through which ecDNA influences gene expression and contributes to the aggressive nature of ecDNA-bearing tumors (Lanctôt et al., 2007).

4.4.4 Interaction with Transcriptional Machinery

The critical role played by ecDNA in regulating gene expression involves its interaction with the transcriptional machinery. Unlike genomic DNA, which is subject to constraints from nucleosome positioning and epigenetic modifications, ecDNA has a more accessible chromatin structure that allows for easier binding of RNA polymerase II and other transcription factors. This is due to its lower density of histones and enrichment of active histone marks such as H3K27ac and H3K4me3, indicating favorable conditions for gene transcription (Miller & Grant, 2013).

This makes ecDNA a hotbed for transcription and leads to the recruitment of co-activators and chromatin remodelers, including BRD4 and the Mediator complex. This recruitment further enhances the efficiency of transcription by facilitating the formation of super-enhancers that drive high-level expression of oncogenes. Ultimately, this physical and functional interaction with the transcriptional machinery gives ecDNA a unique ability to disrupt normal gene regulatory networks, promoting a malignant phenotype that enables cells to adapt and survive under even intense therapeutic pressure (Talukdar & Chatterji, 2023).

4.5 EPIGENETIC LANDSCAPE OF ECDNA

4.5.1 DNA Methylation Patterns

DNA methylation is a crucial epigenetic change in which methyl groups are added to cytosine residues, typically in CpG dinucleotides, to regulate gene expression. However, in the context of extrachromosomal DNA (ecDNA), there are unique methylation patterns that differ significantly from those found in the linear genome. Research has shown that ecDNA tends to have reduced levels of methylation, particularly at regions responsible for promoting and enhancing transcriptional activity. This lowered methylation allows easier access for transcription factors and RNA polymerase complexes to oncogenes, leading to their overexpression (Kiselev et al., 2021; Moore et al., 2013).

It's worth noting that this hypomethylation is not spread evenly across all ecDNA regions; certain areas not involved in active transcription may still exhibit moderate levels of methylation. This selective reduction of methylation on ecDNA highlights its role in increasing permissive transcription in cancer cells. Additionally, the dynamic nature of ecDNA methylation may contribute to cancer cells' ability to quickly adapt to changes or stressors, such as therapeutic treatments (Chandler & Jones, 1988).

4.5.2 Histone Modifications (H3K27ac, H3K4me3)

Histone modifications are crucial for controlling chromatin structure and gene expression in epigenetic regulation. In ecDNA, specific modifications play a significant role in promoting an open and active chromatin state. Of particular importance are two modifications, known as H3K27ac (histone H3 lysine 27 acetylation) and H3K4me3 (histone H3 lysine 4 trimethylation), that are highly concentrated on ecDNA. These markers are well-known indicators of active enhancers and promoters, respectively. The presence of H3K27ac on ecDNA suggests the presence of highly active enhancer elements that contribute to strong transcriptional output. Similarly, the abundance of H3K4me3 at transcription start sites on ecDNA further supports the high expression levels of embedded oncogenes (Bannister & Kouzarides, 2011; Bure et al., 2022).

The modifications to histones found in ecDNA indicate both its transcriptionally active state and ability to attract coactivators like BRD4 and the Mediator complex. These marks promote a chromatin environment that is favorable for the formation of super-enhancers, which are powerful clusters of enhancers that work together to increase gene expression. This specific epigenetic state sets ecDNA apart from neighboring chromosomal DNA, highlighting its distinctive role in promoting oncogenesis (Miller & Grant, 2013).

4.5.3 Chromatin Accessibility and Remodeling

The standout feature of ecDNA is its elevated chromatin accessibility, which sets it apart from the tightly packaged DNA on chromosomes. Rather than being embedded in higher-order structures, ecDNA has a relaxed configuration that enables swift and unimpeded binding by transcription factors, coactivators, and RNA polymerase complexes. This allows for increased expression of oncogenes (Z. Li et al., 2022).

It is believed that chromatin remodeling complexes, such as SWI/SNF, contribute to maintaining this open state by rearranging or removing nucleosomes on ecDNA. The combination of fewer nucleosomes and active histone modifications results in a dynamic and responsive chromatin environment. Not only does this enhance oncogene expression, but it also enables cancer cells to quickly alter their transcriptional programs in response to external stimuli like hypoxia or chemotherapy. The adaptable nature of ecDNA-associated chromatin highlights its crucial role in promoting cancer survival and flexibility (Tang et al., 2010).

4.5.4 Techniques for Epigenetic Profiling of ecDNA

The discovery of the epigenetic characteristics of ecDNA has greatly benefited from the emergence of advanced technologies in genomics and epigenomics. For example, ATAC-seq (Assay for Transposase-Accessible Chromatin using sequencing) has been crucial in analyzing chromatin accessibility on ecDNA, unveiling its significantly more open structure compared to regular DNA. Additionally, ChIP-seq (Chromatin Immunoprecipitation followed by sequencing) has been a

TABLE 4.1
Epigenetic Features of Extrachromosomal DNA (ecDNA)

Epigenetic Feature	Description	Key Observations in ecDNA	References
DNA Methylation Patterns	Addition of methyl groups to CpG sites to repress transcription	ecDNA exhibits hypomethylation, especially near oncogenes. Allows enhanced transcriptional activity. Some non-transcribed regions retain moderate methylation	(Gan et al., 2024)
Histone Modifications (H3K27ac, H3K4me3)	Chemical modifications of histone tails regulate chromatin accessibility and transcription	ecDNA enriched with H3K27ac (enhancers) and H3K4me3 (promoters). Facilitates super-enhancer formation and oncogene activation	(Kumar et al., 2025)
Chromatin Accessibility & Remodeling	Openness of chromatin to allow binding of transcription machinery	ecDNA is highly accessible SWI/SNF complexes remodel chromatin. Promotes rapid transcriptional changes and oncogene overexpression	(Pepin & Schneider, 2024)
Techniques for Epigenetic Profiling	Methods used to study chromatin state, histone modifications, and DNA methylation	ATAC-seq: Chromatin accessibility. ChIP-seq: Histone marks (H3K27ac, H3K4me3). Bisulfite sequencing: DNA methylation. Nanopore/PacBio sequencing: Direct long-read epigenome mapping	(Kumar et al., 2025; Liu & Conesa, 2025)

commonly used method for identifying histone modifications like H3K27ac and H3K4me3, providing evidence of the active transcriptional status of ecDNA (Jiang et al., 2023).

Moreover, the utilization of bisulfite sequencing has facilitated the accurate identification of DNA methylation patterns on ecDNA molecules. A newer development in single-molecule and long-read sequencing techniques, such as nanopore and PacBio sequencing, has eliminated the need for amplification in directly examining both DNA sequence and epigenetic changes on ecDNA. These methods are especially beneficial due to the intricate structure and varying number of copies of ecDNA. In combination, these approaches offer a complete understanding of how epigenetic characteristics impact the functional roles of ecDNA in cancer biology, creating opportunities for diagnosis and treatment interventions (Y. Li & Tollefsbol, 2011). The epigenetic features of extrachromosomal DNA (ecDNA) are summarized in **Table 4.1**.

4.6 FUNCTIONAL IMPLICATIONS IN TUMOR BIOLOGY

4.6.1 Tumor Progression and Metastasis

The existence of extrachromosomal DNA (ecDNA) is a significant contributor to the rapid advancement and spread of cancer. Unlike chromosomal gene amplification, which adheres to strict genomic structure, ecDNA provides an adaptable and ever-changing foundation for overactive oncogenes that play a crucial role in malignancy development. The presence of multiple copies of oncogenes such as MYC, EGFR, and CCND1 on ecDNA allows cancer cells to quickly acquire aggressive characteristics, including heightened growth, resistance to cell death, and enhanced formation of blood vessels (S. Wu et al., 2023).

Moreover, the dynamic nature and self-replicating ability of ecDNA grant cancer cells the ability to rapidly alter gene expression to adapt to specific pressures, including hypoxia and immune surveillance. This serves to aid their survival and invasion capabilities, as evidenced by research showing heightened invasiveness and a greater propensity for distant organ metastasis in tumors containing ecDNA. Additionally, ecDNA promotes epithelial-to-mesenchymal transition (EMT), an important step in metastasis, by promoting increased expression of EMT-regulating transcription factors. Therefore, ecDNA plays a crucial role in both the advancement and dissemination of cancerous cells (Z. Li et al., 2022).

4.6.2 Intra-Tumoral Heterogeneity

The presence of multiple subpopulations with different genetic and physical characteristics within a single tumor, known as intra-tumoral heterogeneity, poses a significant challenge in effectively treating cancer. One factor that contributes to this heterogeneity is ecDNA, which is crucial in creating and sustaining it. Unlike changes in chromosomes that are usually passed down during cell division in a consistent manner, ecDNA does not have centromeres and therefore divides unequally. This uneven distribution results in differing numbers of ecDNA copies among daughter cells and ultimately leads to variability in the levels of oncogene expression (Sun & Yu, 2015).

The unpredictable nature of inheritance leads to a varied tumor environment, where each cellular clone may have its own advantages for growth in specific conditions. ecDNA, with its varying structures and regulations, also adds to the diversity among clones. The flexibility of epigenetics and transcription further increases the ability of tumors to evolve and survive despite targeted treatments (Chen & Song, 2022).

4.6.3 Resistance to Therapy and Relapse

One of the most noteworthy impacts of ecDNA in cancer is its ability to contribute to treatment resistance and recurrence. This is due to the increased expression of drug resistance-associated genes found on ecDNA, which allows cancer cells to withstand various forms of therapy. These genes, including *EGFR, MET,* and *PD-L1*, are often amplified on ecDNA and have been directly associated with acquired resistance against tyrosine kinase inhibitors and immune checkpoint therapies. As a result, ecDNA serves as a rapid source for these resistant traits that can be quickly utilized when needed in response to therapeutic stress (Ashique et al., 2022).

Furthermore, the ever-changing nature of ecDNA enables its quantity to fluctuate in response to drug exposure, allowing for reversible resistance that is highly adaptable. As a result, when treatment is discontinued, ecDNA levels may decrease briefly before rising once more upon re-exposure, contributing to cyclical cycles of resistance and sensitivity. Moreover, the presence of ecDNA-driven heterogeneity guarantees that a subset of cancer cells is typically primed to withstand initial therapy, ultimately resulting in relapse. Hence, targeting mechanisms of resistance mediated by ecDNA is becoming increasingly crucial in conquering therapy failures and enhancing long-term outcomes in cancer treatment (Wei et al., 2020).

4.7 DETECTION AND CHARACTERIZATION TECHNIQUES

4.7.1 Imaging (FISH, 3D Microscopy)

The use of imaging techniques has played a crucial role in the discovery and confirmation of extrachromosomal DNA (ecDNA) in cancer. These methods have provided valuable insights into the spatial organization, abundance, and dynamics of ecDNA within the nucleus. Among these techniques, fluorescence in situ hybridization (FISH) is commonly used to visualize ecDNA by utilizing fluorescent probes that target specific DNA sequences. Through FISH, researchers can differentiate ecDNA from chromosomal DNA through its scattered and punctate distribution within the nucleus. This approach has been instrumental in detecting oncogene-containing ecDNA in tumor samples and cell lines from patients, effectively linking the presence of ecDNA with aggressive tumor phenotypes (Yi et al., 2022).

In addition to traditional FISH, advanced three-dimensional microscopy techniques (such as SIM and confocal microscopy) offer a more comprehensive spatial analysis of ecDNA. Through these high-resolution imaging tools, we can observe the formation of ecDNA hubs: clusters of ecDNA molecules that are closely positioned to transcriptional machinery and active areas of chromatin. These imaging methods provide a unique perspective on how ecDNA is functionally organized within the nucleus, shedding light on its role in amplifying transcription and promoting tumor growth. Furthermore, researchers are currently developing live-cell imaging strategies that can track the inheritance and movements of ecDNA in real time, offering promising opportunities for understanding its impact on intratumoral heterogeneity (Markaki et al., 2012).

4.7.2 Sequencing (WGS, ATAC-seq, ChIP-seq)

The incorporation of sequencing technologies has greatly transformed the examination of ecDNA, allowing for its comprehensive discovery, structural reconstruction, and functional interpretation. Whole genome sequencing (WGS) serves as the primary method for identifying ecDNA in cancer, especially by detecting circular genomic elements with extensive amplification that do not linearly connect to chromosomes. By utilizing sophisticated computational resources on WGS data, the architecture of ecDNA can be reconstructed, exposing intricate rearrangements, breakpoint junctions, and oncogene amplification occurrences (S. Wu et al., 2023).

In addition to revealing the structure of ecDNA, epigenomic sequencing methods provide insights into its regulatory landscape. ATAC-seq has been widely utilized to assess chromatin accessibility on ecDNA, demonstrating that these elements possess a highly open chromatin configuration that is conducive to transcriptional activation. ChIP-seq adds to this by pinpointing specific histone modifications, such as H3K27ac and H3K4me3, which are prevalent on ecDNA and serve as markers for active enhancers and promoters. Together, these sequencing techniques not only uncover the complex structure of ecDNA but also shed light on its regulatory characteristics, offering a deeper understanding of how ecDNA contributes to oncogene overexpression and cancer pathophysiology (Y. Li, 2021).

4.7.3 Computational Tools and Databases

ecDNA data is analyzed using advanced computational tools and curated databases that are specialized in handling sequencing and imaging outputs. For instance, *AmpliconArchitect* reconstructs the ecDNA architecture from WGS data by detecting focal amplifications and assembling circular structures. Moreover, *Amplicon Classifier* further categorizes these amplifications into various structural types, such as ecDNA, breakage–fusion–bridge products, or linear amplifications. These tools enable researchers to distinguish ecDNA from other types of gene amplification and offer thorough maps of ecDNA content and structure (Jiang et al., 2023).

The ecDNA Atlas is an up-and-coming resource that collects information on ecDNA occurrences in different types of cancer. It links the presence of ecDNA to various genomic features, oncogenes, and clinical factors. Additionally, researchers are utilizing databases like TCGA (The Cancer Genome Atlas) and CCLE (Cancer Cell Line Encyclopedia) to study the prevalence and role of ecDNA in a wide range of tumors. To further advance this field, experts are investigating ways to combine imaging data with computerized analyses using machine learning techniques for automated detection of ecDNA in pathology images. As technology progresses, these tools will be crucial for identifying ecDNA as both a biomarker and a potential target for precision oncology treatments (Jiang et al., 2023).

4.8 THERAPEUTIC PERSPECTIVES

4.8.1 Targeting ecDNA Biogenesis

The biogenesis of extrachromosomal DNA (ecDNA) is a key factor in cancer development, making it a promising target for therapeutic intervention. By disrupting processes such as DNA damage repair, chromothripsis, and replication stress, ecDNA production can be limited. This can be achieved through the use of certain therapeutic strategies, including the inhibition of DNA damage response pathways using ATM, ATR, or PARP inhibitors. By targeting these mechanisms, we may be able to prevent the formation of ecDNA and its role in oncogenesis (Dong et al., 2023; Z. Li et al., 2022).

Targeting replication stress through the use of DNA polymerase or topoisomerase inhibitors has shown potential in suppressing breakage–fusion–bridge cycles and blocking erroneous repair mechanisms that result in circular DNA fragments. While these methods are still in the preclinical stage, they provide evidence that hindering ecDNA formation could have a significant impact on slowing cancer progression and increasing the effectiveness of standard therapies (Rodriguez et al., 2023).

4.8.2 Disruption of ecDNA Hubs

The latest findings have brought attention to the presence of ecDNA hubs, microenvironments within the nucleus where multiple ecDNA molecules cluster and interact with transcriptional machinery. These hubs function as super-enhancers, greatly boosting the expression of oncogenes compared to those integrated into chromosomes. Targeting these ecDNA hubs could prove to be a new approach in reducing abnormal gene transcription (Cavalli & Misteli, 2013).

The development of therapeutic interventions for disrupting hub integrity involves targeting key proteins and complexes involved in chromatin looping and nuclear architecture, such as BRD4, CTCF, and cohesion. These interventions have the potential to disrupt transcriptional hotspots and potentially reduce oncogene expression, making existing drugs more effective against tumors. Furthermore, research is being conducted on inhibiting nuclear actin polymerization, which can help disperse and break apart ecDNA clusters that contribute to tumor growth. Although these strategies are still in their early stages, they show promise in combating the effects of hub integrity on cancer cells (Mr et al., n.d.).

4.8.3 Epigenetic Modulators and Inhibitors

The regulation of ecDNA through epigenetic means is a promising avenue for potential therapeutic intervention due to its accessible chromatin structure and unique epigenetic signature. Epigenetic inhibitors, including HDAC inhibitors, BET inhibitors like JQ1, and DNA methyltransferase inhibitors, can silence oncogene expression driven by ecDNA by modifying the chromatin landscape. For instance, BET inhibitors have demonstrated effectiveness in suppressing MYC expression on ecDNA by disrupting interactions between acetylated histones and transcriptional coactivators (S. Wu et al., 2019).

Additionally, the distinguishing histone modification patterns found on ecDNA—including elevated levels of H3K27ac and H3K4me3—make these circular elements especially susceptible to drugs that alter chromatin structure. These medications not only decrease transcriptional activity, but they may also disrupt the maintenance and replication processes of ecDNA. It is worth noting that combining epigenetic therapies with conventional chemotherapy or targeted treatments could effectively combat resistance mechanisms linked to oncogene amplification caused by ecDNA. However, it is crucial to precisely target the ecDNA-specific epigenetic markers without disturbing the overall balance of chromatin in the cell (Jiang et al., 2023).

4.8.4 Future Drug Development

The study of ecDNA is constantly advancing, and upcoming pharmaceutical advancements will likely focus on utilizing the distinct biological weaknesses of ecDNA. A main goal is to create small molecules or biologics that specifically target ecDNA structures, possibly through their replication origins, histone profiles, or related protein complexes. These agents could effectively degrade ecDNA, inhibit its transcriptional capabilities, or hinder its replication process (Z. Li et al., 2022).

As research advances, strategies that target ecDNA-positive tumors through synthetic lethality are becoming increasingly popular. By leveraging the unique characteristics of ecDNA, such as increased replication stress or reliance on certain repair pathways, these approaches specifically target and eliminate cancer cells harboring ecDNA. With further developments, immunotherapeutic methods may also emerge, potentially utilizing ecDNA's role in neoantigen presentation or immune evasion mechanisms (S. Wu et al., 2022).

Advances in personalized medicine have led to the use of ecDNA as a biomarker for predicting prognosis, selecting treatments, and monitoring responses. With the development of liquid biopsy techniques capable of detecting ecDNA fragments in the bloodstream, therapeutic decisions can be made in real time. As our knowledge about ecDNA biology expands, targeted therapy drugs aimed at neutralizing its cancer-causing effects will become more precise and effective, ushering in a new era of cancer treatment (Gambardella et al., 2020).

4.9 CONCLUSION

ecDNA, a circular type of DNA without centromeres and telomeres, has become a significant factor in cancer biology. It has significantly changed our understanding of genomic architecture, the regulation of oncogenes, and the evolution of tumors. In addition to its distinct structural features, ecDNA also differs in its functionality from chromosomal DNA. This remarkable type of DNA can house amplified oncogenes and create active transcriptional hubs. Moreover, it dynamically controls gene expression through enhancer hijacking and epigenetic modifications, making it a potent driver of malignancy. ecDNA is formed through various mechanisms such as breakage–fusion–bridge cycles, chromothripsis, and replication-based errors—all reflections of the chaotic genomic patterns often observed in aggressive tumors.

New technologies, such as high-resolution imaging, next-generation sequencing, and computational modeling, have revolutionized our understanding of ecDNA. Through these tools, we have been able to uncover crucial details about its structure, behavior, and impact in cancer progression. By examining the epigenetic landscape of ecDNA, including distinctive histone marks and chromatin accessibility profiles, we can better comprehend its active regulatory role. With this knowledge in hand, ecDNA presents itself not only as a valuable tumor marker but also as a promising therapeutic target for addressing treatment resistance.

In a clinical setting, disrupting the creation of ecDNA, inhibiting its epigenetic function, or breaking down transcriptional hubs could significantly alter the way we treat ecDNA-related cancers. Researchers are currently investigating the use of epigenetic modifiers, DNA repair blockers, and emerging drugs based on synthetic lethality. Additionally, as immunotherapy advances, comprehending how ecDNA influences the interaction between tumors and the immune system could open up new possibilities for combination treatments such as targeting neoantigens or preventing immune evasion.

In short, ecDNA is a promising area of cancer research that connects basic molecular biology to practical clinical use. As our comprehension of its biology progresses, implementing ecDNA-based knowledge into diagnostic, prognostic, and treatment approaches has the potential to revolutionize cancer treatment, enhance results, and individualize treatment plans.

REFERENCES

Achrem, M., Szućko, I., & Kalinka, A. (2020). The epigenetic regulation of centromeres and telomeres in plants and animals. *Comparative Cytogenetics*, *14*(2), 265–311. https://doi.org/10.3897/CompCytogen.v14i2.51895

Ashique, S., Upadhyay, A., Garg, A., Mishra, N., Hussain, A., Negi, P., Hing, G. B., Bhatt, S., Ali, M. K., Gowthamarajan, K., Singh, S. K., Gupta, G., Chellappan, D. K., & Dua, K. (2022). Impact of ecDNA: A mechanism that directs tumorigenesis in cancer drug resistance-a review. *Chemico-Biological Interactions*, *363*, 110000. https://doi.org/10.1016/j.cbi.2022.110000

Bailey, C., Pich, O., Thol, K., Watkins, T. B. K., Luebeck, J., Rowan, A., Stavrou, G., Weiser, N. E., Dameracharla, B., Bentham, R., Lu, W.-T., Kittel, J., Yang, S. Y. C., Howitt, B. E., Sharma, N., Litovchenko, M., Salgado, R., Hung, K. L., Cornish, A. J., . . . Swanton, C. (2024). Origins and impact of extrachromosomal DNA. *Nature*, *635*(8037), 193–200. https://doi.org/10.1038/s41586-024-08107-3

Bannister, A. J., & Kouzarides, T. (2011). Regulation of chromatin by histone modifications. *Cell Research*, *21*(3), 381–395. https://doi.org/10.1038/cr.2011.22

Bure, I. V., Nemtsova, M. V., & Kuznetsova, E. B. (2022). Histone modifications and non-coding RNAs: Mutual epigenetic regulation and role in pathogenesis. *International Journal of Molecular Sciences*, *23*(10), 5801. https://doi.org/10.3390/ijms23105801

Casas-Delucchi, C. S., Daza-Martin, M., Williams, S. L., & Coster, G. (2022). The mechanism of replication stalling and recovery within repetitive DNA. *Nature Communications*, *13*(1), 3953. https://doi.org/10.1038/s41467-022-31657-x

Cavalli, G., & Misteli, T. (2013). Functional implications of genome topology. *Nature Structural & Molecular Biology*, *20*(3), 290–299. https://doi.org/10.1038/nsmb.2474

Chandler, L. A., & Jones, P. A. (1988). Hypomethylation of DNA in the regulation of gene expression. *Developmental Biology (New York, N.Y.: 1985)*, *5*, 335–349. https://doi.org/10.1007/978-1-4615-6817-9_12

Chen, X., & Song, E. (2022). The theory of tumor ecosystem. *Cancer Communications*, *42*(7), 587–608. https://doi.org/10.1002/cac2.12316

Claringbould, A., & Zaugg, J. B. (2021). Enhancers in disease: Molecular basis and emerging treatment strategies. *Trends in Molecular Medicine*, *27*(11), 1060–1073. https://doi.org/10.1016/j.molmed.2021.07.012

Cooper, L. A., Demicco, E. G., Saltz, J. H., Powell, R. T., Rao, A., & Lazar, A. J. (2018). PanCancer insights from the cancer genome atlas: The pathologist's perspective. *The Journal of Pathology*, *244*(5), 512–524. https://doi.org/10.1002/path.5028

Cortés-Ciriano, I., Lee, J. J.-K., Xi, R., Jain, D., Jung, Y. L., Yang, L., Gordenin, D., Klimczak, L. J., Zhang, C.-Z., Pellman, D. S., & Park, P. J. (2020). Comprehensive analysis of chromothripsis in 2,658 human cancers using whole-genome sequencing. *Nature Genetics*, *52*(3), 331–341. https://doi.org/10.1038/s41588-019-0576-7

de Groot, D., Spanjaard, A., Hogenbirk, M. A., & Jacobs, H. (2023). Chromosomal Rearrangements and Chromothripsis: The alternative end generation model. *International Journal of Molecular Sciences*, *24*(1), 794. https://doi.org/10.3390/ijms24010794

D'haeseleer, P. (2005). How does gene expression clustering work? *Nature Biotechnology*, *23*(12), 1499–1501. https://doi.org/10.1038/nbt1205-1499

Dong, Y., He, Q., Chen, X., Yang, F., He, L., & Zheng, Y. (2023). Extrachromosomal DNA (ecDNA) in cancer: Mechanisms, functions, and clinical implications. *Frontiers in Oncology*, *13*, 1194405. https://doi.org/10.3389/fonc.2023.1194405

Espinosa, E., Bautista, R., Larrosa, R., & Plata, O. (2024). Advancements in long-read genome sequencing technologies and algorithms. *Genomics*, *116*(3), 110842. https://doi.org/10.1016/j.ygeno.2024.110842

Gambardella, V., Tarazona, N., Cejalvo, J. M., Lombardi, P., Huerta, M., Roselló, S., Fleitas, T., Roda, D., & Cervantes, A. (2020). Personalized medicine: Recent progress in cancer therapy. *Cancers*, *12*(4), 1009. https://doi.org/10.3390/cancers12041009

Gan, J., Huang, M., Wang, W., Fu, G., Hu, M., Zhong, H., Ye, X., & Cao, Q. (2024). Novel genome-wide DNA methylation profiling reveals distinct epigenetic landscape, prognostic model and cellular composition of early-stage lung adenocarcinoma. *Journal of Translational Medicine*, *22*(1), 428. https://doi.org/10.1186/s12967-024-05146-2

Hung, K. L., Mischel, P. S., & Chang, H. Y. (2022). Gene regulation on extrachromosomal DNA. *Nature Structural & Molecular Biology*, *29*(8), 736–744. https://doi.org/10.1038/s41594-022-00806-7

Hung, K. L., Yost, K. E., Xie, L., Shi, Q., Helmsauer, K., Luebeck, J., Schöpflin, R., Lange, J. T., Chamorro González, R., Weiser, N. E., Chen, C., Valieva, M. E., Wong, I. T.-L., Wu, S., Dehkordi, S. R., Duffy, C.

V., Kraft, K., Tang, J., Belk, J. A., . . . Chang, H. Y. (2021). ecDNA hubs drive cooperative intermolecular oncogene expression. *Nature*, *600*(7890), 731–736. https://doi.org/10.1038/s41586-021-04116-8

Jiang, R., Yang, M., Zhang, S., & Huang, M. (2023). Advances in sequencing-based studies of microDNA and ecDNA: Databases, identification methods, and integration with single-cell analysis. *Computational and Structural Biotechnology Journal*, *21*, 3073–3080. https://doi.org/10.1016/j.csbj.2023.05.017

Jones-Weinert, C., Mainz, L., & Karlseder, J. (2025). Telomere function and regulation from mouse models to human ageing and disease. *Nature Reviews Molecular Cell Biology*, *26*(4), 297–313. https://doi.org/10.1038/s41580-024-00800-5

Kiselev, I. S., Kulakova, O. G., Boyko, A. N., & Favorova, O. O. (2021). DNA methylation as an epigenetic mechanism in the development of multiple sclerosis. *Acta Naturae*, *13*(2), 45–57. https://doi.org/10.32607/actanaturae.11043

Kumar, B., Navarro, C., Yung, P. Y. K., Lyu, J., Salazar Mantero, A., Katsori, A.-M., Schwämmle, H., Martin, M., & Elsässer, S. J. (2025). Multiplexed chromatin immunoprecipitation sequencing for quantitative study of histone modifications and chromatin factors. *Nature Protocols*, *20*(3), 779–809. https://doi.org/10.1038/s41596-024-01058-z

Lanctôt, C., Cheutin, T., Cremer, M., Cavalli, G., & Cremer, T. (2007). Dynamic genome architecture in the nuclear space: Regulation of gene expression in three dimensions. *Nature Reviews Genetics*, *8*(2), 104–115. https://doi.org/10.1038/nrg2041

Lange, J. T., Rose, J. C., Chen, C. Y., Pichugin, Y., Xie, L., Tang, J., Hung, K. L., Yost, K. E., Shi, Q., Erb, M. L., Rajkumar, U., Wu, S., Taschner-Mandl, S., Bernkopf, M., Swanton, C., Liu, Z., Huang, W., Chang, H. Y., Bafna, V., . . . Mischel, P. S. (2022). The evolutionary dynamics of extrachromosomal DNA in human cancers. *Nature Genetics*, *54*(10), 1527–1533. https://doi.org/10.1038/s41588-022-01177-x

Li, Y. (2021). Modern epigenetics methods in biological research. *Methods (San Diego, Calif.)*, *187*, 104–113. https://doi.org/10.1016/j.ymeth.2020.06.022

Li, Y., & Tollefsbol, T. O. (2011). DNA methylation detection: Bisulfite genomic sequencing analysis. *Methods in Molecular Biology (Clifton, N.J.)*, *791*, 11–21. https://doi.org/10.1007/978-1-61779-316-5_2

Li, Z., Wang, B., Liang, H., & Han, L. (2022). Pioneering insights of extrachromosomal DNA (ecDNA) generation, action and its implications for cancer therapy. *International Journal of Biological Sciences*, *18*(10), 4006–4025. https://doi.org/10.7150/ijbs.73479

Liu, T., & Conesa, A. (2025). Profiling the epigenome using long-read sequencing. *Nature Genetics*, *57*(1), 27–41. https://doi.org/10.1038/s41588-024-02038-5

Lovett, S. T. (2017). Template-switching during replication fork repair in bacteria. *DNA Repair*, *56*, 118–128. https://doi.org/10.1016/j.dnarep.2017.06.014

MacKinnon, R. N., & Campbell, L. J. (2011). The role of dicentric chromosome formation and secondary centromere deletion in the evolution of myeloid malignancy. *Genetics Research International*, *2011*, 643628. https://doi.org/10.4061/2011/643628

Malkova, A., & Ira, G. (2013). Break-induced replication: Functions and molecular mechanism. *Current Opinion in Genetics & Development*, *23*(3), 271–279. https://doi.org/10.1016/j.gde.2013.05.007

Markaki, Y., Smeets, D., Fiedler, S., Schmid, V. J., Schermelleh, L., Cremer, T., & Cremer, M. (2012). The potential of 3D-FISH and super-resolution structured illumination microscopy for studies of 3D nuclear architecture: 3D structured illumination microscopy of defined chromosomal structures visualized by 3D (immuno)-FISH opens new perspectives for studies of nuclear architecture. *BioEssays: News and Reviews in Molecular, Cellular and Developmental Biology*, *34*(5), 412–426. https://doi.org/10.1002/bies.201100176

Miller, J. L., & Grant, P. A. (2013). The role of DNA methylation and histone modifications in transcriptional regulation in humans. *Sub-Cellular Biochemistry*, *61*, 289–317. https://doi.org/10.1007/978-94-007-4525-4_13

Moore, L. D., Le, T., & Fan, G. (2013). DNA methylation and its basic function. *Neuropsychopharmacology*, *38*(1), 23–38. https://doi.org/10.1038/npp.2012.112

Mr, M., Ja, J., M, S., M, M., Mr, M., A, J., N, K., L, A., A, A. M., & Ma, S. (n.d.). Integrated bioinformatics analysis reveals upregulated extracellular matrix hub genes in pancreatic cancer: Implications for diagnosis, prognosis, immune infiltration, and therapeutic strategies. *PubMed*. Retrieved April 21, 2025, from https://pubmed.ncbi.nlm.nih.gov/38639039/

Pepin, A.-S., & Schneider, R. (2024). Emerging toolkits for decoding the co-occurrence of modified histones and chromatin proteins. *EMBO Reports*, *25*(8), 3202–3220. https://doi.org/10.1038/s44319-024-00199-2

Qiu, H., Shao, Z.-Y., Wen, X., & Zhang, L.-Z. (2020). New insights of extrachromosomal DNA in tumorigenesis and therapeutic resistance of cancer. *American Journal of Cancer Research*, *10*(12), 4056–4065.

Rashid, M., & Stingl, U. (2015). Contemporary molecular tools in microbial ecology and their application to advancing biotechnology. *Biotechnology Advances*, *33*(8), 1755–1773. https://doi.org/10.1016/j.biotechadv.2015.09.005

Rodriguez, I., Rossi, N. M., Keskus, A., Xie, Y., Ahmad, T., Bryant, A., Lou, H., Paredes, J. G., Milano, R., Rao, N., Tulsyan, S., Boland, J. F., Luo, W., Liu, J., O'Hanlon, T., Bess, J., Mukhina, V., Gaykalova, D., Yuki, Y., . . . Dean, M. (2023). Insights into the mechanisms and structure of breakage-fusion-bridge cycles in cervical cancer using long-read sequencing. *medRxiv*, 2023.08.21.23294276. https://doi.org/10.1101/2023.08.21.23294276

Singh, M., Raseley, K., Perez, A. M., MacKenzie, D., Kosiyatrakul, S. T., Desai, S., Batista, N., Guru, N., Loomba, K. K., Abid, H. Z., Wang, Y., Udo-Bellner, L., Stout, R. F., Schildkraut, C. L., Xiao, M., & Zhang, D. (2024). Elucidation of the molecular mechanism of the breakage-fusion-bridge (BFB) cycle using a CRISPR-dCas9 cellular model. *Nucleic Acids Research*, *52*(19), 11689–11703. https://doi.org/10.1093/nar/gkae747

Stöber, M. C., Chamorro González, R., Brückner, L., Conrad, T., Wittstruck, N., Szymansky, A., Eggert, A., Schulte, J. H., Koche, R. P., Henssen, A. G., Schwarz, R. F., & Haase, K. (2024). Intercellular extrachromosomal DNA copy-number heterogeneity drives neuroblastoma cell state diversity. *Cell Reports*, *43*(9), 114711. https://doi.org/10.1016/j.celrep.2024.114711

Sun, X., & Yu, Q. (2015). Intra-tumor heterogeneity of cancer cells and its implications for cancer treatment. *Acta Pharmacologica Sinica*, *36*(10), 1219–1227. https://doi.org/10.1038/aps.2015.92

Talukdar, P. D., & Chatterji, U. (2023). Transcriptional co-activators: Emerging roles in signaling pathways and potential therapeutic targets for diseases. *Signal Transduction and Targeted Therapy*, *8*(1), 1–41. https://doi.org/10.1038/s41392-023-01651-w

Tanaka, H., & Watanabe, T. (2020). Mechanisms underlying recurrent genomic amplification in human cancers. *Trends in Cancer*, *6*(6), 462–477. https://doi.org/10.1016/j.trecan.2020.02.019

Tang, L., Nogales, E., & Ciferri, C. (2010). Structure and function of SWI/SNF chromatin remodeling complexes and mechanistic implications for transcription. *Progress in Biophysics and Molecular Biology*, *102*(2–3), 122–128. https://doi.org/10.1016/j.pbiomolbio.2010.05.001

Warburton, P. E., & Sebra, R. P. (2023). Long-read DNA sequencing: Recent advances and remaining challenges. *Annual Review of Genomics and Human Genetics*, *24*, 109–132. https://doi.org/10.1146/annurev-genom-101722-103045

Wei, J., Wu, C., Meng, H., Li, M., Niu, W., Zhan, Y., Jin, L., Duan, Y., Zeng, Z., Xiong, W., Li, G., & Zhou, M. (2020). The biogenesis and roles of extrachromosomal oncogene involved in carcinogenesis and evolution. *American Journal of Cancer Research*, *10*(11), 3532–3550.

Wu, M., & Rai, K. (2022). Demystifying extrachromosomal DNA circles: Categories, biogenesis, and cancer therapeutics. *Computational and Structural Biotechnology Journal*, *20*, 6011–6022. https://doi.org/10.1016/j.csbj.2022.10.033

Wu, S., Bafna, V., Chang, H. Y., & Mischel, P. S. (2022). Extrachromosomal DNA: An emerging hallmark in human cancer. *Annual Review of Pathology*, *17*, 367–386. https://doi.org/10.1146/annurev-pathmechdis-051821-114223

Wu, S., Tao, T., Zhang, L., Zhu, X., & Zhou, X. (2023). Extrachromosomal DNA (ecDNA): Unveiling its role in cancer progression and implications for early detection. *Heliyon*, *9*(11), e21327. https://doi.org/10.1016/j.heliyon.2023.e21327

Wu, S., Turner, K. M., Nguyen, N., Raviram, R., Erb, M., Santini, J., Luebeck, J., Rajkumar, U., Diao, Y., Li, B., Zhang, W., Jameson, N., Corces, M. R., Granja, J. M., Chen, X., Coruh, C., Abnousi, A., Houston, J., Ye, Z., . . . Mischel, P. S. (2019). Circular ecDNA promotes accessible chromatin and high oncogene expression. *Nature*, *575*(7784), 699–703. https://doi.org/10.1038/s41586-019-1763-5

Yang, L., Jia, R., Ge, T., Ge, S., Zhuang, A., Chai, P., & Fan, X. (2022). Extrachromosomal circular DNA: Biogenesis, structure, functions and diseases. *Signal Transduction and Targeted Therapy*, *7*, 342. https://doi.org/10.1038/s41392-022-01176-8

Yang, Q.-L., Xie, Y., Qiao, K., Lim, J. Y. S., & Wu, S. (2025). Modern biology of extrachromosomal DNA: A decade-long voyage of discovery. *Cell Research*, *35*(1), 11–22. https://doi.org/10.1038/s41422-024-01054-8

Yi, E., Gujar, A. D., Guthrie, M., Kim, H., Zhao, D., Johnson, K. C., Amin, S. B., Costa, M. L., Yu, Q., Das, S., Jillette, N., Clow, P. A., Cheng, A. W., & Verhaak, R. G. (2022). Live-cell imaging shows uneven segregation of extrachromosomal DNA elements and transcriptionally active extrachromosomal DNA hubs in cancer. *Cancer Discovery*, *12*(2), 468–483. https://doi.org/10.1158/2159-8290.CD-21-1376

Zhang, H., Liu, B., Cheng, J., Li, Z., Jia, M., Li, M., Zhao, L., Wang, L., & Xi, Y. (2024). Characterization and integrated analysis of extrachromosomal DNA amplification in hematological malignancies. *Neoplasia*, *56*, 101025. https://doi.org/10.1016/j.neo.2024.101025

Zhu, Y., Gong, L., & Wei, C.-L. (2022). Guilt by association: EcDNA as a mobile transactivator in cancer. *Trends in Cancer*, *8*(9), 747–758. https://doi.org/10.1016/j.trecan.2022.04.011

5 ecDNA
Role in Cancer Drug Resistance

Joy Das,[1] Utpal Bhui,[2] Komal Kriti,[3] Yogesh Tiwari,[4] Sachin Raikwar,[5] and Farzad Taghizadeh-Hesary[6]

1 School of Pharmaceutical Sciences
Lovely Professional University
Punjab, India
2 School of Pharmaceutical Sciences
Lovely Professional University
Phagwara, India
3 Department of Pharmaceutical Science
Dhanbad College of Pharmacy and Research Institute
Shaktinagar, India
4 Department of Pharmaceutics
Research Scholar, School of Pharmacy, ITM University
Gwalior, India
5 Department of Pharmacology
Research Scholar, Faculty of Pharmacy, P.K. University
Tehsil-Karera Shivpuri India
6 Breast Cancer Research Center, Iran University
of Medical Sciences, Tehran, Iran

5.1 INTRODUCTION

Extrachromosomal DNA (ecDNA) is circular DNA outside the typical chromosomal genome contained in the nucleus of eukaryotic cells [1]. In contrast to naturally extrachromosomal mitochondrial DNA located in mitochondrial organelles, ecDNA in cancer consists of acentric (not containing centromeres) DNA fragments of circular shape, typically representing sequences of chromosomes, especially oncogene-containing ones [2]. The existence of ecDNA in cancer cells was recognized several decades ago, with early works of the 1960s and 1970s detecting features such as so-called double minutes (DMs) in cytogenetic studies of tumors [3]. These duplicated minutes were subsequently identified as ecDNA, typically present in fast-growing and genetically unstable cancer cells. Yet, the prevalence of ecDNA in tumors, as well as the biological importance of ecDNA in the tumor milieu, came to light only when high-resolution genomic technologies, such as next-generation sequencing and single-cell imaging, were introduced [4]. A defining characteristic of ecDNA compared to chromosomal DNA is that it has neither centromeres nor telomeres and therefore does not participate in Mendelian inheritance following cell division. Rather, ecDNA is randomly distributed into daughter cells, which enhances intratumoral heterogeneity [5]. This feature offers a dynamic platform of rapid genetic evolution, selective growth advantage under environmental stress conditions, such as chemotherapy. ecDNA is highly enriched with oncogenes like MYC, EGFR, CDK4, MDM2, and HER2, which are frequently expressed at amplified levels in tumors [6]. Such ecDNAs are able to induce overexpression of their oncogenic cargo without facing the restrictions of chromatin structure imposed in chromosomes. In addition, they frequently establish ecDNA hubs or

 DOI: 10.1201/9781003655299-5

regions within the cell to enable collaborative interactions between copies of oncogenes, enhancer hijacking, and transcriptional landscape remodeling [7]. These mechanisms provide a deep proliferative and survival advantage to tumor cells.

In recent years, ecDNA has gained substantial attention in oncology for several reasons. First, ecDNA is now recognized as a key driver of tumorigenesis and cancer progression across a broad range of malignancies, including glioblastoma, lung, breast, and colorectal cancers. Second, ecDNA plays a central role in drug resistance [8]. Upon therapeutic pressure, ecDNA can dynamically change in copy number, enabling cancer cells to bypass drug inhibition by amplifying drug-resistance genes or oncogenic pathways. For example, studies have shown increased ecDNA-mediated EGFR amplification in glioblastoma patients relapsing after EGFR-targeted therapies [9]. Third, ecDNA correlates with poor clinical outcomes and aggressive disease phenotypes, such as increased metastasis and shortened survival. Its absence in normal cells makes it an attractive candidate for targeted therapies or diagnostic markers. Technological breakthroughs have further accelerated interest in ecDNA [10]. Tools like AmpliconArchitect, Circle-Seq, and ecTag enable high-resolution detection and characterization of ecDNA in patient samples, allowing researchers to track ecDNA dynamics in real time [11]. Liquid biopsy methods for ecDNA detection are also emerging, offering promising non-invasive diagnostic tools. In light of these developments, ecDNA is now considered not only a hallmark of cancer biology but also a potential therapeutic vulnerability. In this chapter, we aim to explore the multifaceted role of ecDNA in cancer drug resistance, with a focus on its mechanistic underpinnings, oncogenic impact, clinical relevance, and the opportunities it presents for future therapeutic interventions. Table 5.1 describes advances in technologies for extrachromosomal DNA (ecDNA) detection and characterization [12–21].

5.2 BIOGENESIS AND MOLECULAR FEATURES

The formation of ecDNA in cancer is a result of various complex genomic disruptions and faulty repair processes that lead to the production of circular, centromere-lacking DNA fragments. These structures are independent of chromosomes and play a significant role in cancer aggressiveness [22]. Modern genomic technologies, primarily single-cell sequencing and whole-genome mapping, have pointed to the role of ecDNA in oncogene amplification, intratumoral heterogeneity, and therapy resistance. Chromothripsis is one major process explaining the formation of ecDNA, a disorganized chromosomal process in which a stretch of DNA breaks into several parts under replicative tension or mitotic fate [23]. Some of the fragments are reinserted into the genome, whereas some circularize and maintain an extrachromosomal state. Such rings frequently harbor oncogenes as well as regulatory elements that allow them to evade normal chromatin regulation and drive aberrant gene expression. The other extensively documented source is double minutes (DMs), which are small, acentric fragments of DNA seen under the microscope in most cancers [24]. In contrast to the sudden mechanism that characterizes chromothripsis, DMs may occur gradually as a result of extensively acquired genomic instability and are often enriched in oncogenes, including MYC, EGFR, and CDK4. On the same note, incorrect reconnections of DNA fragments occur during repair of the double-strand breaks by error-prone repair pathways such as the non-homologous end joining (NHEJ) pathway or microhomology-mediated end joining (MMEJ), where the edges are reconnected in an inaccurate style, leading to the accidental formation of circular DNA [25]. Telomere loss also leads to processes such as breakage–fusion–bridge (BFB) cycles, which result in repeated breakage and misrepair of chromosomal endings, producing ecDNA gradually. ecDNA is structurally unique because it is acentric and does not have centromeres to segregate equally during mitosis. This leads to uneven segregation of ecDNA into daughter cells during cell division, which increases genomic diversity in tumors and contributes to clonal selection [26]. These are round-shaped DNA molecules whose length differs significantly—a few kilobases to megabases—and where several genes and enhancers can be found [27]. Circle-Seq and AmpliconArchitect technologies have revealed that the ecDNA

TABLE 5.1
Comparative Overview of Modern Tools for ecDNA Detection and Analysis

S.No.	Tool Name	Methodology	Core Features	Benefits	Limitations	Ref
1	ECdetect	Fluorescence Microscopy-Based	Dual adaptive thresholds applied to differentiate DNA and ecDNA in DAPI images	Strong correlation with manual observation	Misses smaller ecDNA due to high background noise	[12]
2	ecSeg	Fluorescence Microscopy-Based	Utilizes U-net neural network to detect ecDNA and FISH probe signals concurrently	Higher F1 score and recall than ECdetect	Poor generalizability; needs dataset-specific tuning	[13]
3	Circle-seq	Sequencing-Based	Uses alkaline lysis and chromatography for ecDNA isolation, followed by sequencing	High specificity and complete ecDNA sequence coverage	Complex protocol and large sample requirement	[14]
4	Circle_finder	Sequencing-Based	ATP-dependent exonuclease digestion followed by random primer amplification	Identifies ecDNA as potential cancer biomarker	Bias toward detection of short-length ecDNA	[15]
5	Circle-Map	Sequencing-Based	Detects ecDNA from short-read sequences crossing circular junctions	High precision even for short soft-clipped reads (>4 nts)	Not suitable for ecDNA containing mutations	[16]
6	ecc_finder	Sequencing-Based	Detects ecDNA from long-read Nanopore sequencing	Identifies ecDNA from repetitive genomic regions	Requires availability of nanopore sequencing data	[17]
7	AmpliconArchitect	Sequencing-Based	Reconstructs ecDNA amplicons using whole-genome sequencing (WGS)	Cost-effective use of short-read WGS data	Manual cutoff setting limits detection precision	[18]
8	AmpliconReconstructor	Sequencing-Based	Integrates next-gen sequencing (NGS) and optical mapping for accurate reconstruction	High-resolution breakpoint detection	Requires additional optical mapping (OM) input	[19]
9	ecTag	CRISPR-Based	Applies CRISPR-Cas and Casilio systems to visualize ecDNA in live cells	Enables live-cell ecDNA visualization	gDNA design is dependent on precise breakpoint identification	[20]
10	CRISPR-CATCH	CRISPR-Based	Profiles genetic and epigenetic properties of large ecDNA molecules	Allows base-resolution ecDNA and epigenetic analysis	Limited to profiling large ecDNA molecules	[21]

often contains distal regulatory regions, including super-enhancers and non-coding RNAs, which contribute to unregulated transcription [28]. One feature of ecDNA is that it exists in an open chromatin state, unlike the tightly packed chromosomal DNA structure. Lacking topological constraints and nuclear lamina anchoring, ecDNA also facilitates free enhancer-promoter interactions. ecDNA clusters tend to become nucleotypic hubs, in which individual ecDNA molecules localize together to increase overall gene expression [29]. These hubs have elevated measures of active histone marks (e.g., H3K27ac, H3K4me3), promoting them as versatile transcription machines. This epigenetic signature enables ecDNA to function not only as gene amplifiers but also as central regulators that re-engineer transcriptional networks to support malignancy and cope with treatment pressure [30].

5.3 ROLE IN ONCOGENE AMPLIFICATION AND TUMOR HETEROGENEITY

ecDNA has become a major factor in the amplification of oncogenes and tumor heterogeneity. In contrast with linear chromosomal DNA, ecDNA fragments are circular without centromeres or telomeres, which is why this ecDNA can be reproduced flexibly and freely and cause oncogenes to be replicated [31]. The increased oncogene expression and the random separation of chromosomes during mitosis that occurs because of this structural advantage result in increased genetic diversity and a rapid clonal selection in the tumor. Some of the most important oncogenes that are commonly amplified by ecDNA are MYC, EGFR, HER2, and MDM2, among others. These genes are involved in the control of cell growth, survival, and tumor suppression resistance [32]. As an example, high copies of MYC present on ecDNA can contribute to its increased transcription as well as permit open chromatin accessibility. In a parallel manner, EGFR amplification on ecDNA is linked with resistance to targeted treatment in glioblastoma and lung cancer [33]. HER2 in ecDNA makes it possible to express it at high levels despite anti-HER2 therapies. An overexpression of MDM2 negates p53 action, favoring cell survival during periods of stress [34]. The loss of centromeres makes ecDNA segregate randomly in cell division, resulting in unequal distribution among daughter cells. This leads to variability of oncogene dosage, which contributes to intratumoral heterogeneity [35]. During treatment pressure, some of the clones acquire a survival advantage, and others become dormant. This promotes drug resistance and relapse. Also, unlike chromosomal mutations, which take time to accumulate, ecDNA-driven adaptation can be rapid and speed up clonal evolution [36]. ecDNA is not stable and can rearrange or fuse with other ecDNA to form hybrid elements with novel regulatory potential. These can be the super-enhancers that enhance the transcription of oncogenes. Moreover, ecDNA activity does not end with the cancer cell, where it potentially impacts the tumor microenvironment by inducing angiogenesis, stromal remodeling, and immune evasion through the secreted vesicles [37]. Clinically, ecDNA is associated with aggressive disease, increased risk of metastasis, treatment resistance, and poor prognosis. Due to its circular architecture and dynamic behavior, it is a potent force in cancer evolution and a promising target of therapeutic intervention and biomarker development [38].

5.4 MECHANISMS OF DRUG RESISTANCE

Whenever drugs are used, drug resistance is the primary obstacle in the successful treatment of cancer, which may subsequently result in recurrence, metastasis, and poor prognosis. Emerging studies have linked ecDNA as a major contributor in conferring treatment resistance of various tumors [39]. The peculiar property of the circular, acentric, and highly dynamic nature of ecDNA helps cancer cells to increase resistance genes, engage and reprogram transcription, and harness swift adjustments in the face of therapeutic pressure. Amplification of therapy-related genes is one of the major resistance strategies [40]. ecDNA may carry copies of various therapy-related genes, including EGFR, HER2, MDM2, and ABC transporters (ABCB1, ABCG2), which mediate drug efflux, antagonize apoptosis, or stimulate DNA repair. For example, glioblastoma cells with ecDNA carrying EGFR show enhanced expression and drug resistance, while breast cancer cells with HER2 on ecDNA resist HER2-targeted therapies [41]. These genes can be rapidly upregulated in response to treatment pressure due to the flexible copy number of ecDNA. Epigenetic reprogramming and transcriptional rewiring further enhance resistance. ecDNA displays an open chromatin conformation enriched with active histone marks (e.g., H3K27ac, H3K4me3), promoting high transcriptional activity [42]. Moreover, ecDNA interacts with distant enhancers and forms nuclear hubs—clusters that concentrate transcriptional machinery and support co-expression of multiple resistance genes. This spatial organization drives alternative transcriptional programs that bypass the effects of drugs such as BET inhibitors, which target super-enhancer-driven oncogenes [43]. A defining feature of ecDNA is its plasticity and adaptability under therapeutic pressure. Post-treatment, cancer cells can increase ecDNA copy number or form new ecDNA elements that provide survival advantages [44]. In glioblastoma and lung cancer models, treatment with EGFR or ALK inhibitors led to the expansion of ecDNA carrying resistant oncogenes. This rapid adaptation enables tumors to escape therapy and evolve quickly compared to chromosomal mutation-based mechanisms. ecDNA can also undergo recombination and fusion, generating hybrid molecules with multiple resistance genes from different loci. These chimeric ecDNAs coordinate the activation of bypass signaling pathways, supporting resistance to combination therapies [45]. Additionally, ecDNA is implicated in immune evasion, with amplified immune checkpoint regulators (e.g., PD-L1) and suppressors of antigen presentation potentially aiding in resistance to immunotherapies. Importantly, ecDNA levels often increase after therapy, even if undetectable beforehand, suggesting a selection or emergence of ecDNA under drug stress [46]. Its detectability in circulating tumor DNA (ctDNA) offers potential for non-invasive monitoring of resistance and relapse.

5.5 CLINICAL IMPLICATIONS AND THERAPEUTIC STRATEGIES

The identification of ecDNA as a critical element in oncogene amplification and therapeutic resistance is reshaping the landscape of cancer diagnostics and treatment. Its firmly established role in enhancing tumor aggressiveness, clonal evolution, and drug resistance has made ecDNA a prospective biomarker and potential therapeutic target [47]. Following the development of high-throughput sequencing technology, spatial transcriptomics, and liquid biopsy platforms, it is now possible to identify and characterize ecDNA with greater accuracy, which has led to the possibility of incorporating ecDNA in clinical oncology [48]. ecDNA is used as a diagnostic and prognostic biomarker, which has become one of the most promising uses. Owing to its prevalence in various cancers and strong correlation with poor outcomes, detecting ecDNA could assist in early diagnosis, tumor classification, and predicting treatment response [49]. Unlike linear chromosomal fragments, ecDNA retains its circular structure and often exists in high copy numbers, which makes it easier to distinguish through technologies like rolling-circle amplification, digital PCR, or advanced sequencing [50]. Importantly, ecDNA can be detected not only in solid tumor biopsies but also in minimally invasive liquid biopsies—such as circulating tumor DNA (ctDNA) and extracellular vesicles in blood—allowing real-time tracking of tumor progression, residual disease, and therapeutic resistance [51]. Longitudinal monitoring of oncogenes such as MYC or EGFR on ecDNA in patient blood samples could enable clinicians to assess treatment efficacy or anticipate relapse before clinical symptoms appear. From a therapeutic perspective, efforts are underway to exploit ecDNA-specific vulnerabilities. Although no drugs currently target ecDNA directly, early research has revealed exploitable pathways [52]. One promising target is CHK1, a checkpoint kinase that mitigates replication stress. Because ecDNA-bearing cells often suffer from elevated replication stress due to unregulated oncogene expression, CHK1 inhibition selectively impairs ecDNA maintenance, reduces copy number, and suppresses tumor growth in preclinical models. Similarly, cancer cells with high ecDNA content heavily depend on DNA damage repair mechanisms, especially ATR and PARP pathways [53]. Inhibiting these pathways may compromise ecDNA stability, offering an effective strategy to weaken ecDNA-driven tumors. Another therapeutic direction involves disrupting ecDNA inheritance.[54] ecDNA has no centromeres and breaks randomly during the

TABLE 5.2
Emerging Clinical Implications of ecDNA in Cancer Diagnosis, Prognosis, Resistance, and Therapy

S.No.	Functional Role	Key Insights	References
1	Diagnostic Indicator	ecDNA fragments have potential as diagnostic biomarkers, detectable in plasma of lung cancer patients.	[58]
2	Prognostic Marker	—Specific ecDNAs correlate with poor outcomes in cancers like glioblastoma and high-grade serous ovarian cancer (HGSOC). —Conversely, some predict favorable prognosis in gastric cardia adenocarcinoma.	[4, 51, 59]
3	Contributor to Drug Resistance	—Amplified oncogenes on ecDNA promote tumor heterogeneity. —ecDNA can act as a hidden source of genetic variation supporting resistance development.	[12, 33, 60]
4	Therapeutic Target	—Hydroxyurea can help overcome drug resistance. —PARP inhibitors may disrupt ecDNA reintegration. —BET inhibitors target ecDNA hubs. —Immune-based therapies could be designed to enhance antigen presentation influenced by ecDNA.	[45, 61]

division of a cell. Therapeutic strategies that disrupt the assembly of spindles or nuclear envelope reformation during mitosis might aid in preventing the retention of ecDNA, thereby affecting the oncogene transmission between daughter cells [55]. Also, there is an emerging interest in targeting the transcriptional product of ecDNA. Chromatin accessibility can be altered by drugs, including histone deacetylase inhibitors (HDACis) or BET inhibitors, and limit the impact of ecDNA-mediated overexpression of resistance proteins [56]. It is possible that by combining these with standard therapies or immune checkpoint blockade, they may exert synergistic effects, particularly in refractory cancers in which ecDNA has a significant presence [57]. Table 5.2 describes the clinical perspectives on ecDNA [58–61].

5.6 EMERGING TECHNOLOGIES AND RESEARCH TOOLS

Recent advances in technology have provided a tremendous increase in the ability to study and understand the role of ecDNA in cancer [62]. The sequencing of single-cell genomes or fluorescence in situ hybridization (FISH) methods, as well as computational pipelines such as AmpliconArchitect, now provide one of the clues to solve the structural and functional mysteries behind ecDNA [63]. These methods enable probing heterogeneity of ecDNA in cancers in more detail, mapping their genetic content, and visualizing their nuclear distribution, which can heavily simplify the task of a scientist. These discoveries are important in revealing how ecDNA contributes to amplifying oncogenes and the progression of cancers and therapy resistance. Single-cell sequencing has become a cornerstone component of tumor heterogeneity studies. Scientists will be able to identify ecDNA copy number variations of thousands of tumor cells with single-cell resolution of DNA [64]. The consequence of such high-resolution profiling is the discovery of rare cellular subsets with unique ecDNA footprints that might underlie the resistance to therapy or their metastatic potential nature. Moreover, together with transcriptomic and epigenomic data, their, so-called single-cell multi-omics, this technology can attain a complete understanding of how ecDNA regulates gene expression and chromatin structure at the single-cell level. FISH, which has been a long-standing approach to cancer cytogenetics, is still one of the most useful methods to discover ecDNA [65]. FISH allows visualizing circular ecDNA structures and their localization in the cell nucleus using gene-specific probes. More recent advances in multiplexed and super-resolution FISH technology have advanced the accuracy of this method, allowing the mapping of ecDNA hubs, dense aggregates of ecDNA copies that have become active transcription units [66]. These imaging breakthroughs have also given us a lot more insight into ecDNA behavior during cell division and indeed how it relates to other aspects of the nucleus, including chromatin and transcription complexes [67]. ecDNA research has been transformed by computational tools such as AmpliconArchitect that can reconstruct complex amplicons through whole-genome sequencing data [68]. The tool also identifies the specific characteristics of ecDNAs, including circular amplicons and breakpoint signatures, which can be used to differentiate them from linear chromosomal amplifications. It is also applied to the detection of enhancer hijacking events, involved in the overexpression of oncogenes, using AmpliconArchitect. It also allows comprehensive ecDNA mapping of large cancer datasets when combined with other bioinformatics tools, providing more comprehensive knowledge about the distribution and functional diversity of ecDNA between different tumor types [69]. New approaches, like live-cell imaging and chromatin interaction assays (e.g., Hi-C and ChIA-PET), have opened up new avenues in ecDNA studies. Live-cell tracking by fluorescent labeling makes it possible to monitor the dynamics of ecDNA mobility, replication dynamics, and segregation in real time. Concurrently, 3D genome mapping technologies have revealed the long-range chromatin interactions that are coordinated by ecDNA hubs and thus depict their contributions to reprogramming nuclear architecture to fuel the expression of oncogenes [70]. These innovative strategies are all projecting our vision of ecDNA as a mobile and vital role in cancer biology.

5.7 CHALLENGES AND FUTURE DIRECTIONS

Despite the rapid development of scientific knowledge about the ecDNA mechanism in cancer, there are still a number of major obstacles in front of its wider translational and clinical use. The necessity to locate and plate precisely and quantify ecDNA is one of the most urgent issues. Due to its cyclic conformation, frequent genomic rearrangements, and variability in copy number, ecDNA can be easily confused with chromosomal amplifications, especially in bulk sequencing data [71]. Modern analysis pipelines, which are mainly developed to handle linear DNA, are sometimes incapable of providing a resolution of the complex ecDNA architecture. Although new methods, including single-cell sequencing and sophisticated computational tools, have better resolutions, they remain limited by high cost, low sensitivity, and scalability constraints to their usability in daily clinical practice. In addition, the dynamic nature of ecDNA through cell divisions and its non-linear spatial structure in the nucleus complicate real-time visualization and tracking. This has a detrimental effect since it is difficult to reliably quantify ecDNA in the typical diagnostic samples, such as tissue biopsies or liquid blood-based diagnostic samples, and this limits the application of ecDNA as a clinical biomarker [72]. Besides the challenges of detection, the mechanisms underlying cancer drug resistance using the ecDNA at the molecular level are not fully understood. Amplification of oncogenes and increase in transcriptional activity are both identified functions, but the downstream pathways, such as epigenetic remodeling, changes in signal transduction, and modifications to the immune or metabolic landscape of the tumor, are all relatively unknown. Learning more about how ecDNA mediates its interaction with the tumor microenvironment, immune evasion, or modulates metabolic states may open new areas to target resistance. Moreover, the details of the biology of the generation, maintenance, and transmission of ecDNA during cell division remain an area of active investigation. These understandings are key to developing approaches that involve destabilization or eradication of ecDNA in cancerous cells [73]. These scientific and technical gaps will need to be handled using multidisciplinary approaches with genomics, imaging, bioinformatics, and clinical oncology. Priority should be given to refining and validating detection tools that are sensitive, scalable, and cost-effective enough for clinical deployment. This could include enhanced single-cell multi-omics workflows, machine learning-powered image analysis for ecDNA visualization, and standardized bioinformatic pipelines tailored to different sample types and tumor contexts. Moreover, longitudinal studies following ecDNA dynamics throughout treatment will be key to understanding how resistance develops and evolves, ultimately guiding more responsive and individualized therapeutic regimens [74].

From a therapeutic standpoint, the next wave of innovation will likely involve combination approaches that simultaneously target ecDNA and canonical oncogenic pathways. As an example, one might combine an inhibitor of proteins involved in DNA replication stress, chromatin remodeling, or the DNA damage response with conventional chemotherapy or immunotherapy to inhibit ecDNA-based drug resistance mechanisms. Additionally, new compounds capable of inhibiting ecDNA clustering or disrupting its segregation during mitosis or specific ecDNA degradation may provide attractive new therapeutic approaches. Personalizing these strategies will involve extensive profiling of ecDNA in individual tumor conditions in order to achieve maximum therapeutic effect and the lowest side effects. Among the most promising avenues is a strategy that seeks to take advantage of synthetic lethality—seeking to disrupt molecular vulnerabilities that are specific to ecDNA-positive cancer cells. As ecDNA-driven tumors can be heavily reliant on particular chromatin regulators or even pathways of the stress response, interfering with those essential mechanisms might specifically kill malignant cells without affecting normal tissue. The other frontier is through the immune system: therapeutics could be designed to detect unique neoantigens or chromatin patterns generated in conflict with neoantigens generated by ecDNA, permitting the selective destruction of resistant cancer clones [75]. The future of treating cancer is ecDNA surveillance integrated into real-time treatment planning, especially with the help of liquid biopsies, which can be transformative. Non-invasively monitoring ecDNA levels and composition changes, clinicians might be able to adjust the treatment in real time and avoid resistant clones' outgrowth. Further, patient-derived organoids and xenograft models with characterized ecDNA profiles will also be valuable in the drug screen and confirmation of ecDNA-specific therapies. Altogether, bridging current gaps in detection, understanding, and treatment of ecDNA will be critical in realizing its full promise as both a biomarker and therapeutic target in precision oncology.

5.8 CONCLUSION

ecDNA has gained considerable attention as a major driver of oncogene amplification and resistance to cancer therapies. Unlike linear chromosomal DNA, ecDNA is circular, acentric fragments, which permit dynamic and tumor-specific expansions of gene copy number, commonly including potent oncogenes contributing to therapy resistance. Its ability to amplify and activate these genes outside the constraints of chromosomal regulation gives cancer cells a powerful advantage, especially under the selective pressure of treatment. Furthermore, because ecDNA is not evenly inherited during cell division, it promotes significant intratumoral diversity—enabling some cell populations to survive, adapt, and repopulate tumors even after initial therapeutic success. Such genomic plasticity has highlighted ecDNA as a mainstay behind tumor aggression, treatment resistance, and tumor relapse. In the future, it will be important to solve the technical challenges with ecDNA detection and quantification before seeing a clinical implementation of ecDNA-based diagnostics. Greater clarity of the mechanisms of ecDNA formation, its persistence in cancer cells, and its relationship with the adjacent tumor microenvironment will provide the essential key to discovering new intervention targets. A novel pathway toward resistance could be through combinatoric therapeutic strategies that use combinations of existing therapeutics with strategies that target the functionality or stability of ecDNA. Simultaneously, the development of new liquid biopsy technologies with the capacity to detect ecDNA levels and composition with a high degree of sensitivity in real time may allow more precise treatment approaches to be tailored to the evolving tumor. Finally, the challenges of ecDNA will be difficult to address without multidisciplinary or cross-functional effort among genomics, oncology, bioinformatics, and molecular biology. Through such synergies, researchers have the opportunity to learn more about how ecDNA influences cancer development and resistance and perhaps discover previously unknown, more effective methods to target drug-resistant tumors using precision medicine.

REFERENCES

1. Wu S, Bafna V, Chang HY, Mischel PS. Extrachromosomal DNA: An emerging hallmark in human cancer. *Annu Rev Pathol*. 2022;17:367–386.
2. Xing J, Ning Q, Tang D, Mo Z, Lei X, Tang S. Progress on the role of extrachromosomal DNA in tumor pathogenesis and evolution. *Clin Genet*. 2021;99(4):503–512.
3. Levan A, Levan G. Have double minutes functioning centromeres? *Hereditas*. 1978;88(1):81–92.
4. Pecorino LT, Verhaak RGW, Henssen A, Mischel PS. Extrachromosomal DNA (ecDNA): An origin of tumor heterogeneity, genomic remodeling, and drug resistance. *Biochem Soc Trans*. 2022;50(6):1911–1920.
5. Kim H, Nguyen NP, Turner K, Wu S, Gujar AD, Luebeck J, et al. Extrachromosomal DNA is associated with oncogene amplification and poor outcome across multiple cancers. *Nat Genet*. 2020;52(9):891–897.
6. Hung KL, Mischel PS, Chang HY. Gene regulation on extrachromosomal DNA. *Nat Struct Mol Biol*. 2022;29(8):736–744.
7. Yi E, Chamorro González R, Henssen AG, Verhaak RGW. Extrachromosomal DNA amplifications in cancer. *Nat Rev Genet*. 2022;23(12):760–771.
8. Zhao L, Jiang Y, Lei X, Yang X. Amazing roles of extrachromosomal DNA in cancer progression. *Biochim Biophys Acta Rev Cancer*. 2023;1878(1):188843.
9. Zhu Y, Gong L, Wei CL. Guilt by association: EcDNA as a mobile transactivator in cancer. *Trends Cancer*. 2022;8(9):747–758.
10. Chen Y, Chen Y, Qiu Q, She J, Yu J. Extrachromosomal circular DNA in colorectal cancer: Biogenesis, function and potential as therapeutic target. *Oncogene*. 2023;42(13):941–951.
11. Noorani I, Mischel PS, Swanton C. Leveraging extrachromosomal DNA to fine-tune trials of targeted therapy for glioblastoma: Opportunities and challenges. *Nat Rev Clin Oncol*. 2022;19(11):733–743.
12. Turner KM, Deshpande V, Beyter D, Koga T, Rusert J, Lee C, et al. Extrachromosomal oncogene amplification drives tumour evolution and genetic heterogeneity. *Nature*. 2017;543(7643):122–125.
13. Rajkumar U, Turner K, Luebeck J, Deshpande V, Chandraker M, Mischel P, et al. EcSeg: Semantic segmentation of metaphase images containing extrachromosomal DNA. *iScience*. 2019;21:428–435.
14. Møller HD, Bojsen RK, Tachibana C, Parsons L, Botstein D, Regenberg B, et al. Genome-wide purification of extrachromosomal circular DNA from eukaryotic cells. *J Vis Exp*. 2016;(110):e54239.
15. Shibata Y, Kumar P, Layer R, Willcox S, Gagan JR, Griffith JD, et al. Extrachromosomal microDNAs and chromosomal microdeletions in normal tissues. *Science*. 2012;336(6077):82–86.
16. Prada-Luengo I, Krogh A, Maretty L, Regenberg B. Sensitive detection of circular DNAs at single-nucleotide resolution using guided realignment of partially aligned reads. *BMC Bioinformatics*. 2019; 20(1):663.
17. Zhang P, Peng H, Llauro C, Bucher E, Mirouze M. Ecc_finder: A robust and accurate tool for detecting extrachromosomal circular DNA from sequencing data. *Front Plant Sci*. 2021;12:743742.
18. Deshpande V, Luebeck J, Nguyen ND, Bakhtiari M, Turner KM, Schwab R, et al. Exploring the landscape of focal amplifications in cancer using AmpliconArchitect. *Nat Commun*. 2019;10(1):392.
19. Luebeck J, Coruh C, Dehkordi SR, Lange JT, Turner KM, Deshpande V, et al. AmpliconReconstructor integrates NGS and optical mapping to resolve the complex structures of focal amplifications. *Nat Commun*. 2020;11(1):4374.
20. Yi E, Gujar AD, Guthrie M, Kim H, Zhao D, Johnson KC, et al. Live-cell imaging shows uneven segregation of extrachromosomal DNA elements and transcriptionally active extrachromosomal DNA hubs in cancer. *Cancer Discov*. 2022;12(2):468–483.
21. Hung KL, Yost KE, Xie L, Shi Q, Helmsauer K, Luebeck J, et al. Targeted profiling of human extrachromosomal DNA by CRISPR-CATCH. *Nat Genet*. 2022;54(11):1746–1754.
22. Bahr G, Gilbert F, Balaban G, Engler W. Homogeneously staining regions and double minutes in a human cell line: Chromatin organization and DNA content. *J Natl Cancer Inst*. 1983;71(4):657–661.
23. Hamkalo BA, Farnham PJ, Johnston R, Schimke RT. Ultrastructural features of minute chromosomes in a methotrexate-resistant mouse 3T3 cell line. *Proc Natl Acad Sci U S A*. 1985;82(4):1126–1130.
24. Rattner JB, Lin CC. Ultrastructural organization of double minute chromosomes and HSR regions in human colon carcinoma cells. *Cytogenet Cell Genet*. 1984;38(3):176–181.
25. Wu S, Turner KM, Nguyen N, Raviram R, Erb M, Santini J, et al. Circular ecDNA promotes accessible chromatin and high oncogene expression. *Nature*. 2019;575(7784):699–703.
26. Korbel JO, Campbell PJ. Criteria for inference of chromothripsis in cancer genomes. *Cell*. 2013; 152(6):1226–1236.
27. Stephens PJ, Greenman CD, Fu B, Yang F, Bignell GR, Mudie LJ, et al. Massive genomic rearrangement acquired in a single catastrophic event during cancer development. *Cell*. 2011;144(1):27–40.

28. Rausch T, Jones DT, Zapatka M, Stütz AM, Zichner T, Weischenfeldt J, et al. Genome sequencing of pediatric medulloblastoma links catastrophic DNA rearrangements with TP53 mutations. *Cell.* 2012;148(1–2):59–71.
29. Dillon LW, Kumar P, Shibata Y, Wang YH, Willcox S, Griffith JD, et al. Production of extrachromosomal MicroDNAs is linked to mismatch repair pathways and transcriptional activity. *Cell Rep.* 2015;11(11):1749–1759.
30. Koche RP, Rodriguez-Fos E, Helmsauer K, Burkert M, MacArthur IC, Maag J, et al. Extrachromosomal circular DNA drives oncogenic genome remodeling in neuroblastoma. *Nat Genet.* 2020;52(1):29–34.
31. Nikolaev S, Santoni F, Garieri M, Makrythanasis P, Falconnet E, Guipponi M, et al. Extrachromosomal driver mutations in glioblastoma and low-grade glioma. *Nat Commun.* 2014;5:5690.
32. Morton AR, Dogan-Artun N, Faber ZJ, MacLeod G, Bartels CF, Piazza MS, et al. Functional enhancers shape extrachromosomal oncogene amplifications. *Cell.* 2019;179(6):1330–41.e13.
33. Yu L, Zhao Y, Quan C, Ji W, Zhu J, Huang Y, et al. Gemcitabine eliminates double minute chromosomes from human ovarian cancer cells. *PLoS One.* 2013;8(8):e71988.
34. Møller HD, Mohiyuddin M, Prada-Luengo I, Sailani MR, Halling JF, Plomgaard P, et al. CRISPR-c: Circularization of genes and chromosome by CRISPR in human cells. *Nucleic Acids Res.* 2018;46(22):e131.
35. Zuberi L, Adeyinka A, Kuriakose P. Rapid response to induction in a case of acute promyelocytic leukemia with MYC amplification on double minutes at diagnosis. *Cancer Genet Cytogenet.* 2010;198(2):170–172.
36. Gibaud A, Vogt N, Hadj-Hamou NS, Meyniel JP, Hupé P, Debatisse M, et al. Extrachromosomal amplification mechanisms in a glioma with amplified sequences from multiple chromosome loci. *Hum Mol Genet.* 2010;19(7):1276–1285.
37. McClintock B. Chromosome organization and genic expression. *Cold Spring Harb Symp Quant Biol.* 1951;16:13–47.
38. McClintock B. The stability of broken ends of chromosomes in Zea mays. *Genetics.* 1941;26(2):234–282.
39. Boukamp P, Popp S, Krunic D. Telomere-dependent chromosomal instability. *J Investig Dermatol Symp Proc.* 2005;10(2):89–94.
40. Zhang CZ, Spektor A, Cornils H, Francis JM, Jackson EK, Liu S, et al. Chromothripsis from DNA damage in micronuclei. *Nature.* 2015;522(7555):179–184.
41. Zakov S, Kinsella M, Bafna V. An algorithmic approach for breakage-fusion-bridge detection in tumor genomes. *Proc Natl Acad Sci U S A.* 2013;110(14):5546–5551.
42. Toledo F, Buttin G, Debatisse M. The origin of chromosome rearrangements at early stages of AMPD2 gene amplification in Chinese hamster cells. *Curr Biol.* 1993;3(5):255–264.
43. Shoshani O, Brunner SF, Yaeger R, Ly P, Nechemia-Arbely Y, Kim DH, et al. Chromothripsis drives the evolution of gene amplification in cancer. *Nature.* 2021;591(7848):137–141.
44. Umbreit NT, Zhang CZ, Lynch LD, Blaine LJ, Cheng AM, Tourdot R, et al. Mechanisms generating cancer genome complexity from a single cell division error. *Science.* 2020;368(6488).
45. Yang L, Luquette LJ, Gehlenborg N, Xi R, Haseley PS, Hsieh CH, et al. Diverse mechanisms of somatic structural variations in human cancer genomes. *Cell.* 2013;153(4):919–929.
46. Lee JA, Carvalho CM, Lupski JR. A DNA replication mechanism for generating nonrecurrent rearrangements associated with genomic disorders. *Cell.* 2007;131(7):1235–1247.
47. Bao Y, Liu J, You J, Wu D, Yu Y, Liu C, et al. Met promotes the formation of double minute chromosomes induced by sei-1 in NIH-3T3 murine fibroblasts. *Oncotarget.* 2016;7(35):56664–56675.
48. Utani K, Fu H, Jang SM, Marks AB, Smith OK, Zhang Y, et al. Phosphorylated SIRT1 associates with replication origins to prevent excess replication initiation and preserve genomic stability. *Nucleic Acids Res.* 2017;45(13):7807–7824.
49. Xu K, Ding L, Chang TC, Shao Y, Chiang J, Mulder H, et al. Structure and evolution of double minutes in diagnosis and relapse brain tumors. *Acta Neuropathol.* 2019;137(1):123–137.
50. Bailey MH, Tokheim C, Porta-Pardo E, Sengupta S, Bertrand D, Weerasinghe A. Comprehensive characterization of cancer driver genes and mutations. *Cell.* 2018;173(2):371–385.e18.
51. deCarvalho AC, Kim H, Poisson LM, Winn ME, Mueller C, Cherba D, et al. Discordant inheritance of chromosomal and extrachromosomal DNA elements contributes to dynamic disease evolution in glioblastoma. *Nat Genet.* 2018;50(5):708–717.
52. Helmsauer K, Valieva ME, Ali S, Chamorro González R, Schöpflin R, Röefzaad C, et al. Enhancer hijacking determines extrachromosomal circular MYCN amplicon architecture in neuroblastoma. *Nat Commun.* 2020;11(1):5823.
53. Zhu Y, Gujar AD, Wong CH, Tjong H, Ngan CY, Gong L, et al. Oncogenic extrachromosomal DNA functions as mobile enhancers to globally amplify chromosomal transcription. *Cancer Cell.* 2021;39(5):694–707.e7.

54. Hung KL, Yost KE, Xie L, Shi Q, Helmsauer K, Luebeck J, et al. ecDNA hubs drive cooperative intermolecular oncogene expression. *Nature*. 2021;600(7890):731–736.
55. Carroll SM, DeRose ML, Gaudray P, Moore CM, Needham-Vandevanter DR, Von Hoff DD, et al. Double minute chromosomes can be produced from precursors derived from a chromosomal deletion. *Mol Cell Biol*. 1988;8(4):1525–1533.
56. Windle B, Draper BW, Yin YX, O'Gorman S, Wahl GM. A central role for chromosome breakage in gene amplification, deletion formation, and amplicon integration. *Genes Dev*. 1991;5(2):160–174.
57. Ronneberger O, Fischer P, Brox T. U-Net: Convolutional networks for biomedical image segmentation. *ArXiv*. 2015.
58. Kumar P, Dillon LW, Shibata Y, Jazaeri AA, Jones DR, Dutta A. Normal and cancerous tissues release extrachromosomal circular DNA (eccDNA) into the circulation. *Mol Cancer Res*. 2017;15(9):1197–1205.
59. Møller HD, Mohiyuddin M, Prada-Luengo I, Sailani MR, Halling JF, Plomgaard P, et al. Circular DNA elements of chromosomal origin are common in healthy human somatic tissue. *Nat Commun*. 2018;9(1):1069.
60. Jiang W, Zhao X, Gabrieli T, Lou C, Ebenstein Y, Zhu TF, et al. Cas9-assisted targeting of CHromosome segments CATCH enables one-step targeted cloning of large gene clusters. *Nat Commun*. 2015;6:8101.
61. Amler LC, Schwab M. Amplified n-myc in human neuroblastoma cells is often arranged as clustered tandem repeats of differently recombined DNA. *Mol Cell Biol*. 1989;9(11):4903–4913.
62. VanDevanter DR, Piaskowski VD, Casper JT, Douglass EC, Von Hoff DD. Ability of circular extrachromosomal DNA molecules to carry amplified MYCN proto-oncogenes in human neuroblastomas in vivo. *J Natl Cancer Inst*. 1990;82(23):1815–1821.
63. Cen Y, Fang Y, Ren Y, Hong S, Lu W, Xu J. Global characterization of extrachromosomal circular DNAs in advanced high grade serous ovarian cancer. *Cell Death Dis*. 2022;13(4):342.
64. Von Hoff DD, Forseth B, Clare CN, Hansen KL, VanDevanter D. Double minutes arise from circular extrachromosomal DNA intermediates which integrate into chromosomal sites in human HL-60 leukemia cells. *J Clin Invest*. 1990;85(6):1887–1895.
65. Ng AWT, Contino G, Killcoyne S, Devonshire G, Hsu R, Abbas S, et al. Rearrangement processes and structural variations show evidence of selection in oesophageal adenocarcinomas. *Commun Biol*. 2022;5(1):335.
66. Zhao XK, Xing P, Song X, Zhao M, Zhao L, Dang Y, et al. Focal amplifications are associated with chromothripsis events and diverse prognoses in gastric cardia adenocarcinoma. *Nat Commun*. 2021;12(1):6489.
67. Morales C, García MJ, Ribas M, Miró R, Muñoz M, Caldas C, et al. Dihydrofolate reductase amplification and sensitization to methotrexate of methotrexate-resistant colon cancer cells. *Mol Cancer Ther*. 2009;8(2):424–432.
68. Singer MJ, Mesner LD, Friedman CL, Trask BJ, Hamlin JL. Amplification of the human dihydrofolate reductase gene via double minutes is initiated by chromosome breaks. *Proc Natl Acad Sci U S A*. 2000;97(14):7921–7926.
69. Tyler LC, Le AT, Chen N, Nijmeh H, Bao L, Wilson TR, et al. MET gene amplification is a mechanism of resistance to entrectinib in ROS1+ NSCLC. *Thorac Cancer*. 2022;13(21):3032–3041.
70. Xu G, Shi W, Ling L, Li C, Shao F, Chen J, et al. Differential expression and analysis of extrachromosomal circular DNAs as serum biomarkers in lung adenocarcinoma. *J Clin Lab Anal*. 2022;36(6):e24425.
71. Cristofanilli M, Budd GT, Ellis MJ, Stopeck A, Matera J, Miller MC, et al. Circulating tumor cells, disease progression, and survival in metastatic breast cancer. *N Engl J Med*. 2004;351(8):781–791.
72. Dawson SJ, Tsui DW, Murtaza M, Biggs H, Rueda OM, Chin SF, et al. Analysis of circulating tumor DNA to monitor metastatic breast cancer. *N Engl J Med*. 2013;368(13):1199–1209.
73. Becker A, Thakur BK, Weiss JM, Kim HS, Peinado H, Lyden D, et al. Extracellular vesicles in cancer: Cell-to-cell mediators of metastasis. *Cancer Cell*. 2016;30(6):836–848.
74. von Groll A, Levin Y, Barbosa MC, Ravazzolo AP. Linear DNA low efficiency transfection by liposome can be improved by the use of cationic lipid as charge neutralizer. *Biotechnol Prog*. 2006;22(4):1220–1224.
75. Raymond E, Faivre S, Weiss G, McGill J, Davidson K, Izbicka E, et al. Effects of hydroxyurea on extrachromosomal DNA in patients with advanced ovarian carcinomas. *Clin Cancer Res*. 2001;7(5):1171–1180.

6 Imaging, Visualization, and Characterization of ecDNA in Cancer Cells

Navjot Kaur Sandhu,[1] *Issac V. Cherian,*[1]
Abhishek VijuKumar,[1] *Sheikh Mohammad Faisal,*[1]
and Chowdhury Mobaswar Hossain[2]

1 Department of Quality Assurance
ISF College of Pharmacy
Moga, India
2 Department of Pharmaceutical Technology
Maulana Abul Kalam Azad University of Technology
Harnghata, India

6.1 INTRODUCTION

Research has confirmed that Extrachromosomal DNA (ecDNA) is a vital genetic element that exerts significant biological impacts on cancer progression. The circular configuration of extrachromosomal DNA enables it to operate as acentric, decompartmentalized genetic material due to its absence of chromosomal linkage during mitotic cell division. The lack of centromeres allows for uneven DNA distribution among daughter cells. Studies revealed ecDNA as a genetic abnormality over several decades ago, but many experts believed it was unusual [1]. Recent technical improvements in high-resolution imaging as well as next-generation sequencing have demonstrated that ecDNA appears widely across various types of human malignancies and helps drive oncogene amplification while guiding tumor growth. Software-generated extrachromosomal elements combine oncogenes and regulatory elements coupled with functional genomic domains, which assist tumor cell evolution. Their specific design and operating properties make them different from chromosomal rearrangements or localized amplifications while making them significant clinical biomarkers for Oncology therapeutic targets [2]. The fundamental defining feature of ecDNA comprises its ability to maintain enhanced oncogene duplication, which promotes elevated gene dosage, resulting in increased oncogene expression levels. ecDNA elements exist as free-floating circular entities that can become present hundreds of times per cell yet stay independent from chromosomal amplifications by linear genomic structure. The circular form of ecDNA facilitates fast duplications and unequal cellular distribution throughout cell division, which causes intratumoral heterogeneity that defines aggressive cancer presentation [3]. The oncogenes EGFR, MYC, MDM2, CDK4, and ERBB2 have been discovered on ecDNA in studies of glioblastoma neuroblastoma, along with lung cancer cells. Researchers show that where whole genome sequencing partners up with cytogenetic analysis, ecDNA amplification leads to considerable overexpression of oncogenes beyond chromosomal levels, thus causing uncontrolled cell proliferation and genome instability accompanying treatment resistance. The cancer progression processes of ecDNA offer functions above simple oncogene amplification patterns because the mobile DNA structures work as an adaptable regulatory center that reorganizes gene expression activities through different regulatory channels [4]. For ecDNA, the fundamental trait is its ability to provide

DOI: 10.1201/9781003655299-6

transcriptional machinery access. Oncogenes implanted in ecDNA gain greater expression levels due to ecDNA's loosened chromatin structure mixed with higher transcriptional activity. The mobile super-enhancers within ecDNA form ecDNA bodies, which concentrate in the nucleus to establish "ecDNA bodies" that prolong transcriptional outputs [5]. Through spatial clustering, ecDNA hubs produce stronger collaborative interactions between oncogenes and enhancers that boost signaling networks, which favor cancer cell growth and survival. Cancer cells receive evolutionary benefits through ecDNA because this DNA structure permits quick adaptability to hypoxic stress and nutritional shortages in addition to therapeutic obstacles. Tumor evolution and genetic plasticity have made great advances through ecDNA biology study. During mitosis, ecDNA is distributed unevenly, leading to daughter cells obtaining variable quantities of ecDNA, which form groupings of tumor subclones that carry unique oncogene expressions. The diverse character of tumor cells increases both clonal selection and tumor development notably during treatment-driven stress [6].

Today's research indicates that ecDNA continuously adjusts both its genetic material counts and its structural components following outside stimuli exposure because cancer cells use this power to briefly raise resistance gene levels and control elements. Cancer cell adaptability facilitated by ecDNA helps tumors grow by emphasizing drug resistance development, leading to targeted therapy failure and tumor recurrence. It has been discovered that ecDNA interacts actively in nuclear three-dimensional architecture redesign processes. Research reveals that ecDNA prefers to accumulate inside unique nuclear subdomains where it contacts chromatin components to generate broad transcriptional pathways. Distant proto-oncogenes and tumor suppressor genes gain activation or suppression as a consequence of ecDNA-mediated spatial organizing. The presence of extrachromosomal DNA participates in heightened genomic instability through its stimulation of recombination events, as well as chromothripsis and DNA damage responses to create additional genomic changes accountable for cancer aggressiveness. Research reveals ecDNA serves as a crucial cancer-driving component that proves its significance for both prognostic biomarker discovery and therapeutic intervention development. Studying ecDNA's structural components plus its functional processes and regulatory features in cancer cells lays the basis for uncovering cancer's genetic pathways and improving present diagnosis and therapy methods. The considerable role that ecDNA plays in oncogene amplification, as well as intratumoral heterogeneity and medication resistance, puts this genetic entity as a vital research target in fundamental oncology and for clinical applications.

6.2 IMAGING TECHNIQUES FOR VISUALIZING ECDNA

Observing extrachromosomal DNA (ecDNA) in cancer cell architecture is crucial for understanding its structural characteristics, spatial organization, and functional implications. The advancement of imaging techniques facilitated the enhancement of ecDNA analysis, enabling researchers to understand its function in oncogenesis, tumor heterogeneity, and chemoresistance [7]. This section delineates the principal imaging modalities employed in ecDNA research, examining their methodology, applications, and substantial contributions to the field.

6.2.1 Fluorescence in Situ Hybridization (FISH)

Fluorescence in situ hybridization (FISH) is a molecular cytogenetic method that employs fluorescent probes to identify and map specific DNA sequences on chromosomes. Concerning ecDNA, FISH has proven effective in identifying and characterizing the extrachromosomal elements found in cancer cells [8].

The procedure involves the hybridization of fluorescently labeled DNA probes with the corresponding locations inside the target ecDNA. Following hybridization, the samples are examined under a fluorescence microscope, revealing ecDNA and unique fluorescent signals that differ from chromosomal signals. This method has demonstrated significant efficacy in identifying oncogene amplifications present on ecDNA, including Myc, Egfr, and Mdm2 [9].

One of the differentiating advantages of FISH in comparison to other technologies is the contribution given by FISH to the spatial context, demonstrating the localization of ecDNA within the nucleus and its aggregation with chromosomal areas. Additionally, FISH can differentiate ecDNA and homogeneously staining regions (HSRs), chromosomal amplifications, on the basis of a dissimilar hybridization pattern. This is crucial to note since ecDNA and HSRs may have diverse effects with regard to gene expression and tumor activity [10].

But FISH has its limitations. The proposed methodology needs to know the target sequence first before designing specialized probes, which may not be reachable for novel or uncharacterized ecDNA components. Further, FISH is performed on fixed cells, thereby foreclosing the discovery of dynamic ecDNA activity within live cells. Although there are limitations, FISH is still significant in ecDNA investigations and in the study of existence and structure of ecDNA inside cancer cells.

6.2.2 Super-Resolution Microscopy

Super-resolution microscopy encompasses a collection of sophisticated imaging techniques that surpass the diffraction limit of conventional light microscopy, achieving resolutions between 20 and 100 nanometers. These approaches have revolutionized the visualization of subcellular organelles, such as ecDNA, by enabling meticulous examination of their shape and spatial overlap within the nuclear domain [11].

Techniques like as Stochastic Optical Reconstruction Microscopy (STORM), Photoactivated Localization Microscopy (PALM), and Stimulated Emission Depletion (STED) microscopy have been employed to examine ecDNA. The methods described focus on the direct localization of individual fluorescent molecules, producing high-resolution images that depict the nanoscale architecture of ecDNA [12].

Super-resolution microscopy has explained the clustering characteristic of ecDNA in the nucleus, indicating that ecDNA can form transcriptionally active hubs. These hubs are implicated in dysregulation of oncogenic pathways within cancer cells due to elevated gene expression. Moreover, the dynamic interaction of ecDNA with nuclear structure, including nuclear lamina and transcription factories, has been revealed by super-resolution imaging, thus revealing the regulatory mechanisms underpinning the regulation of ecDNA function [13].

Multivalent as it may be, super-resolution microscopy demands specific equipment and abilities, and it is not always easy to prepare a given material. In addition, the high-intensity light utilized in some super-resolution techniques may cause phototoxicity, hence restricting the application of these techniques to live-cell imaging [14]. However, the remarkable level of detail given by these approaches makes them a crucial resource in dismantling the complex architecture of the ecDNA.

6.2.3 Confocal and Electron Microscopy

Confocal microscopy is an optical imaging approach whereby a spatial pinhole is used to exclude out-of-focus light to boost resolution and contrast. Through this approach, one is able to obtain clear, high-resolution photographs and reconstruction of three-dimensional structures in the cells. In the field of ecDNA, confocal microscopy has been utilized to see the location of ecDNA in the nucleus, which is frequently applied in combination with fluorescent probe directed to selected DNA sequences [15]. Confocal microscopy has provided information on the localization of ecDNA in connection to chromosomal areas and nuclear markers, including nucleoli and nuclear pores. Such studies have shed some light on the various processes through which ecDNA may impact the level of gene expression and nuclear architecture. Further, time-lapse confocal imaging has facilitated the tracking of ecDNA dynamics throughout the cell cycle and the identification of patterns of segregation and replication that drive tumor heterogeneity [16].

Electron microscopy (EM) (functioning as transmission electron microscopy (TEM) and scanning electron microscopy (SEM), for example) even allows much higher resolution imaging down to the level of nanometers. EM has been utilized to directly visualize the ultrastructure of ecDNA to get information on the size, physical appearance, and interaction with other nuclear components [17]. Owing to these examples, for example, SEM has been used to study metaphase spreads in which ecDNA was found as independent circular DNA elements from chromosomal components. The EM pictures of high resolution have shown that ecDNA is circular in nature and have indicated its connection with chromatin and nuclear membranes. Still, EM techniques sometimes entail a lot of sample preparation, which involves fixation, dehydration, and sectioning. Hence, artifacts may be produced, and the dynamic processes cannot be observed. Furthermore, EM does not give molecular specificity without the cooperation with the immunogold labeling or other methods of contrast enhancement [18]. However, the pairing of the confocal and electron microscopes has made it a robust approach to the comprehensive characterization of ecDNA. Through a combination of molecular specificity associated with fluorescence imaging and ultrastructural detail connected with EM, scientists will be able to acquire a holistic understanding of the significance of ecDNA in cancer biology [19].

6.3 MOLECULAR AND COMPUTATIONAL APPROACHES FOR ECDNA CHARACTERIZATION

The understanding and characterization of extrachromosomal DNA (ecDNA) in cancer have significantly advanced due to molecular and computational tools. These methodologies facilitate the identification, structural analysis, and functional interpretation of ecDNA, hence enhancing the comprehension of their oncogenic potential and therapeutic applications. The application of high-throughput sequencing methods, specialized computational software, and single-cell multivariate approaches has collectively introduced novel avenues for exploring the intricacies of ecDNA biology [20, 21].

6.3.1 Next-Generation and Whole-Genome Sequencing (NGS/WGS)

Next-generation sequencing (NGS) and whole-genome sequencing (WGS) have been essential in the detection and characterization of ecDNA in cancer cells. These technologies enable researchers to analyze genomic alterations on an unprecedented scale and resolution, therefore acquiring circular DNA elements independent of chromosomal DNA [22].

Initially, ecDNA was not easily differentiated from high-level chromosomal amplifications. However, the improvement in the sequencing depth and read length has made it possible to predict circular structures based on particular properties such as the junctional sequences and copy number patterns. Suitable in this scenario are techniques like the paired-end and mate-pair sequencing, which can detect the aberrant read alignments, demonstrating that certain DNA structures are circular. Whole-genome sequencing reveals a comprehensive landscape of genomic alterations and may disclose complicated rearrangements that are typical of ecDNA [23]. Research employing WGS has found that ecDNAs are significantly rearranged and have oncogenes, enhancer components, and regulatory sequences that lead to aggressive cancers. For example, there have been several examples of amplified oncogenes such as MYC, EGFR, and CDK4 being identified within ecDNA, in which circumstances the expression of the genes is usually boosted substantially compared to those on linear chromosomal copies [24].

Further, the arrival of the long-read sequencing technologies, such as PacBio and Oxford Nanopore, has improved the direct sequencing of ecDNA molecules, permitting correct mapping of the molecules' circular structure. These approaches are able to overcome the restrictions of the short-read sequencing, which is not often capable of resolving complicated repetitive areas in ecDNA [24]. However, there are still obstacles that the typical NGS and WGS approaches have to overcome, such as identifying ecDNA from HSRs in the chromosomes. Complementary experimental procedures, including, but not limited to, circle-seq, that pre-enrich circular DNA prior to sequencing, have been devised for maximizing the detection and analysis of ecDNA [25].

6.3.2 Computational Tools and Algorithms for ecDNA Detection

ecDNA's increasing association with cancer has inspired the construction of computer algorithms that can recognize and categorize ecDNA from the sequencing data. These tools exploit the particular qualities of the ecDNA, including high copy number amplifications, circular topology, and complicated breakpoint patterns. Among the most prominent tools available is AmpliconArchitect (AA), which builds the structure of amplified genomic fragments based on WGS data [26]. AmpliconArchitect recognizes highly amplified sections and evaluates the amplification structure type (whether it is linear or circular). Research with the use of AA has found that the ecDNA-amplification of oncogenes is more widespread than estimated in several forms of cancer [27]. Another computer method is Circle-Map, which is a program that discovers circular DNA using split alignment paired-end methodology and discordant read-end data from sequencing data. Circle-Map was created to locate tiny circular fragments of DNA (eccDNA) but has been updated for bigger carcinogenic ecDNAs as well [28]. Other techniques like ECdetect and HolistIC are produced not only for detecting ecDNA but also for illustrating the structural complexities and putative gene regulation activities of the ecDNA [29]. Such algorithms integrate read depth information, breakpoint analysis, and structural variant calling in rebuilding ecDNA structures.

Additionally, the machine learning approaches are being welcomed in boosting the specificity and sensitivity of ecDNA detection. Trained algorithms are also capable of predicting the existence of circular DNA from complex genomic markers through the usage of known ecDNA datasets, hence fostering computing capacity [30]. Notwithstanding these improvements, there are still issues regarding how to accurately differentiate ecDNA from chromosomal amplifications, as well as recreating the entire structure of ecDNAs that are severely rearranged. The coupling of computational predictions with experimental validation approaches, such as FISH and optical mapping, is still important to validate the presence and pattern of ecDNA [4].

6.3.3 Single-Cell Multiomics and Spatial Transcriptomics

Single-cell technologies have revolutionized the concept of tumor heterogeneity and have made a tremendous contribution to the grasp of ecDNA dynamics in a single cell. Conventional bulk sequencing hides the distinctions between tumor cells, while single-cell multiomics allows the unpicking of the ecDNA presence, structure, and function on the scale of an individual cell. It is using single-cell DNA sequencing (scDNA-seq) that ecDNA is discovered as focal amplifications not aligned linearly across chromosomes [31]. scDNA-seq studies have demonstrated that ecDNA distribution is highly variable between tumor cells, leading to varying oncogene copy numbers and gene expression profiles within the same tumor mass. This heterogeneity is the primary backbone of tumor evolution, treatment resistance, as well as metastatic potential [32].

Furthermore, single-cell RNA sequencing (scRNA-seq) has been utilized to assess for existence of ecDNA being connected to transcriptional modifications at a single cell level. Cells containing ecDNA frequently demonstrate significant levels of overexpression of oncogenes on the ecDNA, revealing the functional consequence of ecDNA-based amplification. Recent breakthroughs in multiomics analysis at the single-cell level now allow concurrent profiling of DNA, RNA, and epigenetic marks, painting a united picture of how ecDNA alters transcriptional and regulatory landscapes. Advanced technologies such as scATAC-seq have shown that ecDNA may work in a transcriptionally mobile fashion to make the oncogenes come in contact with active enhancers and facilitate their activation [33].

The knowledge of ecDNA has been further advanced by spatial transcriptomics, which has kept spatial information at the tissue level. With the help of linking spatial gene expression with the presence of ecDNAs, researchers can map the ecDNA-driven oncogene expression throughout diverse locations of tumors. Spatial transcriptomics investigations have shown that the locations with high ecDNA content are likely to have particular transcriptional programs that are important in proliferation, survival, and invasion [34]. Remarkably, the merging of single-cell and geographical data has indicated that the ecDNA distribution is not random but might be influenced by the microenvironment in the tumor. Such locations as hypoxic ones, for instance, can include a larger burden of ecDNA-positive cells, reflecting favorable benefits of ecDNA under particular stress situations [35].

Taken together, single-cell and spatial technologies have enabled precise insights into the contribution of ecDNA to the intratumoral heterogeneity, cancer progression, and resistance to therapy. Some further improvements developed in single-cell multiomics, and spatial studies are likely to give new insights into the functional biology of ecDNA in cancer.

6.4 FUNCTIONAL AND CLINICAL RELEVANCE

Extrachromosomal DNA is responsible for the pathophysiology, tumor progression, and prognosis for many malignancies. Although the understanding of its structural and imaging properties is essential, exposing the functional and clinical value of ecDNA became central in contemporary cancer biology. Its capacity for oncogene amplification and its participation in medication resistance and therapy response are two of the most essential features of the ecDNA biology that have an impact on clinical outcomes [36]. Such features of the ecDNA make it an intriguing object for newly created diagnostic tools, prognosis indicators, and therapeutic approaches, as shown in **Table 6.1**.

6.4.1 Oncogene Amplification

Perhaps one of the most characterized and the most significant roles of ecDNA is its stimulation of amplification of oncogenes. As opposed to chromosomal DNA, which undergoes strictly controlled replication and segregation during the mitotic cell division, ecDNA replicates in an autonomous way, and its distribution is not balanced between the daughter cells. This results in great variance of the copy number, allowing a significant and broad amplification of genes that provide a selective advantage to a cancer cell. There is a lot of evidence that indicates that frequently, ecDNA carries driver oncogenes, including but not limited to MYC, EGFR, CDK4, CCND1, and MDM2 [37].

TABLE 6.1
Comparison of Current Therapeutic Strategies, Their Mechanisms, and Clinical Considerations in Cancer Treatment

Category	Examples/Approaches	Description
Conventional Therapies	Chemotherapy, radiation therapy	Kill rapidly dividing cells; nonspecific and affect healthy tissues too
Targeted Therapy	EGFR inhibitors, BRAF inhibitors	Targets specific molecular changes in cancer cells
Hormonal Therapy	Tamoxifen, aromatase inhibitors	Used in hormone-sensitive cancers (e.g., breast, prostate)
Immunotherapy	Immune checkpoint inhibitors, CAR-T cell therapy	Stimulates body's immune system to attack cancer
Precision Medicine	Genetic profiling, personalized treatment plans	Tailors treatment based on individual patient's genetic/molecular profile
Resistance Mechanisms	Mutation, drug efflux, tumor heterogeneity	Cancer cells develop ways to evade treatment
Combination Therapies	Chemo + targeted therapy, immunotherapy + radiation	Improve efficacy and overcome resistance
Challenges and Future Directions	Biomarker identification, drug delivery, cost-effectiveness	Critical areas needing advancement for better therapeutic outcomes

Amplification of these genes resulted in an augmentation of transcriptional levels because of gene dosage and ecDNA epigenetic environment. Unlike the chromosomal loci, ecDNA does not have the same level of structural inhibition from the compaction of chromatin. This relates to the fact that genes that are present on ecDNA are defined by the presence of open chromatin configurations and are accompanied by super-enhancers, which provide a transcriptionally active state [38].

The circularity of ecDNA enables amplified oncogenes to be put near enhancers, into which they were potentially not typically attached, which is dubbed "enhancer hijacking." This is the potential of resulting in the development of highly active transcriptional hubs that highly overexpress oncogenes. Particularly, the presence of no centromeres on ecDNA leads to random segregation of these structures during mitosis and enhances intratumoral heterogeneity and clonal development under the stresses of natural selection [7]. From a clinical perspective, the presence of oncogene-bearing ecDNA bears a tight relation to aggressive tumor characteristics, poor differentiation, and high-grade malignancies. Tumors rich in ecDNA indicate an enhanced rate of proliferation, high metastatic competence, and stronger expression of oncogenic pathways. For instance, in glioblastoma, an EGFR amplification in high levels on ecDNA has been associated with poorer prognosis and lower survival. In a similar fashion, in neuroblastoma, MYCN amplification through the ecDNA has been found to lead to the aggressive nature of the tumor and treatment resistance to the traditional therapy [19].

Notably, ecDNA is also able to dynamically develop in the tumor microenvironment. When under treatment pressure or environmental stress, structures of ecDNA may reorganize or fuse, or the ecDNA can be destroyed altogether, leading to adaptive alterations in dosage of oncogenes. This adaptability shows the potential of ecDNA in not only tumor genesis but also the development of the disease and recurrence.

6.4.2 Drug Resistance and Therapy Response

Apart from encouraging oncogene proliferation, ecDNA is increasingly recognized as a key arbiter of medication resistance and response modulator. This resistance develops by both direct and indirect mechanisms, some of which are dependent on the idiosyncrasies of ecDNA biology. One key process takes place in the form of gene amplification of drug resistance mediators [18]. For example, ecDNA may contain genes, like ABCB1, which encode P-glycoprotein, which actively transfers chemotherapeutic medication out of cancer cells. Amplification of such genes by ecDNA leads to the multidrug resistance, especially in malignancies such as ovarian, breast, and colon cancers [39]. ecDNA also supports adaptive transcriptional responses that allow the survival of the tumor in the context of targeted therapy. Inhibitors that target oncogenes like EGFR or ALK may, at first, limit tumor growth; however, the existence of ecDNA allows for quick re-amplification or redistribution of copies of oncogenes, leading to resistance re-emergence. This dynamic activity of ecDNA under strain of therapy can give rise to resistant subclones, which later on become dominant over the cancer population [40].

In addition, unequal segregation of ecDNA during mitosis gives rise to the diversity of cells in the malignancies. Some of the cells might have such large ecDNA copy numbers that bestow drug resistance, while others do not. The intratumoral heterogeneity adds to Darwinian selection since drug-resistant clones advance during the therapy, while the susceptible ones are lost. Consequently, chemo, rad, and targeted therapy campaigns are associated with poor responses to malignancies of high ecDNA levels [41]. One further point of ecDNA-driven resistance is the epigenetic remodeling. Genes on ecDNA are in accessible chromatin areas and are more sensitive to transcriptional reprogramming. This permits cancer cells to quickly adjust their gene expression programs in order to adapt to numerous stimuli, such as exposure to medicines, among others. According to some comments, the mobile character of ecDNA makes it feasible to interact with new regulatory environment, affecting the output of the adjacent genes and helping in setting up non-genetic defense mechanisms [42].

The consequences of the ecDNA-mediated resistance in terms of the clinical situation have important significance. Patients with malignancies, having significant quantities of ecDNA, are usually characterized by shorter progression-free survival and greater rates of relapse. For instance, when it comes to non-small cell lung cancer (NSCLC), patients with ecDNA mutant EGFR had poor responses to EGFR inhibitors [43]. In the instance of pediatric glioblastoma, the resistance to the radiation therapy, mediated by ecDNA was also discovered, wherein the ecDNA helps the survival of cells, as well as the restoration of the DNA damage in the post-irradiational environment. Significantly, ecDNA possesses a dynamic and plastic feature, which makes it challenging to target but one that can be potentially potent [4]. There is ongoing research for the tactics that target the disruption of ecDNA creation or inhibit ecDNA replication or aim to disrupt the ecDNA-related epigenetic regulators. Also, the discovery of ecDNA in circulating tumor DNA (ctDNA) and tumor biopsies may be utilized as a predictor of response to therapy and a marker of resistance to it. As scientists are exploring the part played by ecDNA in therapeutic resistance, developing techniques such as single-cell sequencing, long-read genome analysis, and live-cell imaging are allowing real-time tracking of ecDNA dynamics [44]. These approaches are anticipated to deliver a better knowledge of the treatment failure and employ it toward the development of ecDNA-connected precision medicines.

6.5 EMERGING TECHNOLOGIES AND FUTURE DIRECTIONS

The progress in comprehending extrachromosomal DNA (ecDNA) has unveiled its crucial function in oncogenesis, treatment resistance, and intratumoral heterogeneity. As research increasingly highlights the diagnostic and prognostic capabilities of ecDNA, the integration of advanced technologies is important. The advancement of tools in artificial intelligence (AI), machine learning (ML), and translational medicine is advancing ecDNA research from fundamental science to clinical application [45]. These developments are poised to transform cancer diagnosis and treatment planning from a transient association to an integral component of cancer management. Emerging tools like AI are expected to optimize each stage shown in Figure 6.1.

6.5.1 ML/AI in ecDNA Analysis

Machine learning (ML) and artificial intelligence (AI) are transforming the interpretation of complex biological datasets. With the large multi-dimensional data generated from ecDNA investigations (e.g., sequencing, imaging, and transcriptomics), AI/ML technologies are particularly suited to recognizing patterns that are not easily identified by the standard techniques of bioinformatics [46]. One of the initial uses of AI in ecDNA research is the detection and classification of ecDNA structures utilizing genomic data, namely whole genome sequencing (WGS). ML systems may be taught to recognize particular ecDNA properties, including circular genome topology, breakpoint junctions, and high-copy number signals when looking at novel cancer genomes, on datasets for which these features are known. Compounds like AmpliconArchitect and Circle-Map have utilized machine learning to improve the accuracy of circular DNA reconstruction and address intricate genomic rearrangements [10]. AI is also utilized for analysis of ecDNA from imaging data, such as fluorescence in situ hybridization (FISH), super-resolution microscopy, or 3D confocal microscope. Computer-assisted recognition of images with deep learning helps to discriminate chromosomal and extrachromosomal DNA signals in terms of size, position, and intensity. This significantly reduces the manual component of ecDNA measurement and facilitates scalable, high-throughput visualization in research and diagnostic applications [10].

Apart from structure detection, ML models are already emerging to correlate the ecDNA properties with outcomes of the disease, response to treatment, or the tumor. For instance, prediction models that are trained on sets of patients' tumors can detect the relationships between ecDNA ONA levels and survival results. Such models are also being attempted to infer the elevation in drug resistant-prone cancers from ecDNA copy number dynamics or chromatin accessibility. One of the emerging domains of inquiry is the use of a multi-modal AI platform using genomic, epigenomic, transcriptomic, and geographical data to generate holistic ecDNA "portraits" of individual tumors [30]. With the increase in data, the predictive capacity of these models is set to rise, and hence it will be possible to make real-time judgments with respect to clinical oncology. However, issues remain. The quality and diversity of AI/ML models rely on a restricted pool of ecDNA-characterized patient samples due to scarcity. Development of public ecDNA archives, standardization of data formats, and enlargement of training sets among different types of cancer will be the core components of harnessing AI potential in this field.

6.5.2 Translational Research

ecDNA studies have presently turned toward more translational applications—that of translating laboratory findings onto clinical practice. This encompasses the characterization of ecDNA as a biomarker, a therapeutic target, and an indicator of the predictability of response to treatment and disease progression. ecDNA has a great promise as a biomarker. The existence and constitution, as well as the amount of ecDNA, can indicate tumor aggressiveness, stage, and potential for being resistant [18]. A number of studies recommend adopting circulating ecDNA (circulating tumor ecDNA) in liquid biopsies as a non-invasive approach to diagnosing cancer early or tracking recurrence. This method has the edge over the usual circulating tumor DNA (ctDNA) on account of the unique structural features of ecDNA that allow it to be selectively amplified and detected. At the same time, there is a rising interest in therapeutically targeting ecDNA. As ecDNA is lacking centromeres and replicates independently, researchers are studying the inhibitors that interfere with its replication, conservation, or transcriptional regulation. One of the ways is the use of epigenetic inhibitors, including BET inhibitors, which can hinder transcriptional activity of oncogenes associated with ecDNA, thanks to super-enhancers [47]. Another developing concept is to impair the ecDNA packaging or anchoring to nuclear structures, therefore reducing its abilities for transcription or aiding its destruction.

In addition, one can aim at the mechanisms inherent in the ecDNA biogenesis (e.g., chromothripsis, breakage–fusion–bridge cycles, or replication stress) and thus effectively prevent the production of the ecDNA. Blocking these pathways should minimize the formation of ecDNA-driven resistance clones, notably in therapy-resistant tumors [48]. In clinical trials, attempts are being carried out to stratify patients according to ecDNA presence or burden. Such a method may be effective in identifying those individuals who are less likely to respond to traditional therapies because of ecDNA-mediated resistance and steering them toward various combinations and therapeutic regimens targeting the inhibition of ecDNA. Moreover, clinical processes integration of such profiling is likely to assist with the monitoring of minimal residual disease since abundance or configuration change in ecDNA might foreshadow radiological or symptomatic recurrence [4]. Another possible path is the use of spatial transcriptomics and single-cell technologies to define the microenvironmental interaction of ecDNA-positive cells [33]. These technologies allow researchers to explore the way in which the ecDNA-driven gene expression affects the immune evasion, angiogenesis, or stromal remodeling, all of which are significant for immunotherapy and anti-angiogenic treatments [49]. One of these findings in translation to clinical decisions may open up new opportunities for ecDNA-focused combo treatments.

From a drug development point of view, ecDNA is starting to be recognized as a drug development biomarker by pharmaceutical companies. ecDNA status might be utilized to create patient selection criteria, stratify responses to investigational medicines, or serve as a surrogate endpoint in early-phase clinical investigations [4]. Preclinical investigations based on targets of ecDNA-associated vulnerabilities have already started at glioblastoma, neuroblastoma, and lung cancer models [1]. The shift of ecDNA from a research curiosity to a clinical value is currently undergoing. As further innovation continues to occur in AI, single-cell analytics, and non-invasive detection approaches, ecDNA is poised to redefine the face of precision oncology as a new axis of molecular stratification for cancer patients [10].

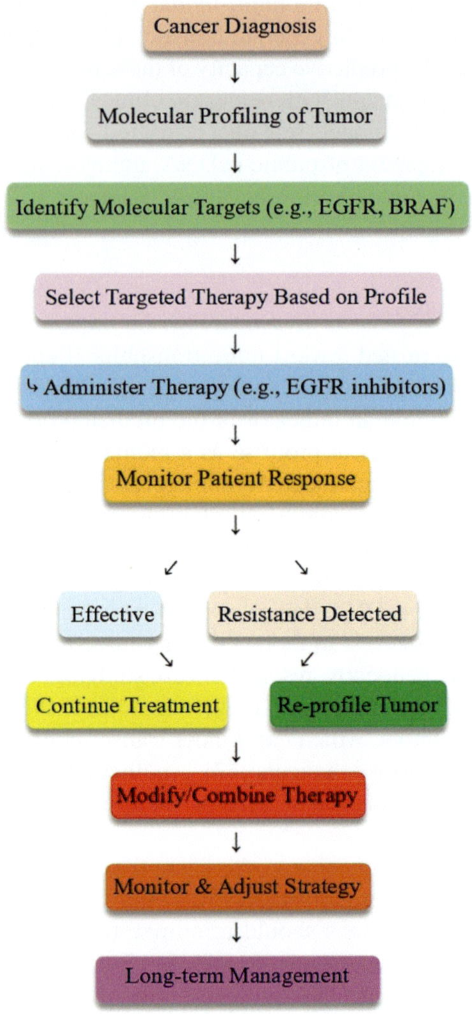

FIGURE 6.1 Flowchart summarizing targeted cancer therapy strategy from diagnosis to long-term management.

6.6 CONCLUSION

The research of extrachromosomal DNA (ecDNA) has started a revolutionary age in cancer biology, providing key clues concerning heterogeneity of tumors, amplification of oncogenes, and resistance that is blurred outside the typical chromosomal limits. The progress gained in the previous decade in the development of experimental as well as computational approaches has substantially improved our potential to identify, visualize, and comprehend the function of ecDNA in promoting cancer. It is out of such imaging methods as fluorescence in situ hybridization (FISH), super-resolution microscopy, and electron microscopy, as well as genomic methods, such as the next generation and whole-genome sequencing, that the technology tool that researchers have at their disposal has become highly sophisticated. These technologies not only enhanced the precision of mapping ecDNA of tumors but also uncovered its dynamic behavior in tumor cells, such as its unequal segregation in mitosis, independent replication, and the ability to host oncogenes that drive aggressive tumor phenotypes. The interaction between molecular biology and computer science has broadened the scope of ecDNA studies. Now, machine learning methods and AI models for analysis of large-scale sequencing and imaging datasets for automation of ecDNA detection and further predictive information about clinical outcomes are being applied. In the realm of personalized medicine, these findings are especially promising, as ecDNA profiles may be used to inform therapy choices, to monitor the course of disease, or to foresee therapeutic resistance. The advent of the new multiomics of single-cell and spatial transcriptomics has introduced another aspect of resolution in which the influences on the spatial and cellular anatomy of tumors may be mapped. This multi-dimensional approach is vital to comprehending how ecDNA rewires the tumor environment, affects gene expression, and plays a role in immune escape.

Clinically, the impacts of ecDNA are immense. Its function in multiplying oncogenes, promoting a rapid growth of tumors, and developing resistance toward targeted therapies makes it one of the main indicators in the field of cancer. The possibility of disclosing ecDNA utilizing liquid biopsies opens a new sphere for non-invasive diagnostics and monitoring the disease in real time. Moreover, addressing the functional machinery that serves the purpose of maintaining and expressing ecDNA, which comprises the transcriptional regulators or the replication enzymes, introduces a new category of therapy options. During further research, ecDNA can also be a vital stratifier in clinical trials to better characterize patient populations and examine the efficacy of medications with higher molecular specificity. In the future, a new age where ecDNA is an intrinsic part of precision oncology frameworks is on the horizon for the field. However, there are still various hurdles, such as the need for standardized methods, a bigger ecDNA-specific database, and longitudinal studies that examine ecDNA evolution during the course of treatment. Further multidisciplinary cooperation among molecular biologists, doctors, data scientists, and pharmaceutical developers will be a vital aspect for harnessing the complete translational potential of ecDNA. Packed with the integration of developing technologies and optimization of detection techniques, ecDNA research is on pace to change cancer diagnostics, prognostics, and treatments in the foreseeable future. The road ahead is full of possibilities, and gives the hope for more tailored, adaptive, and successful cancer care.

REFERENCES

[1] Z. Li, B. Wang, H. Liang, and L. Han, "Pioneering insights of extrachromosomal DNA (ecDNA) generation, action and its implications for cancer therapy," *Int. J. Biol. Sci.*, vol. 18, no. 10, p. 4006, 2022.

[2] V. Bafna and P. S. Mischel, "Extrachromosomal DNA in cancer," *Annu. Rev. Genomics Hum. Genet.*, vol. 23, no. 1, pp. 29–52, 2022.

[3] Y. Dong, Q. He, X. Chen, F. Yang, L. He, and Y. Zheng, "Extrachromosomal DNA (ecDNA) in cancer: Mechanisms, functions, and clinical implications," *Front. Oncol.*, vol. 13, p. 1194405, 2023.

[4] I. Noorani, P. S. Mischel, and C. Swanton, "Leveraging extrachromosomal DNA to fine-tune trials of targeted therapy for glioblastoma: Opportunities and challenges," *Nat. Rev. Clin. Oncol.*, vol. 19, no. 11, pp. 733–743, 2022.

[5] K. L. Hung et al., "ecDNA hubs drive cooperative intermolecular oncogene expression," *Nature*, vol. 600, no. 7890, pp. 731–736, 2021.

[6] S. Wu, T. Tao, L. Zhang, X. Zhu, and X. Zhou, "Extrachromosomal DNA (ecDNA): Unveiling its role in cancer progression and implications for early detection," *Heliyon*, vol. 9, no. 11, 2023.

[7] K. L. Hung, P. S. Mischel, and H. Y. Chang, "Gene regulation on extrachromosomal DNA," *Nat. Struct. Mol. Biol.*, vol. 29, no. 8, pp. 736–744, 2022.

[8] A. R. Shakoori, "Fluorescence in situ hybridization (FISH) and its applications," in *Chromosome Structure and Aberrations*, Springer India, 2017, pp. 343–367.

[9] J. Tang et al., "Enhancing transcription–replication conflict targets ecDNA-positive cancers," *Nature*, vol. 635, no. 8037, pp. 210–218, 2024.

[10] A. Arshadi, D. Tolomeo, S. Venuto, and C. T. Storlazzi, "Advancements in focal amplification detection in tumor/liquid biopsies and emerging clinical applications," *Genes (Basel)*, vol. 14, no. 6, p. 1304, 2023.

[11] J. Valli, A. Garcia-Burgos, L. M. Rooney, B. V. de M. e Oliveira, R. R. Duncan, and C. Rickman, "Seeing beyond the limit: A guide to choosing the right super-resolution microscopy technique," *J. Biol. Chem.*, vol. 297, no. 1, 2021.

[12] A. Carsten, M. Wolters, and M. Aepfelbacher, "Super-resolution fluorescence microscopy for investigating bacterial cell biology," *Mol. Microbiol.*, vol. 121, no. 4, pp. 646–658, 2024.
[13] K. Purshouse, S. M. Pollard, and W. A. Bickmore, "Imaging extrachromosomal DNA (ecDNA) in cancer," *Histochem. Cell Biol.*, vol. 162, no. 1, pp. 53–64, 2024.
[14] K. L. Tosheva, "Defining the limits of live-cell microscopy." UCL (University College London), 2024.
[15] E. Yi et al., "Live-cell imaging shows uneven segregation of extrachromosomal DNA elements and transcriptionally active extrachromosomal DNA hubs in cancer," *Cancer Discov.*, vol. 12, no. 2, pp. 468–483, 2022.
[16] A. N. Lacen and H.-T. Lee, "Tracing the chromatin: From 3C to live-cell imaging," *Chem. Biomed. Imaging*, vol. 2, no. 10, pp. 659–682, 2024.
[17] J. Madren, J. Chen, W. Dennis, C. Ford, K. White, and E. Brunk, "A standardized protocol for sample preparation for scanning electron microscopy (SEM) to visualize extrachromosomal DNA (ecDNA)," *bioRxiv*, pp. 2008–2023, 2023.
[18] S. Ashique et al., "Impact of ecDNA: A mechanism that directs tumorigenesis in cancer drug resistance-a review," *Chem. Biol. Interact.*, vol. 363, p. 110000, 2022.
[19] M. Karami Fath et al., "Revisiting characteristics of oncogenic extrachromosomal DNA as mobile enhancers on neuroblastoma and glioma cancers," *Cancer Cell Int.*, vol. 22, no. 1, p. 200, 2022.
[20] R. Jiang, M. Yang, S. Zhang, and M. Huang, "Advances in sequencing-based studies of microDNA and ecDNA: Databases, identification methods, and integration with single-cell analysis," *Comput. Struct. Biotechnol. J.*, vol. 21, pp. 3073–3080, 2023.
[21] S. Satya, S. Sharma, G. Choudhary, and G. Kaushik, "Advances in environmental microbiology: A multi-omic perspective," in *Microbial Omics in Environment and Health*, Springer, 2024, pp. 175–204.
[22] L. Chang et al., "Single-cell third-generation sequencing-based multi-omics uncovers gene expression changes governed by ecDNA and structural variants in cancer cells," *Clin. Transl. Med.*, vol. 13, no. 8, p. e1351, 2023.
[23] N. D. Olson et al., "Variant calling and benchmarking in an era of complete human genome sequences," *Nat. Rev. Genet.*, vol. 24, no. 7, pp. 464–483, 2023.
[24] A. Hamdan and A. Ewing, "Unravelling the tumour genome: The evolutionary and clinical impacts of structural variants in tumourigenesis," *J. Pathol.*, vol. 257, no. 4, pp. 479–493, 2022.
[25] N. E. Weiser, T. B. K. Watkins, H. Y. Chang, and P. S. Mischel, "A guide to extrachromosomal DNA: Cancer's dynamic circular genome," *Cancer Discov.*, pp. OF1–OF10, 2025.
[26] D. Schreyer, "A comprehensive study of extrachromosomal circular DNA in pancreatic ductal adenocarcinoma." University of Glasgow, 2024.
[27] J. T. Lange et al., "The evolutionary dynamics of extrachromosomal DNA in human cancers," *Nat. Genet.*, vol. 54, no. 10, pp. 1527–1533, 2022.
[28] I. Prada-Luengo, A. Krogh, L. Maretty, and B. Regenberg, "Sensitive detection of circular DNAs at single-nucleotide resolution using guided realignment of partially aligned reads," *BMC Bioinformatics*, vol. 20, no. 1, p. 663, Dec. 2019, https://doi.org/10.1186/s12859-019-3160-3.
[29] L. Yang et al., "Extrachromosomal circular DNA: Biogenesis, structure, functions and diseases," *Signal Transduct. Target. Ther.*, vol. 7, no. 1, p. 342, 2022.
[30] S. Wang et al., "Machine learning-based extrachromosomal DNA identification in large-scale cohorts reveals its clinical implications in cancer," *Nat. Commun.*, vol. 15, no. 1, p. 1515, 2024.
[31] K. L. Hung et al., "Coordinated inheritance of extrachromosomal DNAs in cancer cells," *Nature*, vol. 635, no. 8037, pp. 201–209, 2024.
[32] X. Tang et al., "Single-cell sequencing: High-resolution analysis of cellular heterogeneity in autoimmune diseases," *Clin. Rev. Allergy Immunol.*, vol. 66, no. 3, pp. 376–400, 2024.
[33] W. Lv et al., "Spatial–temporal diversity of extrachromosomal DNA shapes urothelial carcinoma evolution and the tumor immune microenvironment," *Cancer Discov.*, pp. OF1–OF22, 2025.
[34] I. Noorani et al., "Extrachromosomal DNA driven oncogene spatial heterogeneity and evolution in glioblastoma," *bioRxiv*, pp. 2010–2024, 2024.
[35] Q. Zhu et al., "Single cell multi-omics reveal intra-cell-line heterogeneity across human cancer cell lines," *Nat. Commun.*, vol. 14, no. 1, p. 8170, 2023.
[36] J. Xing, Q. Ning, D. Tang, Z. Mo, X. Lei, and S. Tang, "Progress on the role of extrachromosomal DNA in tumor pathogenesis and evolution," *Clin. Genet.*, vol. 99, no. 4, pp. 503–512, 2021.
[37] Y. Chen, Q. Qiu, J. She, and J. Yu, "Extrachromosomal circular DNA in colorectal cancer: Biogenesis, function and potential as therapeutic target," *Oncogene*, vol. 42, no. 13, pp. 941–951, 2023.
[38] G. Wright, "A conserved DNA rereplication response in adaptation to environmental stress." University of Colorado Denver, Anschutz Medical Campus, 2023.

[39] M. Reitberger, "Regulation of ABCB1 expression is a potential therapeutic target in drug resistant pancreatic cancer." 2022.
[40] K. Kobayashi and A. C. Tan, "Unraveling the impact of intratumoral heterogeneity on EGFR tyrosine kinase inhibitor resistance in EGFR-mutated NSCLC," *Int. J. Mol. Sci.*, vol. 24, no. 4, p. 4126, 2023.
[41] M. Gerlinger and C. Swanton, "How Darwinian models inform therapeutic failure initiated by clonal heterogeneity in cancer medicine," *Br. J. Cancer*, vol. 103, no. 8, pp. 1139–1143, 2010.
[42] B. Terrasa, S. Ramírez-Amaro, A. Picornell, and C. Ramon, "State of the art review of environmental DNA genomics," 2022.
[43] A. Desai et al., "ctDNA for the evaluation and management of EGFR-mutant non-small cell lung cancer," *Cancers (Basel)*, vol. 16, no. 5, p. 940, 2024.
[44] E. Beddowes, S. J. Sammut, M. Gao, and C. Caldas, "Predicting treatment resistance and relapse through circulating DNA," *The Breast*, vol. 34, pp. S31–S35, 2017.
[45] L. Zhang and S. Li, "Artificial intelligence/machine learning tools for single molecule data analysis in biomedicine," in *Machine Learning and Artificial Intelligence in Chemical and Biological Sensing*, Elsevier, 2024, pp. 341–368.
[46] D. N. Dwivedi, G. Mahanty, and V. Nath Dwivedi, "Intelligent conservation: A comprehensive study on AI-enhanced environmental monitoring and preservation," in *The Convergence of Self-Sustaining Systems with AI and IoT*, IGI Global, 2024, pp. 215–226.
[47] P. Tsiakanikas et al., "Beyond the chromosome: Recent developments in decoding the significance of extrachromosomal circular DNA (eccDNA) in human malignancies," *Life*, vol. 14, no. 8, p. 922, 2024.
[48] A. Arshadi, "Amplificazione genomica nel cancro: Circolarizzazione degli ampliconi tramite CRISPR-C, conseguenze trascrizionali e implicazioni terapeutiche," 2024.
[49] Q.-L. Yang, Y. Xie, K. Qiao, J. Y. S. Lim, and S. Wu, "Modern biology of extrachromosomal DNA: A decade-long voyage of discovery," *Cell Res.*, pp. 1–12, 2025.

7 Extrachromosomal DNA in Non-Invasive Cancer Diagnostics

Prabir Manna,[1] *Suravi Saha,*[1] *Pritha Dey,*[2]
Bijoy Paria,[3] *and Payal Chakraborty*[1]

1 Department of Pharmaceutical Chemistry
Calcutta Institute of Pharmaceutical Technology
and Allied Health Sciences
Uluberia, Howrah, India
2 Department of Pharmaceutical Chemistry Derozio Pharma Institute
Gopal Chak
Moyna, Dakshin Ankha, India
3 Department of Pharmaceutical Chemistry Derozio Pharma Institute
Dakshin Ankha, India

7.1 INTRODUCTION

Extrachromosomal DNA (ecDNA) has emerged as a potent driver involved in metastatic transformations in aggressive cancer phenotypes (Ur Rehman, 2025). In cases for non-invasive diagnostic strategies, ecDNA's presence in body fluids, particularly blood plasma, provides a critical advantage for liquid biopsy approaches (Ur Rehman, 2025; Luo et al., 2023) whereas liquid biopsy emerged as a promising, minimally invasive alternative, enabling the analysis of circulating biomarkers such as cell-free DNA (cfDNA) and circulating tumor DNA (ctDNA). ecDNA entities are stable and express cancer-specific genomic signatures allowing detection of tumor-derived genetic material without invasive tissue sampling (Luo et al., 2023). Recent studies demonstrate that small ecDNA profiles in plasma mirror tumor tissue patterns, achieving AUC (Area Under Curve) values >0.9 for multi-cancer detection through biomarkers like ETV6-derived fragments (i.e., translocation-Ets-leukemia virus) protein—a transcription factor in humans. The ETV-6 protein regulates the development and growth of diverse cell types, particularly those of hematological tissues. This approach surpasses the conventional tumor marker performances of the carcinoembryonic antigen (CEA) test or cancer antigen (CA) 19–9 test, offering real-time monitoring of treatment response and relapse (Ur Rehman, 2025; Wu et al., 2023). In contrast to chromosomal DNA, ecDNA molecules in circulation undergo oncogene amplification and tumor heterogeneity by evading typical regulatory mechanisms (Dong et al., 2023; Ur Rehman, 2025). They have structural uniqueness that enhances chromatin accessibility and ability to hijack enhancers, for example, enabling elevation of oncogene expression such as MYCN—a member of the Myc family of oncogenes and epidermal growth factor receptor (EGFR), a membrane-spanning glycoprotein. These genes play important roles in regulating cell growth and division (proliferation), invasiveness, metastasis (pro-survival effects), and the self-destruction of cells (apoptosis) (Dong et al., 2023; Wu et al., 2023) (Figure 7.1).

The dynamic nature of ecDNA evolves with the tumor mechanistic pathways, including therapy resistance (Wan et al., 2017; Koche et al., 2020). Their characteristic inheritable phenomenal patterns

DOI: 10.1201/9781003655299-7

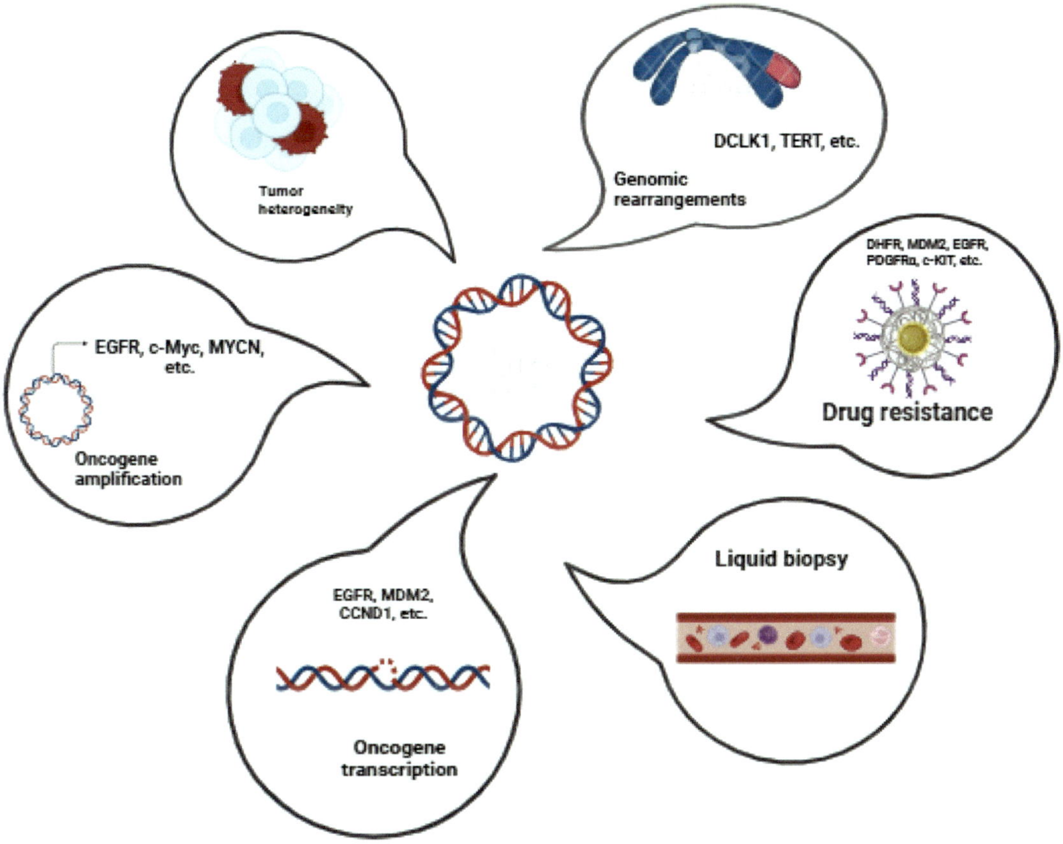

FIGURE 7.1 Biological functions and diagnostic relevance of ecDNA in cancer.

and rapid copy number changes create identifiable molecular footprints during the critical phases of disease progression (Koche et al., 2020; Turner et al., 2017). Advancements in the underlying molecular domain, ecDNA profiling, promise to revolutionize precision oncology by enabling early interception of treatment-resistant clones and personalized therapeutic strategies (Dong et al., 2023; Wan et al., 2017).

7.1.1 Correlation between the ecDNA Profile Pattern and Tumor Progression

Evidence suggests that ecDNA is linked to more aggressive tumor behavior and poor clinical outcomes, which highlight the fact that "the clinical potential extends beyond detection." The key factors of ecDNA enhance tumor progression through (1) extensive genomic remodeling, (2) enhanced oncogene expression, and (3) rapid adaptation to therapeutic pressures. These circular DNA fragments have the capacity to bypass the rules of chromosomal inheritance and amplify cancer-promoting genes.

Main Mechanisms Leading to Tumor Progressions Are as Follows:

(A) **Oncogene Amplification and Expression**
As ecDNA harbors amplified oncogenes, such as EGFR/MYCN, leading to their overexpression, it allows for higher levels of transcriptional activity. Partially, ecDNA can hijack both the local and distal enhancers, forming ultra-long chromatin contacts that further boost oncogene expression. For example, in glioblastoma, ecDNA containing EGFR and its enhancers

results in elevated EGFR activity, while in neuroblastoma, MYCN is co-amplified with a distal enhancer, enhancing its expression and promoting tumor cell viability (Dong et al., 2023).

(B) **Genetic Heterogeneity and Tumor Evolution**
The lack of centromeres in ecDNA leads to its random and unequal segregation during cell division. This non-Mendelian inheritance increases genetic diversity within tumor cell populations, fueling intratumoral heterogeneity and enabling rapid adaptation to environmental and therapeutic pressures. As a result, tumor cells with advantageous ecDNA amplifications can quickly become dominant clones, contributing to cancer progression and therapy resistance (Kim et al., 2024; Wu et al., 2024).

(C) **Association with Malignant Progression and Metastasis**
Clinical studies have shown that ecDNA is detected at higher frequencies in advanced, metastatic, and treatment-resistant tumors compared to newly diagnosed cancers. Tumors from patients who have undergone chemotherapy or targeted therapies often exhibit increased ecDNA content, indicating a role in treatment response and resistance. Furthermore, ecDNAs are more likely to be retained during tumor evolution than chromosomal amplifications and often harbor localized hypermutations, which may further drive metastatic potential and poor prognosis (Turner et al., 2017; Wu et al., 2022a).

7.2 NON-INVASIVE DIAGNOSTIC POTENTIAL AND CURRENT CHALLENGES OF THE ECDNA PROFILING IN DIAGNOSIS AND TREATMENT

Early in the 1980s, studies were carried out through transmission electron microscopy (TEM) and demonstrated extrachromosomal DNA (ecDNA) as circular in shape and made up of chromosomal fibers incorporating nucleosomes (Wu et al., 2022a). Using ultrasound imaging, optical imaging, and whole genome sequencing (WGS) data analysis, Wu et al. (2019) tried to establish more information about the structure of ecDNA. The major groundbreaking findings concluded that ecDNA accessibility led to increased expression of ecDNA loci (Decarvalho et al., 2018), presence of excess copies leads to overexpression of genes, mostly including oncogenes (Wu et al., 2019). Moreover, WGS research demonstrated that oncogene expression was higher when ecDNA molecules were amplified in comparison to the linear chromosomal DNA (Kim et al., 2020). The source of ecDNA generation extrapolates to genomic instability, chromothripsis—term linked to chromosomal damage, DNA replication, breakage–fusion–bridge (BFB) cycles, fork stalling, and template swapping, etc. (Vendramin et al., 2021; Yi et al., 2022; Morton et al., 2019). The current emergence of the CRISPR-Cas9 system has evolved to create ecDNA from DNA fragments, which can be removed from chromosomes in human cell lines (Wu et al., 2022a).

The utmost potential for the ecDNA to serve as a non-invasive diagnostic marker lies in the fact that it acts as a reservoir for DNA recombination by reintegrating into and being excised from chromosomes, which are the two ways contributing to the formation and evolution of cancer phenotypes. Oncogene amplification on ecDNA was frequently observed in a variety of cancer types. For instance, a number of oncogenes, such as MYC (myelocytomatosis oncogene), MYCN (Neuroblastoma MYC oncogene or N-Myc), EGFR (epidermal growth factor receptor), PDGFRA (Platelet-derived growth factor receptor alpha), MET (Mesenchymal-Epithelial Transition), the MECOM-PIK3CA-SOX2 (M doublet 1 oncogene-Phosphatidylinositol-4,5-bisphosphate 3-kinase catalytic subunit alpha-SRY-box transcription factor 2) gene cluster, and the CDK4-MDM2 (cyclin-dependent kinase-4, murine double minute-2) gene cluster, were discovered on ecDNA in a study of glioblastoma cells. Furthermore, it was shown that 22 out of the 25 ecDNAs identified carried an oncogene in several tumor samples from various individuals (Decarvalho et al., 2018). According to a prior study, oncogenes carried by ecDNA may account for the localized amplification in roughly half of the WGS samples obtained from patient cell lines and tumor tissues (Marusyk et al., 2020). The bulk of localized amplifications (85.7%) found by WGS in a neuroblastoma cell line consistently correlated with oncogenes linked to ecDNA. Moreover, higher oncogene expression was linked to a higher copy number on ecDNA. For example, examination of the same cell line's allele-specific messenger RNA expression (ASE) verified that ecDNA was the source of the elevated gene expression (Yan et al., 2024). Similarly, examination of clinical samples and cancer cell lines revealed that a high copy of ecDNA was found in cancer and that higher oncogene expression was associated with a higher copy number of ecDNA (Zhang et al., 2024).

Second, local or distal enhancer hijacking, ecDNA can form ultra-long chromatin connections that can stimulate oncogene expression. As a rule for the local enhancers, oncogenes and local upstream enhancers have the ability to circularize and create ecDNA, which favors enhancer-oncogene interaction to promote expression. For instance, in glioma samples from TCGA, EGFR and two active enhancers were co-amplified and preserved on circular ecDNA. Through ChlP-Seq analysis, ecDNA synthesis may be able to unite non-adjoining segments and facilitate communication between EGFR promoters and upstream enhancers that were initially divided by insulators. Subsequent analysis revealed that distal enhancer co-amplification was present in 30 out of 43 amplified oncogenes in different tumor types (Zhang et al., 2024).

In the case of distal enhancer hijacking model, oncogenes and distal enhancers on the same or different chromosomes circularize to produce ecDNA (Wu et al., 2022b). MYCN and distal enhancer e4 were co-amplified in neuroblastoma cell lines, and chromatin capture analysis showed that e4 interacted with the MYCN promoter to increase MYCN expression (Helmsauer et al., 2020). Furthermore, as shown in the cases of EGFR in glioblastoma and MYCN in neuroblastoma, hijacking enhancers by creating ecDNA may alter cell survival by controlling oncogene expression (Helmsauer et al., 2020).

ecDNAs intervene with chromosomal DNA or form ecDNA hubs to regulate the expression of oncogenes. Interacting with chromosomal DNA, ecDNA acts as a mobile trans-acting component to improve transcription. Here, we cite a few examples where two of the three cell lines obtained from glioblastoma patients tested positive for ecDNA, according to Hi-C analysis, and ecDNA interacted with chromosomal non-coding regions that were abundant in H3K27ac signals. Furthermore, transcriptional activity was linked to the connection between ecDNA and chromatin, and chromosomal oncogene expression was elevated in ecDNA-negative cells by synthetic circular DNA that mimicked ecDNA. Thus, the aforementioned data demonstrated that ecDNA interacted with chromosomal DNA to function as a strong trans-acting regulator of oncogene expression (Dong et al., 2023).

Creation of ecDNA hubs, where 10–100 ecDNAs might combine to form the ecDNA hubs. The protein known as the bromodomain and extra-terminal domain (BET) was found to be crucial for preserving the hubs' structural integrity. Enhancers in ecDNA hubs may be regulated in MYC ecDNA, indicating that the development of ecDNA hubs increased interactions between enhancers and oncogenes to increase oncogene expression (Yan et al., 2024). Intra- and inter-chromosomal rearrangements may result from ecDNA's ability to reintegrate into linear chromosomes. Koche et al. discovered evidence that ecDNA causes neuroblastoma cells to undergo oncogenic genome remodeling by reintegrating into nuclear chromosomes; this mechanism was linked to worse clinical outcomes; for example, the EGFRvIII mutant allele is borne in high copy number on ecDNA with a consistent circular shape of 1.29 Mb in a glioblastoma multiforme (GBM) cell line. These findings again reemphasize the fact that "the clinical potential extends beyond detection," as referred to an example depicted in Table 7.1. (Yang et al., 2023; Yi et al., 2022).

7.3 MACHINE LEARNING-BASED COMPUTATIONAL FRAMEWORK AND IMPLEMENTATION FOR IDENTIFYING ECDNA AMPLIFICATION

Extrachromosomal DNA segments are variably located in the exteriors of chromosomes in a cell's nucleus (Fessler et al., 2024; Wang et al., 2025). They are mostly linked to tumor proliferation and amplified oncogenes in cancer investigation. Approximately 5% of human cell lines, such as MCF-7, MDA-MB-231, PPT2, and NCIH2170, were experimentally validated, which demonstrated a crucial knowledge gap restricting the basic cancer therapeutics in a diverse manner (Fessler et al., 2024; Bailey et al., 2020). Some of the most widely utilized cancer cell line models have undergone comprehensive profiling conducted by the Cancer Cell Line Encyclopedia (CCLE) and

TABLE 7.1
Indicating ecDNA Amplification as a Prognostic Marker for Overall Survival (OS) Rate in an Advanced Gastric Cancer (AGC) Cohort Study as an Example with the Anti-PD-1 Drug Toripalimab

AGC Cohort (anti-PD-1 drug toripalimab)			
Variables	**Attributes**	**N (Sample Size)**	**p-value**
PDL1_status (Immunotherapy)	< 1%	43	0.720
	>= 1%	8	
Tumor Mutation Burden (TMB)_ status	Low	26	0.035
	High	25	
Liver_Metastasis	No	34	0.165
	Yes	17	
Tumor_purity		51	0.004
Aneuploidy	Low	27	0.163
	High	24	

the Dependency Map (DepMap) initiatives, which maintain a large, extensive, and data-intensive population-scale repositories and offer multiomic data for nearly 2000 distinct cancer cell lines (Yang et al., 2023). High-end state-of-the-art facilities like fluorescence in situ hybridization (FISH) and Scanning Electron Microscopy (SEM) can detect the heterogeneity in the ecDNA variants and continue to replicate independently and partition unevenly into the daughter cells. As a result, individual cells show up with significant quantities of disparate ecDNA copies (Wu et al., 2022a). There are evidence contributing to drug resistance through the heterogeneous cells by augmenting the quantity of ecDNA within cells, depleting ecDNA from cells, or reintegrating ecDNA within homogenous staining regions (HSRs) on chromosomes and restoring them following the cessation of drug treatment. Detailed identification methodologies by the supplementary cell lines harboring extrachromosomal DNA (ecDNA) are of high expense and low-yield throughput by the conventional karyotyping, such as FISH and G-banding. These limitations were overcome by the computational algorithmic methodologies, which result in the quantification of the ecDNA amplification utilizing next-generation sequencing data (Fessler et al., 2024; Miroshnychenko et al., 2022).

A thorough computational approach for ecDNA amplification is collected from many cohorts of non-invasive cancer individuals and associated with clinical and genetic information (Zhu et al., 2024). Cohort studies are not only epidemiological but also statistical research endeavors that track patient categories with varying exposures over time to identify relevant outcomes for each exposure group or to look at how one or more participant variables affect predicted outcomes (Wang et al., 2024). Recently, software algorithms and machine-learning-based strategies have been developed for recognizing and determining the total number of ecDNAs present in the massive cell populations.

Utilizing computer vision-based techniques (ECdetect), semi-automatically detects and measures ecDNA in 4',6-diamidino-2-phenylindole (DAPI)-stained micrographs. These process protocols are designed with high precision but low sensitivity in comparison to the human eye, which results in an undercounting of ecDNA, especially ecDNA particles near chromosomes, despite their purpose to reduce false positives (Wu et al., 2023). Moreover, the ecSeg software (available on https://github.com/UCRajkumar/ecSeg), which was developed more recently, not only utilizes a deep neural network methodology to enhance the specificity of DAPI-stained ecDNA detection in the stage of metaphase cells but also incorporates FISH signals (Wu et al., 2022a).

The limitation of low sensitivity of visual identification is enhanced by FISH, as certain ecDNAs could be too tiny to see with DNA dyes. Under the proof-of-concept, imaging-based machine learning approaches successfully unveiled the conformational arrangements, content, size, copy number, and variation of ecDNA in several non-invasive cancer detection phenotypes, using sequencing

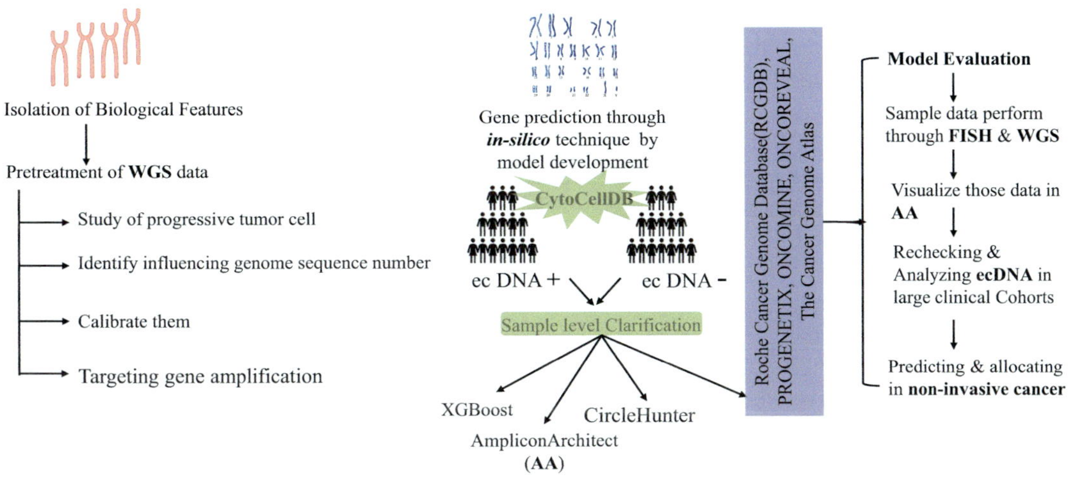

FIGURE 7.2 Algorithm depicting the biological extractive features to build a predictive model for the ecDNA quantitative identification and further processing and performance validation of the sequencing and computational method.

and computational tools. Though there are pros and cons in the sequencing-based computation approach, nevertheless, analyzing and sequencing of ecDNA is publicly available from several cancer genome databases such as Roche Cancer Genome Database (RCGDB), Progenetix, Oncomine, Oncoreveal, The Cancer Genome Atlas (Bergstrom et al., 2022; Wu et al., 2022a, 2023).

Here we like to highlight the most popular and widely used sequencing and computational tools such as AmpliconArchitect (generic amplicon model), CCDA-seq (investigate the variety of ecDNA found in open chromatin zones), ECCsplorer (Linking treatment strategies for cancer-associated ecDNA to active transposable elements), Circle-Map (posterior likelihood of re-aligning the soft splice segment with the genome of reference), Circle-finder (detect the ecDNA fragment as well as analyzing and exploring next generation sequencing data to illustrate non-invasive cancer), which are streamlined for the quantitative identification and evaluation of ecDNAs (Tao et al., 2020; Yang et al., 2023). Finally, for analyzing the ecDNA data, several cluster methods (such as Gaussian mixture clustering, k-means clustering, density-based clustering, etc.) are followed before building the model, but predominantly, good as well as standard gold value is required for the experimental karyotype cell line model-derived data. Analyzing the "true positive" (TP) and "true negative" (TN) numerically represents expressing ecDNA (TP) and not expressing ecDNA (TN), respectively (Wang et al., 2024). Statistically, quantitative models were evaluated with R^2, Pearson, Spearman correlation test, Python 3.8+ using scikit-learn, XGBoost, PyTorch, joblib, etc. methodologies, which take a pivotal role in performing best, with random seeds for reproducibility. In the future, ecDNA analysis would be more reliable, robust, and useful if the sample sizes for "true positive" and "true negative" were increased. This would also boost the power of high-end computational analyses (Wang et al., 2024; Bergstrom et al., 2022) (Figure 7.2).

7.4 FUTURE OF ECDNA IN CLINICAL DIAGNOSIS: AS A POTENTIAL NON-INVASIVE PERSONALIZED CANCER DIAGNOSTIC BIOMARKER

Experimentally, we have several pieces of evidence where ecDNA can be depleted in cell cultures using CRISPR cutting. As ecDNA-directed clinical trials are imminent, and it specifies that ecDNA can function as a predictive biomarker for clinical outcomes, including malignant transformation, patient survival, and therapy response, as previously discussed, there is an urgent need to develop a more affordable and advanced ecDNA detection tool (Zhao et al., 2022).

The domain of ecDNA biology is experiencing rapid expansion. Numerous aspects can be explored utilizing existing methodologies. The initial focus pertains to fundamental ecDNA underlying molecular biology. There is a pressing requirement to elucidate the mechanisms responsible for chromosomal instability associated with ecDNA formation, along with a comprehensive examination of the roles played by the double-stranded breaks and non-homologous end joining (NHEJ) pathways in this context (Ma et al., 2021). Additionally, the implications of ecDNA reintegration in genomic remodeling and its relocation as a heterochromatin structure are yet to be fully elucidated. Emerging citations revealed that the formation of heterochromatin structure is linked to drug resistance; however, the mechanisms, motivations, and specific locations of this phenomenon are still largely unexplored. We are beginning to comprehend the significance of ecDNA in tumor evolution. Nonetheless, numerous areas within this domain require further investigation, including the influence of ecDNA on clonal competition or cooperation, the timing of ecDNA formation—particularly in relation to phenomena such as whole-genome doubling—and whether occurrences such as aneuploidy or whole-genome doubling predispose cells to ecDNA formation (Zahid et al., 2013). The identification of ecDNA within circulating tumor DNA could provide critical insights into prognosis and treatment strategies. This understanding will be enriched if we are able to delineate the landscape of ecDNA-mediated drug resistance and ascertain the feasibility of therapeutically targeting ecDNA formation. Investigating the role of ecDNA in neoantigen generation and the immune microenvironment may also reveal new therapeutic opportunities tailored to personalized medication. This encompasses the refinement of existing sequencing and purification techniques to isolate circular amplicons, as well as the development of computational tools that leverage technologies such as long-read and single-cell sequencing to enhance the reconstruction of circular amplicons (Ling et al., 2021). Employing these technologies to address the temporal aspects of ecDNA formation and integration poses significant challenges as a real-time diagnostic non-invasive biomarker in diverse cancer and tumor phenotypes. Recent findings have highlighted ecDNA as a crucial factor in tumor evolution. Unconstrained by the traditional and hereditary limitations, its presence confers rapid selection advantages that promote aggressive tumor behavior. A deeper understanding of this process will undoubtedly enhance our treatment strategies for future perspectives (Cheng et al., 2024).

REFERENCES

Bailey C, Shoura MJ, Mischel PS, Swanton C. Extrachromosomal DNA—relieving heredity constraints, accelerating tumour evolution. Annals of Oncology. 2020 Jul 1;31(7):884–893.

Bergstrom EN, Luebeck J, Petljak M, Khandekar A, Barnes M, Zhang T, et al. Mapping clustered mutations in cancer reveals APOBEC3 mutagenesis of ecDNA. Nature. 2022 Feb 17;602(7897):510–517.

Cheng X, Bae J. Next-generation DNA-enhanced electrochemical energy storage: Recent advances and perspectives. Current Applied Physics. 2024 Jul 14; 67:1–17.

Decarvalho AC, Kim H, Poisson LM, Winn ME, Mueller C, Cherba D, et al. Discordant inheritance of chromosomal and extrachromosomal DNA elements contributes to dynamic disease evolution in glioblastoma. Nature Genetics. 2018 May;50(5):708–717.

Dong Y, He Q, Chen X, Yang F, He L, Zheng Y. Extrachromosomal DNA (ecDNA) in cancer: Mechanisms, functions, and clinical implications. Frontiers in Oncology [Internet]. 2023 [cited 2025 Jun 9];13:1194405. Available from: https://pmc.ncbi.nlm.nih.gov/articles/PMC10338009/.

Fessler J, Ting S, Yi H, Haase S, Chen J, Gulec S, et al. CytoCellDB: A comprehensive resource for exploring extrachromosomal DNA in cancer cell lines. NAR Cancer. 2024 Sep;6(3):zcae035.

Helmsauer K, Valieva ME, Ali S, Chamorro González R, Schöpflin R, Röefzaad C, et al. Enhancer hijacking determines extrachromosomal circular MYCN amplicon architecture in neuroblastoma. Nature Communications. 2020 Nov 16;11(1):5823.

Kim H, Kim S, Wade T, Yeo E, Lipsa A, Golebiewska A, et al. Mapping extrachromosomal DNA amplifications during cancer progression. Nature Genetics. 2024 Nov 1;56:2447–2454.

Kim H, Nguyen NP, Turner K, Wu S, Gujar AD, Luebeck J, et al. Extrachromosomal DNA is associated with oncogene amplification and poor outcome across multiple cancers. Nature Genetics. 2020 Sep;52(9):891–897

Koche RP, Rodriguez-Fos E, Helmsauer K, Burkert M, MacArthur IC, Maag J, et al. Extrachromosomal circular DNA drives oncogenic genome remodeling in neuroblastoma. Nature Genetics [Internet]. 2020 Jan 1 [cited 2025 Jun 9];52(1):29–34. Available from: https://www.nature.com/articles/s41588-019-0547-z.

Ling X, Han Y, Meng J, Zhong B, Chen J, Zhang H, et al. Small extrachromosomal circular DNA (eccDNA): Major functions in evolution and cancer. Molecular Cancer. 2021 Dec; 20:1–5.

Luo X, Zhang L, Cui J, An Q, Li H, Zhang Z, et al. Small extrachromosomal circular DNAs as biomarkers for multi-cancer diagnosis and monitoring. Clinical and Translational Medicine [Internet]. 2023 Sep [cited 2025 Jun 9];13(9). Available from: https://pubmed.ncbi.nlm.nih.gov/37649244/.

Ma W, Zhan Y, Zhang Y, Mao C, Xie X, Lin Y. The biological applications of DNA nanomaterials: Current challenges and future directions. Signal Transduction and Targeted Therapy. 2021 Oct 8;6(1):351.

Marusyk A, Janiszewska M, Polyak K. Intratumor heterogeneity: The rosetta stone of therapy resistance. Cancer Cell. 2020 Apr 13;37(4):471–484.

Miroshnychenko A, Zeraatkar D, Phillips MR, Bakri SJ, Thabane L, Bhandari M, et al. Cohort studies investigating the effects of exposures: Key principles that impact the credibility of the results. Eye. 2022 May;36(5):905–906.

Morton AR, Dogan-Artun N, Faber ZJ, MacLeod G, Bartels CF, Piazza MS, et al. Functional enhancers shape extrachromosomal oncogene amplifications. Cell. 2019 Nov 27;179(6):1330–1341.

Tao K, Bian Z, Zhang Q, Guo X, Yin C, Wang Y, et al. Machine learning-based genome-wide interrogation of somatic copy number aberrations in circulating tumor DNA for early detection of hepatocellular carcinoma. EBioMedicine. 2020 Jun 1;56.

Turner KM, Deshpande V, Beyter D, Koga T, Rusert J, Lee C, et al. Extrachromosomal oncogene amplification drives tumour evolution and genetic heterogeneity. Nature [Internet]. 2017 Mar 2 [cited 2025 Jun 9];543(7643):122–125. Available from: https://www.nature.com/articles/nature21356.

Ur Rehman, S. The multifaceted role of extrachromosomal DNA in cancer: Functions and clinical implications. Bangladesh Journal of Medicine [Internet]. 2025 May 5;36(2):82–91. Available from: https://www.banglajol.info/index.php/BJMED/article/view/79673.

Vendramin R, Litchfield K, Swanton C. Cancer evolution: Darwin and beyond. The EMBO Journal. 2021 Sep 15;40(18):e108389.

Wan JCM, Massie C, Garcia-Corbacho J, Mouliere F, Brenton JD, Caldas C, et al. Liquid biopsies come of age: Towards implementation of circulating tumour DNA. Nature Reviews Cancer [Internet]. 2017 Feb 24 [cited 2025 Jun 9];17(4):223–238. Available from: https://www.nature.com/articles/nrc.2017.7.

Wang S, Wu CY, He MM, Yong JX, Chen YX, Qian LM, et al. Machine learning-based extrachromosomal DNA identification in large-scale cohorts reveals its clinical implications in cancer. Nature Communications. 2024 Feb 19;15(1):1515.

Wang Y, Cope O, Chen J, Mehta A, Fleifel D, Ford C, et al. Extrachromosomal DNA gives cancer a new evolutionary pathway. bioRxiv. 2025:2025-04.

Wu H, Liu S, Wu D, Zhou H, Wu G. Tumor extrachromosomal DNA: Biogenesis and recent advances in the field. Biomedicine and Pharmacotherapy. 2024;174.

Wu P, Liu Y, Zhou R, Liu L, Zeng H, Xiong F, et al. Extrachromosomal circular DNA: A new target in cancer. Frontiers in Oncology. 2022a;12. Frontiers Media S.A.

Wu S, Bafna V, Chang HY, Mischel PS. Extrachromosomal DNA: An Emerging Hallmark in Human Cancer. Annu Rev Pathol. 2022b Jan 24; 17:367–386. doi: 10.1146/annurev-pathmechdis-051821-114223. Epub 2021 Nov 9. PMID: 34752712; PMCID: PMC9125980.

Wu S, Tao T, Zhang L, Zhu X, Zhou X. Extrachromosomal DNA (ecDNA): Unveiling its role in cancer progression and implications for early detection. Heliyon. 2023 Nov 1;9(11).

Wu S, Turner KM, Nguyen N, Raviram R, Erb M, Santini J, et al. Circular ecDNA promotes accessible chromatin and high oncogene expression. Nature. 2019 Nov 28;575(7784):699–703.

Yan X, Mischel P, Chang H. Extrachromosomal DNA in cancer. Nature Reviews Cancer. 2024 Apr;24(4):261–273.

Yang M, Zhang S, Jiang R, Chen S, Huang M. Circlehunter: A tool to identify extrachromosomal circular DNA from ATAC-Seq data. Oncogenesis. 2023 May22;12(1):28.

Yi E, Chamorro González R, Henssen AG, Verhaak RG. Extrachromosomal DNA amplifications in cancer. Nature Reviews Genetics. 2022 Dec;23(12):760–771.

Zahid M, Kim B, Hussain R, Amin R, Park SH. DNA nanotechnology: A future perspective. Nanoscale Research Letters. 2013 Dec;8:1–3.

Zhang S, Xiao X, Yi Y, Wang X, Zhu L, Shen Y, et al. Tumor initiation and early tumorigenesis: Molecular mechanisms and interventional targets. Signal Transduction and Targeted Therapy. 2024 Jun 19;9(1):149.

Zhao Y, Yu L, Zhang S, Su X, Zhou X. Extrachromosomal circular DNA: Current status and future prospects. Elife. 2022 Oct 18;11:e81412.

Zhu K, Jones MG, Luebeck J, Bu X, Yi H, Hung KL, et al. CoRAL accurately resolves extrachromosomal DNA genome structures with long-read sequencing. Genome Research. 2024 Sep 1;34(9):1344–1354.

8 Extrachromosomal DNA (ecDNA) as Diagnostic Biomarkers for Cancers

Riffat Maqsood,[1] *Muhammad Usman,*[1]
Muhammad Nasir Bhaya,[2] *Ayush Madan,*[3]
and Benedict Terkula Iber[4]

1 Faculty of Veterinary Science
University of Agriculture Faisalabad
Pakistan
2 Department of Pathology
Faculty of Veterinary Science
University of Agriculture Faisalabad
Pakistan
3 Department of Biotechnology
School of Research & Technology, People's University
Bhanpur, Bhopal, India
4 Institute of Tropical Aquaculture and Fisheries
Universiti Malaysia Terengganu
Terengganu, Malaysia

8.1 INTRODUCTION TO EXTRACHROMOSOMAL DNA (ECDNA)

It is a double-stranded circular structure present outside chromosomes, constructed from nucleosomal chromatin, and present in pairs. It lacks a centromere and is inherited in a non-Mendelian pattern during mitosis. It is stable in the bloodstream and resistant to enzymatic degradation because there are no free ends for DNA exonucleases to work. Extrachromosomal circular DNA (eccDNA) was called double minutes (DMs) in 1964 when observed in pig sperm and wheat embryo by Hotta and Bassel, and is chromosome independent [1]. These DMs carry oncogenes such as epidermal growth factor receptor (EGFR), but later on, it was discovered that DMs make up only 30% of eccDNA. Thus, it was renamed as eccDNA. Extrachromosomal DNA (ecDNA) was found in childhood malignancies. Then, its correlation with aging was indicated when eccDNA was amplified in human skin fibroblasts with Werner syndrome [2]. In 2018, Møller et al. identified tens of thousands of eccDNAs in normal human muscle and blood cells, containing full genes or gene fragments [3]. This discovery highlighted the widespread presence of eccDNA in healthy tissues and sparked growing interest in its diverse biological roles, particularly in disease processes. Later on, its cancerous and tumor-associated roles were discovered. This eccDNA contains promoters and enhancers, responsible for turning genes off or on. These are active in making RNA (transcriptional activity) and too much protein, so when eccDNA containing oncogenes is overexpressed, it grows uncontrollably.

ecDNA is crucial for some necessary and mandatory functions in the cell. It plays a critical role in gene expression and thus amplifies specific genes. When these genes are overexpressed, they contribute to cancer development. Gene expression is enhanced by transcriptional activity when ecDNA forms transcriptional hubs. It is also used for gene therapy, such as oncogenes are stable and

 DOI: 10.1201/9781003655299-8

resistant to chemotherapy [4]. ecDNA can stimulate the cGAS-STING pathway, a sensor of cytosolic DNA that is responsible for the production of cytokines, interferons, and anti-tumor immune response. It may accumulate in non-cancerous cells due to failed degradation pathways and DNA repair errors. This ecDNA induces cellular aging and causes age-related inflammation and diseases.

8.2 BIOGENESIS OF ECCDNA

Biogenesis is not so clear, but some biological steps are as follows:

8.2.1 DNA Damage Initiation

DNA damage occurs due to replication stress, radiation, chemicals, or apoptosis. This damage includes double-strand breaks (DSBs), which are crucial triggers for ecDNA formation. DNA damage occurred between replication forks from replication bubble. This forms small fragments after breaking, called episomes.

8.2.2 DNA Repair Pathway Activation

The cell activates DNA damage repair pathways by homologous recombination (HR), non-homologous end joining (NHEJ), and micro-homology-mediated end joining (MMEJ) [5]. These pathways attempt to repair broken DNA, especially in areas rich in repetitive sequences or short homologies.

8.2.3 Chromothripsis or BFB Cycles

Chromothripsis is a catastrophic genomic event involving massive chromosome fragmentation and rearrangement in a single step. During this process, DNA fragments can circularize to form ecDNA. PARP and DNA-PKcs enzymes are essential in mediating chromothriptic events [6]. Chromothripsis has been associated with increased ecDNA carrying oncogenes like MYC and ERBB2, contributing to tumor heterogeneity and drug resistance.

8.2.4 Circularization of DNA Fragments

The DNA fragments formed from damage or rearrangement are ligated into circular DNA using DNA Ligase 3 (Lig3). This ligation may be facilitated by Micro-homology between ends (MMEJ), Direct tandem repeats, or Enzymes like PARP and DNA-PKcs.

8.2.5 Formation of ecDNA

Circular DNA fragments without centromeres form ecDNA. ecDNA formation may be induced by apoptosis using enzymes like CAD or EndoG for DNA fragmentation [7]. Furthermore, episomes can undergo recombination and circularization, forming ecDNA. These may carry oncogenes (e.g., MYC, KRAS, and EGFR), including enhancer elements, or avoid regular chromosome segregation during cell division.

8.2.6 Amplification and Evolution of ecDNA

ecDNA is randomly inherited during mitosis. It can undergo self-replication, leading to high copy numbers, fusion with other ecDNAs to form larger ecDNA structures, and accumulation of mutations due to poor repair efficiency. ecDNA evolves rapidly under selective pressures such as drug treatment, resulting in tumor adaptability, oncogene amplification, and resistance.

8.3 CLASSIFICATION

The classification of eccDNA is described next.

8.3.1 Small Poly-Dispersed DNA (spcDNA)

It is a variant of eccDNA characterized by its size, between 100 base pairs and 10 kilobases, with a diameter ranging from 0.1 to 2 μm. It arises from direct tandem repeat regions within chromosomes [8]. The biogenesis of spcDNA is associated with the DNA ligase-IV-dependent repair mechanism. spcDNA is present in low abundance in normal eukaryotic cells. However, it is more frequently detected in genetically unstable cells, such as in tumors or those undergoing carcinogenic stress. Its presence in HeLa cells, colon cancer cells, and cells exposed to mutagens indicates a strong link between spcDNA and genomic instability.

8.3.2 Telomeric Circles (T-Circles)

T-circles are a subtype of eccDNA composed of repetitive telomeric sequences (TTAGGG/CCCTAA), often existing as single-stranded DNA in units of 738 bp. They form through recombination or telomere loop cleavage as discovered in mouse and human cell nuclei, and can replicate via a rolling-circle mechanism [9]. This contributes to alternative lengthening of telomeres (ALT), a telomere maintenance pathway active in ~15% of cancers, aiding in tumor cell survival.

8.3.3 MicroDNA

It is formed from non-repetitive genomic regions with high GC content, with a length ranging from 100 to 400 bp. They may exist as single-stranded circular form that arises from replication or ligation errors or double-stranded form as a result of circularized DNA break fragments [10]. MicroDNA is much longer and can be detected through plasma and serum in cancer patients.

8.3.4 Extrachromosomal DNA (ecDNA)

ecDNA is a circular DNA molecule, 1–3 Mb in size, lacking telomeres and centromeres. It often arises during DNA repair and is primarily found in cancer cells—present in nearly 46% of cases but absent in normal cells. ecDNA may carry multiple genes and regulatory elements and is linked to gene amplification, particularly of oncogenes. This amplification occurs as either ecDNA (also called double minutes) or chromosomal HSRs (show uniform staining with trypsin-Giemsa) [11]. Being euchromatic, ecDNA undergoes rapid genomic changes, contributing to tumor heterogeneity, treatment resistance, and enhanced malignancy.

8.4 CHARACTERISTICS OF ECDNA

8.4.1 Kyklonic Hypermutation in ecDNA

In various cancers, ecDNA often undergoes kyklonic hypermutation, particularly in genes like TP53, ARNT, and MDM2. This mutation pattern is highly associated with the APOBEC3 enzyme, which normally restricts retroviruses by mutating viral genomes [12]. Since ecDNA mimics viral DNA, it can trigger APOBEC3, resulting in extensive mutations. Studies show about 30% of ecDNAs are hypermutated, with 97.8% of these mutations linked to APOBEC3 activity. This process likely contributes to tumor evolution and therapy resistance.

8.4.2 Nucleosome Structure of ecDNA

ecDNA shows higher gene expression than linear DNA due to its looser nucleosome structure. Assays like ATAC-seq reveal that ecDNA is significantly more chromatin-accessible than chromosomal DNA—up to twice as accessible, with about 80% of ecDNA regions being actively transcribed [13, 14]. This open chromatin state enhances its transcriptional activity.

8.4.3 ecDNA Hubs

ecDNA molecules can cluster into ecDNA hubs, comprising 10–100 ecDNA units in a single nucleus. These hubs form via interactions between ecDNAs, enhancers, and linear chromosomes [15]. The BRD4 protein facilitates hub formation, and its inhibition can reduce oncogene expression. ecDNA hubs also interact with RNA polymerase II, forming transcriptionally active zones that promote high-level gene expression.

8.4.4 Amplified Oncogenes

ecDNA commonly carries oncogenes like MYC, CDK4, and EGFR, which are frequently amplified in cancers such as glioblastoma and lung cancer, contributing to tumor growth and aggressiveness [16]. This amplification occurs because of oncogenes and transcriptional activity that boost gene expression.

8.4.5 Enhanced Transcriptional Activity

Due to open chromatin and absence of histone compaction, ecDNA shows higher transcriptional activity than chromosomal DNA. It also recruits enhancers and transcriptional machinery more effectively [17]. Furthermore, ecDNA can hijack enhancers and form hubs, thus augmenting gene expression.

8.4.6 Drug Resistance and Genetic Evolution

ecDNA contributes to drug resistance by carrying genes that can be rapidly amplified in response to therapeutic pressures. Its random segregation during cell division leads to genetic heterogeneity, enabling tumors to adapt quickly to treatments and evolve resistance mechanisms. Shortly, ecDNA enables rapid gene amplification and unequal segregation during mitosis, increasing tumor heterogeneity and contributing to drug resistance.

8.5 ECDNA IN CANCER PATHOGENESIS AND DIAGNOSIS

8.5.1 Mechanism of Tumorigenesis Mediated by ecDNA

Extrachromosomal circular DNA (eccDNA) has emerged as a pivotal driver of cancer development due to its ability to amplify genes, remodel transcription, and promote genomic instability. Unlike chromosomal DNA, eccDNA exists as independent circular DNA elements that lack centromeres and telomeres. This structural independence provides tumor cells with genetic flexibility that supports aggressive behavior and therapy resistance.

8.5.1.1 Oncogene Amplification and Drug Resistance

One of the basic roles of eccDNA in tumorigenesis is its ability to carry multiple copies of oncogenes such as DHFR, MYC, and EGFR [18]. These circular structures allow for uncontrolled amplification. For instance, amplification of the DHFR gene via eccDNA enables cancer cells to overproduce dihydrofolate reductase, diminishing the effect of methotrexate (MTX)—chemotherapy drug that targets this enzyme. As a result, tumor cells continue to synthesize DNA and proliferate even under the drug pressure.

8.5.1.2 Formation of Transcriptional Hubs

EccDNA molecules often aggregate into well-organized structures called "eccDNA hubs." These hubs concentrate transcriptional machinery, including proteins like BRD4 and RNA polymerase II, at specific genomic loci [19]. This leads to an amplified transcriptional output. Experimental studies show that disrupting these hubs results in reduced gene expression and induced cancer cell apoptosis, making them a target for therapeutic intervention.

8.5.1.3 Enhancer Hijacking and Long-Range Regulation

Beyond gene amplification, eccDNA can enhance gene expression by recruiting distant enhancer elements—a process known as enhancer hijacking. This occurs when enhancers are located far from a gene and become close due to circularization. These eccDNAs then act as mobile regulatory platforms, enhancing oncogene transcription in a way that is independent of normal chromosomal architecture.

8.5.1.4 Genomic Instability and Structural Rearrangements

EccDNA also contributes to chromosomal instability. These elements can reintegrate into linear chromosomes, causing structural mutations such as duplications, insertions, deletions, or translocations [20]. Such rearrangements promote genetic diversity within the tumor, allowing cells to rapidly evolve and adapt to environmental stress, including chemotherapy. This ability to remodel the genome makes eccDNA a major driver of tumor evolution and progression.

8.5.1.5 Accumulation in Senescent Cells

In normal cells, eccDNA is typically removed from the nucleus by nuclear pore complexes (NPCs). However, in aged cells, these removal mechanisms become less effective, leading to eccDNA accumulation. This buildup may hinder gene regulation and DNA stability, contributing to age-related cancer risk. It also reflects defective cellular maintenance, a hallmark of both aging and tumorigenesis [21].

8.5.1.6 Activation of Innate Immunity

eccDNA can leak into the cytoplasm, where it acts as a danger-associated molecular pattern (DAMP). It activates the cyclic GMP-AMP synthase (cGAS)-STING pathway, which in turn stimulates the production of type I interferons and pro-inflammatory cytokines. While this response can enhance immune surveillance of tumors, chronic activation may support a pro-tumor inflammatory environment, helping cancer cells evade immune control and continue to proliferate.

8.5.1.7 Promotion of Tumor Heterogeneity

Due to the absence of centromeres, eccDNAs are not uniformly distributed during mitosis. This causes unequal inheritance of oncogene copies among daughter cells, leading to functional diversity within the tumor. Some cells may acquire higher oncogene levels and become more aggressive or drug-resistant, while others remain more sensitive. This heterogeneity complicates treatment and is a significant barrier to effective, long-term cancer control.

8.5.2 ecDNA as an Early Diagnostic Biomarker

Extrachromosomal circular DNA (eccDNA) is emerging as a promising diagnostic biomarker in cancer due to several unique characteristics that distinguish it from traditional biomarkers like circulating tumor DNA (ctDNA), proteins, or mRNA fragments [22] (Table 8.1). It is used as a valuable biomarker for diagnosis because of the following reasons:

The abundant production of ecDNA during the early stages of the tumor may prove to be alarming. eccDNA is often generated early during tumorigenesis and accumulates in large quantities in

TABLE 8.1
Comparison of ecDNA and Traditional Cancer Biomarkers

Type of Biomarker	Source	Tumor Heterogeneity	Detectability in Early Cancer	Tumor specificity	Methods of Detection	References
ecDNA	Circular, tumor-derived DNA elements with oncogene amplification	Yes	Abundant release during early stages	Very high	Liquid biopsy methylation-seq; Circle-seq; FISH + DAPI + AI tools (ecSeg)	[23]
ctDNA	Fragmented linear tumor DNA	Partially	Limited release during early stages	Normal	ddPCR, CAPP-Seq, NGS, digital PCR assays	[24]
Exosomal DNA/RNA	Protected within extracellular vesicles	Limited	Moderate release during early stages	Normal	ddPCR, CRISPR/Cas-based assays (e.g., SHERLOCK)	[25]

tumor tissues. Unlike chromosomal DNA, eccDNA can be highly amplified, often carrying multiple copies of oncogenes, such as EGFR, MYC, or DHFR. This amplification results in a robust signal that can be more easily detected in blood or other bodily fluids compared to low-abundance traditional biomarkers. Additionally, eccDNA is more stable than linear DNA in circulation due to its circular structure, which protects it from exonucleases digestion. It is found in body fluids as cell-free eccDNA (cf-eccDNA), making it accessible via liquid biopsy for non-invasive, real-time cancer monitoring. Furthermore, eccDNA carries tumor-specific mutations, oncogenes, and regulatory elements; its presence in circulation can reflect the genomic status of the tumor with high specificity. It can provide a broader snapshot of tumor evolution and heterogeneity than traditional biopsy methods. The use of eccDNA as a biomarker, combined with next-generation sequencing (NGS) and advanced liquid biopsy technologies, allows for the detection of cancer with higher sensitivity and specificity. This is particularly useful for early detection, even when tumors are too small for imaging techniques to detect (Table 8.1).

8.6 FUNCTIONS OF ECDNA

eccDNA (extrachromosomal circular DNA) is a small, circular piece of DNA found outside the chromosomes. eccDNA plays a major role in how cancers grow and survive by supporting several key cancer traits known as the hallmarks of cancer.

8.6.1 Boosts Oncogene Copies and Activity

eccDNA holds extra copies of genes that drive cancer, like MYC, EGFR, and DHFR. These extra copies lead to overproduction of proteins, helping tumors grow faster and resist treatment. Example: More DHFR copies make cancer cells resist the chemotherapy drug methotrexate [26]. ecDNA splits randomly during division, leading to increased copies and resulting in variety.

8.6.2 Promotes Genetic Instability and Increases Gene Expression through Hubs

eccDNA can cause DNA changes, such as rearrangements and mutations. It will become unstable. This makes cancer cells more adaptable and can lead to drug resistance and aggressive behavior, leading to survival even during treatment. eccDNA can group together to form transcription hubs. These hubs make it easier for the cell to turn on genes, especially cancer-promoting ones. Disrupting these hubs could be a new way to slow down the growth of cancer.

8.6.3 Leads to Tumor Cell Diversity and Builds Up in Aging Cells

eccDNA gets passed on because it doesn't follow normal cell division rules. This creates genetic differences between tumor cells, making it harder to treat all of them. This feature leads to tumor progression, heterogeneity, and drug resistance. In old or damaged cells, eccDNA isn't removed properly and starts to accumulate. This build-up can either slow cancer (by stopping cell division) or sometimes promote it, depending on the situation [27]. Sometimes, they interfere with cell division and DNA replication, slowing down the proliferation of cancerous cells. On the other hand, some carry oncogenes that lead to tumor progression.

8.6.4 Activates the Immune System

eccDNA released into the cytoplasm can stimulate immune responses by activating the cGAS-STING pathway, which alerts the body with interferons and inflammation [28]. It acts like a warning signal, alerting the immune system to DNA damage, infection, or malignancy. This body's surveillance mechanism helps to detect abnormal cells, such as tumor cells or virus-infected cells. This feature makes eccDNA a preferred immunological biomarker and a therapeutic target to increase anti-tumor immunity.

8.6.5 ecDNA Hijacks Enhancers to Boost Cancer Genes

ecDNA can enhance the activity of oncogenes by hijacking enhancers—DNA elements that boost gene expression. Local and distal Enhancer Hijacking are the two kinds of ecDNA hijacks enhancers. In Local Enhancer Hijacking, enhancers and oncogenes are circularized together into ecDNA. They get closer, thus increasing gene activation. For instance, in glioblastoma, EGFR and its enhancers are co-amplified on ecDNA, boosting EGFR expression [29]. In Distal Enhancer Hijacking, distant enhancers (even from other chromosomes) join with oncogenes on ecDNA. This allows far-away enhancers to activate oncogenes. For instance, in neuroblastoma, MYCN is co-amplified with a distant enhancer (e4), increasing its expression [30].

8.6.6 ecDNA as a Source of DNA Recombination

ecDNA not only carries extra genes but also acts as a reservoir for DNA rearrangements, reshaping the cancer genome. It can form from DNA fragments or other ecDNAs through cutting or ligation and reinsert into chromosomes, causing genetic teratologies, driving cancer progression by remodeling the genome. It is seen in cancers like neuroblastoma and glioblastoma, where ecDNA reintegration alters gene structure and function [31]. In glioblastoma, EGFRvIII ecDNA reintegrates into chromosomes, creating tandem duplications.

8.7 DETECTION METHODS

Various detection methods playing a crucial role in identifying the levels of ecDNA in body fluids are described next.

8.7.1 Scanning Electron Microscopy

Scanning electron microscopy (SEM) enables high-resolution imaging of the eccDNA's surface structure and size. It uses electron beams to detect surface morphology and composition, requiring samples to be coated with conductive metals like gold or platinum [32]. Low-voltage settings (3–5 kV) enhance resolution for chromosomal imaging. In breast cancer cell lines, SEM has revealed eccDNA in both single forms and as double minutes (DMs). These DMs can cluster into ecDNA hubs, networks of 10–100 ecDNA molecules within a single nucleus, formed through interactions with other ecDNA, enhancers, or chromosomes. Additionally, optical microscopy of tumor tissue karyotypes has identified high-molecular-weight eccDNA as small, double-stained bodies in metaphase cells [33]. Now, atomic force microscopy and structured illumination microscopy are widely used. These provide detailed information about morphology, size, and distribution of ecDNA.

First, isolate the ecDNA from cells or tissues using DNA extraction kits. Then fix these samples on glass coverslip using paraformaldehyde. Dehydrate the samples using a series of alcohol concentrations and coat the samples with a thin layer of conductive material to enhance imaging. Use low voltage to enhance the resolution of image [34]. Modern SEMs have a resolution of about 1–5nm. A smaller electron beam diameter provides more resolution. Some challenges, such as high-energy electron beams, can damage ecDNA, limiting resolution. The drying process can cause ecDNA molecules to flatten, affecting their apparent structure. Insufficient coating or charging can affect image quality and resolution. ecDNA's small size and fragility require careful handling and preparation. ecDNA's low atomic number and small size can result in low contrast images, making it difficult to distinguish from the substrate.

8.7.2 Fluorescence In Situ Hybridization (FISH)

Fluorescence microscopy is a key method for visualizing extrachromosomal DNA (ecDNA) to determine its spatial distribution within cells. ecDNA can be labeled using DAPI (4',6-diamidino-2-phenylindole), a fluorescent dye that binds to adenine-thymine-abundant regions of DNA and detected using FISH (fluorescence in situ hybridization) probes targeting specific oncogenes and centromeres. In interphase cells, both chromosomal and extrachromosomal oncogene signals may appear, whereas metaphase cells, due to their well-organized chromosomes, allow more precise localization of these genetic structures.

However, the combined use of DAPI and FISH does not allow accurate quantification of ecDNA copy numbers, which is necessary to assess variation between individual cells. To overcome this limitation, a tool was developed called ECdetect, a semi-automated software that quantifies ecDNA in DAPI-stained metaphase spreads [35]. ECdetect uses two adaptive thresholds: the first filters out small background elements to identify chromosome regions, and the second, applied to the remaining area, detects structures within the size range of ecDNA (3–75 pixels). It has high consistency, but it may be challenging due to a high signal-to-noise ratio, resulting in the loss of small ecDNA.

Despite ECdetect's effectiveness, the detection of ecDNA remains challenging due to its irregular shape, small size, and interference from background noise such as nuclei and chromosome signals. To improve upon this, researchers developed ecSeg, a deep learning-based image analysis platform utilizing the U-Net convolutional neural network architecture, widely used in biomedical image segmentation [36]. ecSeg integrates DAPI and FISH data to classify each pixel in an image as ecDNA, chromosome, nucleus, or cytoplasm. Detected ecDNA components are then grouped and counted, and a final post-processing step associates them with chromosomal regions using FISH signals. Although it has a high F1 score or recall rate, it is limited because of the requirement of fine-tuning for every input data. Performance-wise, ecSeg demonstrates significant improvements over ECdetect, achieving an average precision and recall of 82%, while ECdetect typically shows a recall below 50%. ecSeg also yields a higher F1 score, reflecting its enhanced accuracy in ecDNA detection.

8.7.3 Sequencing-Based Tools

With the advancements in studies, many other tools are used to detect ecDNA. Circle-seq includes cell lysis using proteinase K, column chromatography, removal of linear DNA using plasmid-safe DNase, rolling circle amplification with φ29 DNA polymerase, and sequencing for isolating ecDNA [37]. It is completely readable because of high specificity and enrichment, but a high sample size is challenging. Circle-finder isolates ecDNA using ATP-dependent exonuclease, which is amplified by any primer. It works as a biomarker for tumor detection. ecc-finder detects ecDNA from nanopore reads but is somewhat demanding. Amplicon Reconstructor reconstructs ecDNA by uniting NGS and OM data. It has high resolution, but extra input of the optical mapping (OM) of long DNA fragments is required and is time-consuming [38].

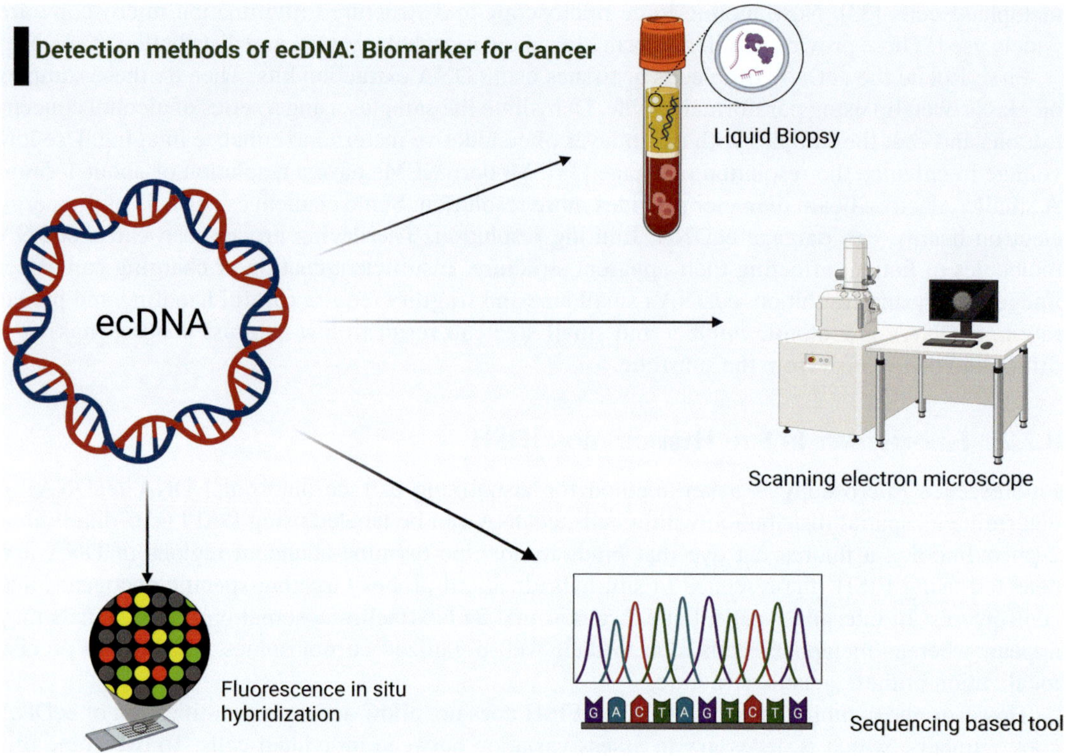

FIGURE 8.1 Techniques and methods used to detect the levels of ecDNA in body fluids [42].

8.7.4 CRISPR-Based Tools

The CRISPR-Cas techniques like ecTag and CRISPR-CATCH enable precise ecDNA visualization by targeting specific sequences with guide RNAs. The ecTag platform identifies ecDNA breakpoints using AmpliconArchitect and then uses dead Cas9 and the Casilio system to fluorescently label ecDNA in live cells. This allows real-time tracking, revealing unequal ecDNA segregation during mitosis [39]. CRISPR-CATCH, originally designed for cloning large DNA fragments, has been adapted to enrich intact, megabase-scale ecDNA. It preserves the circular structure and epigenetic features, enabling detailed sequence and epigenomic analysis [40].

8.7.5 Liquid Biopsy

Traditional tumor biopsies are invasive and offer only limited insight. Liquid biopsies, using biomarkers like circulating tumor cells (CTCs), circulating tumor DNA (ctDNA), and extracellular vesicles (EVs), are emerging as a less invasive alternative [41]. ecDNA, which circulates more readily than linear DNA, can be released from tumors and serve as a biomarker. It is linked to oncogene amplification, drug resistance, and tumor progression (Figure 8.1). In lung adenocarcinoma, specific ecDNAs are enriched in patient plasma and outperform conventional markers like CEA and CYFRA21-1 in early detection.

8.8 APPLICATIONS OF ECCDNA IN CANCER

8.8.1 Early Diagnosis

eccDNA is generated early during tumorigenesis. It reflects active gene expression and epigenetic changes. eccDNA works as a biomarker, released into blood and fluids, and serves as a non-invasive marker in liquid biopsies such as NGS (next-generation sequencing) [43]. It stimulates immune response, making eccDNA a common early indicator across cancer types. It improves detection, where other biomarkers fail or precancerous lesions are unidentifiable. It is highly specific and sensitive as compared to other biomarkers such as KRAS mutations.

8.8.2 Cancer Treatment

eccDNA amplifies drug-resistance genes (e.g., DHFR, RAB3B), helping tumors adapt [26]. If its production is blocked by inhibiting eccDNA production pathways like Chromothripsis and DNA repair mechanisms, it can reduce gene overexpression. Proteins encoded by eccDNA may serve as therapy targets. Tumor cells use eccDNA to respond quickly to environmental stress (e.g., methotrexate exposure).

8.8.3 Prognosis

Prognostic effects vary by eccDNA structure, frequency, and cancer type. The increased levels of eccDNA like MYC and EGFR may lead to poor survival in many cancers. For instance, high eccDNA levels of EGFR and MYC in pancreatic cancer result in shorter survival. eccDNA is prevalent in high-grade malignancies like glioblastoma, sarcoma, and esophageal cancer. Techniques like WGS, Circle-Seq, and AmpliconArchitect allow accurate detection and prognosis prediction [44].

8.9 CHALLENGES AND LIMITATIONS

Extrachromosomal DNA (ecDNA) plays a significant role in cancer development through oncogene amplification, tumor evolution, and treatment resistance. However, several barriers limit the study and clinical translation of ecDNA, falling under technical/methodological and standardization-related challenges.

8.9.1 Sequencing Limitations

Traditional short-read sequencing technologies, such as Illumina platforms, often fall short in detecting complex ecDNA structures, including rearranged breakpoints, tandem duplications, and circular topologies, due to their complex circular structures and repetitive sequences. These limitations hinder accurate reconstruction of ecDNA, essential for understanding its genomic architecture and its role in oncogenesis [45]. While long-read sequencing platforms, like Oxford Nanopore, have benefits in resolving complex genomic regions, they also have some demerits. They can suffer from higher error rates, which may influence the accurate detection of ecDNA.

8.9.2 Imaging and Detection Issues

Visualizing ecDNA within cells is difficult due to its small, irregular shape and dynamic nature. Standard microscopy lacks sufficient resolution to clearly distinguish ecDNA from chromosomal DNA. Advanced tools like FISH (fluorescence in situ hybridization) and DAPI staining provide some insights but are limited by their resolution and quantification capabilities. Preparing samples for high-resolution imaging techniques, such as scanning electron microscopy (SEM), requires meticulous protocols. Variations in sample preparation can introduce artifacts or damage delicate ecDNA structures, leading to misinterpretation of imaging data.

8.9.3 Sample Preparation Variability

The integrity of ecDNA can be affected by differences in sample extraction and preparation techniques. Fragmentation, contamination from environmental DNA or cross-sample interference, or low abundance of ecDNA in biofluids can compromise detection. Especially in liquid biopsies, inconsistent processing affects result accuracy and reproducibility [46, 47].

8.9.4 Quantitative Analysis Constraints

Reliable quantification of ecDNA copy number and abundance remains a challenge. Many current methods lack the sensitivity to detect low-frequency ecDNA molecules, particularly in early-stage cancers or minimal residual disease. This hampers efforts to use ecDNA as a consistent biomarker.

8.9.5 Inadequate Quality Controls

Few ecDNA studies incorporate rigorous controls or validation steps. Some studies include negative controls (cells not containing eccDNA) or positive controls. Artifacts from amplification bias or sequencing errors cause problems for real eccDNA detection. Techniques like electron microscopy or Sanger sequencing are rarely used. Without benchmarks or consensus standards, distinguishing true ecDNA signals from artifacts remains problematic.

8.10 FUTURE PROSPECTS

8.10.1 Advances in ecDNA Detection

eccDNA was first discovered in 1965 and later linked to gene amplification and drug resistance in cancer cells (e.g., DHFR in mice) [48, 49]. Whole-genome studies have confirmed eccDNA presence across multiple tumor types, often carrying oncogenes like MYC, EGFR, and MYCN [50]. Recent findings include microDNA and enhancer-only eccDNAs, suggesting roles in gene regulation and oncogene overexpression. Cell-free eccDNA in plasma shows promise as a cancer biomarker. Additionally, "kyklonic hypermutation" on eccDNA may drive tumor heterogeneity [51].

Emerging technologies are significantly enhancing the ability to detect and analyze extrachromosomal DNA (ecDNA), especially in cancer diagnostics. Liquid biopsy approaches enable the identification of circulating ecDNA in blood, offering a non-invasive method for monitoring tumor dynamics and treatment response. High-throughput sequencing techniques, such as CAPP-Seq and long-read sequencing, have improved the resolution of ecDNA structural mapping, helping to identify oncogene amplifications and complex rearrangements.

In imaging, artificial intelligence tools like ECdetect and ecSeg have shown promise for automating the detection of ecDNA in metaphase spreads. These tools apply adaptive learning (e.g., U-Net architectures) to distinguish ecDNA from chromosomal DNA, even in noisy cellular environments.

8.10.2 Applications in Personalized Medicine

ecDNA plays a critical role in tumor evolution by driving oncogene amplification, contributing to intratumoral heterogeneity, and promoting drug resistance [4, 52]. Because of these properties, ecDNA content and distribution are being explored as predictive biomarkers for treatment response and disease progression.

Studies suggest that ecDNA-specific profiles can guide personalized treatment strategies. For example, tumors with high ecDNA levels may respond differently to checkpoint inhibitors or targeted therapies. The dynamic nature of ecDNA also makes it a promising target for real-time disease monitoring through liquid biopsies, potentially enabling adaptive therapy approaches tailored to the patient's molecular profile.

8.11 CONCLUSION

The growing body of research on extrachromosomal circular DNA (ecDNA) underscores its pivotal role in offering new insights into tumor development, progression, diagnosis, and resistance mechanisms. Its unique characteristics, such as carrying amplified oncogenes, enhancing transcription through chromatin accessibility, and promoting genomic instability, make ecDNA a powerful driver of tumorigenesis. Its stability, abundance in tumor cells, and release into circulation allow for real-time cancer detection through liquid biopsy. Despite technical and methodological challenges, recent advances in sequencing technologies, imaging, and computational tools are rapidly improving our ability to detect and analyze ecDNA. As research progresses, ecDNA is expected to play a central role in personalized oncology, not only as a diagnostic biomarker but also as a therapeutic target. Its integration into clinical workflows could significantly improve early detection, treatment monitoring, and prognosis of various cancers.

REFERENCES

1. Hotta, Y. and A. Bassel, *Molecular size and circularity of DNA in cells of mammals and higher plants*. Proceedings of the National Academy of Sciences, 1965. **53**(2): p. 356–362.
2. Kunisada, T., et al., *Appearance of extrachromosomal circular DNAs during in vivo and in vitro ageing of mammalian cells*. Mechanisms of Ageing and Development, 1985. **29**(1): p. 89–99.
3. Møller, H.D., et al., *Circular DNA elements of chromosomal origin are common in healthy human somatic tissue*. Nature Communications, 2018. **9**(1): p. 1069.
4. Turner, K.M., et al., *Extrachromosomal oncogene amplification drives tumour evolution and genetic heterogeneity*. Nature, 2017. **543**(7643): p. 122–125.
5. Paulsen, T., et al., *Micro DNA levels are dependent on MMEJ, repressed by c-NHEJ pathway, and stimulated by DNA damage*. Nucleic Acids Research, 2021. **49**(20): p. 11787–11799.
6. Shoshani, O., et al., *Chromothripsis drives the evolution of gene amplification in cancer*. Nature, 2021. **591**(7848): p. 137–141.
7. Li, L.Y., X. Luo, and X. Wang, *Endonuclease G is an apoptotic DNase when released from mitochondria*. Nature, 2001. **412**(6842): p. 95–99.
8. Liao, Z., et al., *Classification of extrachromosomal circular DNA with a focus on the role of extrachromosomal DNA (ecDNA) in tumor heterogeneity and progression*. Biochimica et Biophysica Acta (BBA)-Reviews on Cancer, 2020. **1874**(1): p. 188392.
9. Griffith, J.D., et al., *Mammalian telomeres end in a large duplex loop*. Cell, 1999. **97**(4): p. 503–514.
10. Moqrich, A., *Peripheral pain-sensing neurons: From molecular diversity to functional specialization*. Cell Reports, 2014. **6**(2): p. 245–246.
11. Shimizu, N., *Gene amplification and the extrachromosomal circular DNA*. Genes, 2021. **12**(10): p. 1533.
12. Bergstrom, E.N., et al., *Mapping clustered mutations in cancer reveals APOBEC3 mutagenesis of ecDNA*. Nature, 2022. **602**(7897): p. 510–517.
13. Chen, W., et al., *Sequencing of methylase-accessible regions in integral circular extrachromosomal DNA reveals differences in chromatin structure*. Epigenetics & Chromatin, 2021. **14**: p. 1–11.
14. Chapman, O.S., et al., *The landscape of extrachromosomal circular DNA in medulloblastoma*. bioRxiv, 2021: p. 2021.10.18.464907.
15. Hung, K.L., P.S. Mischel, and H.Y. Chang, *Gene regulation on extrachromosomal DNA*. Nature Structural & Molecular Biology, 2022. **29**(8): p. 736–744.
16. Brennan, C.W., et al., *The somatic genomic landscape of glioblastoma*. Cell, 2014. **157**(3): p. 753.
17. Cho, W.-K., et al., *Mediator and RNA polymerase II clusters associate in transcription-dependent condensates*. Science, 2018. **361**(6400): p. 412–415.
18. Wu, S., et al., *Circular ecDNA promotes accessible chromatin and high oncogene expression*. Nature, 2019. **575**(7784): p. 699–703.
19. Koo, D.-H., et al., *Extrachromosomal circular DNA-based amplification and transmission of herbicide resistance in crop weed Amaranthus palmeri*. Proceedings of the National Academy of Sciences, 2018. **115**(13): p. 3332–3337.
20. Zheng, H. and W. Xie, *The role of 3D genome organization in development and cell differentiation*. Nature Reviews Molecular Cell Biology, 2019. **20**(9): p. 535–550.
21. Hanahan, D., *Hallmarks of cancer: New dimensions*. Cancer Discovery, 2022. **12**(1): p. 31–46.
22. Li, R., et al., *Extrachromosomal circular DNA (eccDNA): An emerging star in cancer*. Biomarker Research, 2022. **10**(1): p. 53.
23. Wu, S., et al., *Extrachromosomal DNA (ecDNA): Unveiling its role in cancer progression and implications for early detection*. Heliyon, 2023. **9**(11).
24. Wang, W., Z. Zheng, and J. Lei, *CTC, ctDNA, and exosome in thyroid cancers: A review*. International Journal of Molecular Sciences, 2023. **24**(18): p. 13767.
25. Tang, H., et al., *The new advance of exosome-based liquid biopsy for cancer diagnosis*. Journal of Nanobiotechnology, 2024. **22**(1): p. 610.
26. Verhaak, R.G., V. Bafna, and P.S. Mischel, *Extrachromosomal oncogene amplification in tumour pathogenesis and evolution*. Nature Reviews Cancer, 2019. **19**(5): p. 283–288.

27. Calcinotto, A., et al., *Cellular senescence: Aging, cancer, and injury*. Physiological Reviews, 2019. **99**(2): p. 1047–1078.
28. de Oliveira Mann, C.C. and P.J. Kranzusch, *cGAS conducts micronuclei DNA surveillance*. Trends in Cell Biology, 2017. **27**(10): p. 697–698.
29. Møller, H.D., et al., *CRISPR-C: Circularization of genes and chromosome by CRISPR in human cells*. Nucleic Acids Research, 2018. **46**(22): p. e131.
30. Helmsauer, K., et al., *Enhancer hijacking determines extrachromosomal circular MYCN amplicon architecture in neuroblastoma*. Nature Communications, 2020. **11**(1): p. 5823.
31. Morton, A.R., et al., *Functional enhancers shape extrachromosomal oncogene amplifications*. Cell, 2019. **179**(6): p. 1330–1341.e13.
32. Sumner, A.T., A.R. Ross, and E. Graham. *Preparation of chromosomes for scanning electron microscopy*. In Chromosome Analysis Protocols, 1994, Humana Press. p. 41–50.
33. Cox, D., C. Yuncken, and A. Spriggs, *Minute chromatin bodies in malignant tumours of childhood*. The Lancet, 1965. **286**(7402): p. 55–58.
34. Frey, M.D., *Low kv scanning electron microscopy*. In Scanning Microscopy for Nanotechnology: Techniques and Applications. 2007, Springer. p. 101–119.
35. Purshouse, K., S.M. Pollard, and W.A. Bickmore, *Imaging extrachromosomal DNA (ecDNA) in cancer*. Histochemistry and Cell Biology, 2024. **162**(1): p. 53–64.
36. Ronneberger, O., P. Fischer, and T. Brox. *U-net: Convolutional networks for biomedical image segmentation*. In Medical Image Computing and Computer-Assisted Intervention–MICCAI 2015: 18th International Conference, Munich, Germany, October 5–9, 2015, Proceedings, Part III 18. 2015. Springer.
37. Henssen, A., I. MacArthur, R. Koche, and H. Dorado García. *Purification and sequencing of large circular DNA from human cells*. Protocol Exchange, 2019. doi:10.1038/protex.2019.006
38. Luebeck, J., et al., *Amplicon Reconstructor integrates NGS and optical mapping to resolve the complex structures of focal amplifications*. Nature Communications, 2020. **11**(1): p. 4374.
39. Yi, E., et al., *Live-cell imaging shows uneven segregation of extrachromosomal DNA elements and transcriptionally active extrachromosomal DNA hubs in cancer*. Cancer Discovery, 2022. **12**(2): p. 468–483.
40. Hung, K.L., et al., *Targeted profiling of human extrachromosomal DNA by CRISPR-CATCH*. Nature Genetics, 2022. **54**(11): p. 1746–1754.
41. Becker, A., et al., *Extracellular vesicles in cancer: Cell-to-cell mediators of metastasis*. Cancer Cell, 2016. **30**(6): p. 836–848.
42. Irdianto, S.A., A. Dwiranti, and A. Bowolaksono, *Extrachromosomal circular DNA: A double-edged sword in cancer progression and age-related diseases*. Human Cell, 2025. **38**(2): p. 58.
43. Poulet, G., J. Massias, and V. Taly, *Liquid biopsy: General concepts*. Acta Cytologica, 2019. **63**(6): p. 449–455.
44. Zhao, X.-K., et al., *Focal amplifications are associated with chromothripsis events and diverse prognoses in gastric cardia adenocarcinoma*. Nature Communications, 2021. **12**(1): p. 6489.
45. Giurgiu, M., et al., *Decoil: Reconstructing extrachromosomal DNA structural heterogeneity from long-read sequencing data*. In International Conference on Research in Computational Molecular Biology. 2024. Springer.
46. Comaills, V. and M. Castellano-Pozo, *Chromosomal instability in genome evolution: From cancer to macroevolution*. Biology, 2023. **12**(5): p. 671.
47. Luebeck, J., et al., *Amplicon suite: An end-to-end workflow for analyzing focal amplifications in cancer genomes*. bioRxiv, 2024: p. 2024.05.06.592768.
48. Alt, F.W., et al., *Selective multiplication of dihydrofolate reductase genes in methotrexate-resistant variants of cultured murine cells*. Journal of Biological Chemistry, 1978. **253**(5): p. 1357–1370.
49. Koche, R.P., et al., *Extrachromosomal circular DNA drives oncogenic genome remodeling in neuroblastoma*. Nature Genetics, 2020. **52**(1): p. 29–34.
50. Kohl, N.E., et al., *Transposition and amplification of oncogene-related sequences in human neuroblastomas*. Cell, 1983. **35**(2): p. 359–367.
51. Li, Z., et al., *Pioneering insights of extrachromosomal DNA (ecDNA) generation, action and its implications for cancer therapy*. International Journal of Biological Sciences, 2022. **18**(10): p. 4006.
52. Kim, C., et al., *Chemoresistance evolution in triple-negative breast cancer delineated by single-cell sequencing*. Cell, 2018. **173**(4): p. 879–893.e13.

9 ecDNA Knowledge for Personalized Medicines

Navpreet Kaur,[1] Omji Porwal,[2] Narendra Kumar Pandey,[1] Shubham Kumar,[1] Sumel Ashique,[3] and Bimlesh Kumar[1]

1 School of Pharmaceutical Sciences
Lovely Professional University
Punjab, India
2 Department of Pharmacy
Qaiwan International University
Sulaymaniyah, Iraq
3 Department of Pharmaceutical Technology
Bharat Technology
Uluberia, Howrah, India

9.1 INTRODUCTION

Extrachromosomal DNA is defined as the DNA outside the chromosomes of the cell and found in tumor cells [1]. Extrachromosomal DNA (ecDNA) was initially found in wheat embryo and boat sperm, which is absent from the chromosomal genome. Extrachromosomal DNA (ecDNA) is a key factor for diagnosis, treatment, and discovery of a variety of cancer treatments. ecDNA is also involved in the formation and progression of cancer cells and is even used for rapid advances in detection techniques [2]. Oncogenes and the regulatory elements that go along with them are found in extrachromosomal DNAs (ecDNAs), a ubiquitous characteristic of cancer. Their distinct structural characteristics change chromatin accessibility and enable the quick amplification of oncogenes, which promotes tumorigenesis and the development of cancer. Uneven ecDNA segregation during cell division increases intercellular genetic heterogeneity, which aids in tumor evolution and may lead to chemotherapy tolerance and drug resistance. Furthermore, ecDNA can attach itself to or separate from chromosomal DNA; this development causes structural changes and genomic rearrangements in cancer cells [3].

Medical and pharmaceutical treatments that impact cancer cells at several locations, including nearby and far-off areas, are the foundation of the systemic anti-cancer strategy. Among the interventions are biological drugs that restore the function of damaged proteins in mutated tumors, hormonal interventions that change the tumor environment and impair its ability to survive and grow, cytotoxic chemotherapy agents that directly destroy proliferating cells, and immunological drugs that promote efficient immune recognition of tumor cells and the corresponding immune response [4, 5]. The goal of "personalized medicine in oncology" is to reduce side effects and increase the effectiveness of anti-cancer treatments. Potential candidates are determined by pathological-molecular evaluation of distinct biological alterations in the tumor cells and/or their immediate surroundings, as well as clinically based on indication for therapy and tolerance to it. Treatment that targets the predominant biological changes in a safe and efficient manner [6]. By elaboration on the molecular alterations linked to ecDNA, treatments can be developed to target these changes specifically, enhancing treatment results and reducing adverse effects [7].

A key feature of these cancer-associated ecDNAs is their ability to carry and amplify significant portions of the genome, including entire genes and their regulatory elements. The exact

DOI: 10.1201/9781003655299-9

mechanisms for ecDNA formation are under active investigation, but several pathways have been identified that often involve genomic instability and DNA repair processes. Double-strand DNA breaks (DSBs) are a critical factor in ecDNA formation. When DSBs occur, particularly in fragile or amplification-prone areas (like common fragile sites or oncogene regions), the cell's repair processes may incorrectly join the broken ends in a circular fashion, detaching a piece of DNA from the chromosome [8, 9]. During cell division, asymmetric chromosome segregation can lead to an uneven distribution of genetic material. Mis-segregation can cause chromosome fragments or entire chromosomes to enter micronuclei, tiny nuclear bodies separate from the main nucleus. These micronuclei are highly unstable and likely to break apart and circularize, leading to the formation of ecDNAs [10–12].

Extrachromosomal DNAs (ecDNAs), a type of circular DNA element that was initially discovered in tumor cells and termed "minute chromatin bodies." Oncogene-enriched ecDNA or transcriptional enhancers-carrying ecDNA can drive gene co-amplification, which causes the formation of specialized nuclear cluster particles that leads to intra-tumoral genetic heterogeneity. ecDNA can be reintegrated into the genome, potentially disrupting gene regulation network in tumor cells [13, 14]. Researchers found ecDNA in numerous cancer cells and not in healthy cells. However, recent studies reported its presence in precancerous cells and may initiate cellular carcinogenesis. Additionally, ecDNA has emerged as a focus in cancer research, following the debut of chemotherapy, targeted therapy, and immunotherapy, with the potential to trigger the development in the field of cancer therapy [3, 15].

9.2 THE BIOLOGY OF ECDNA: STRUCTURE, FORMATION, AND DYNAMICS

Extrachromosomal DNA (ecDNA) has become an important factor in cancer growth, evolution, and drug resistance. Understanding how it works, including its structure, formation, and behavior, is essential for creating personalized cancer treatments [16]. The structure of ecDNA is mainly circular, which is important for amplifying and regulating oncogenes in cancer [15]. Recent studies show that ecDNA has unique structural features that improve chromatin accessibility and oncogene expression, helping tumors develop. The following sections outline the key aspects of ecDNA structure. ecDNA is mostly circular double-stranded DNA, with sizes ranging from hundreds of base pairs to more than 150 kilobases [9, 15]. It can exist in multiple copies within a single cell, with estimates suggesting several hundred to a thousand molecules per cell. ecDNA can contain rearranged sequences from different chromosomal locations, resulting in structural variation. Researchers have created advanced computational tools, like Decoil, to reconstruct complex ecDNA structures from long-read sequencing data, uncovering diverse shapes in tumors. Studies indicate that ecDNA may exist in high copy numbers, which can lead to significant overexpression of genes, including oncogenes. Other studies have found that amplifying ecDNA molecules results in higher oncogene expression compared to linear DNA. Therefore, ecDNA is defined as circular molecules with relatively loose chromatin, resulting in increased gene expression levels [17].

9.2.1 Formation of ecDNA

The exact ways ecDNA forms are still under active investigation, but researchers have proposed several pathways. Excisional ecDNA formation is a common process. It involves cutting segments of chromosomal DNA from the chromosome, usually due to double-strand breaks (DSBs), which are then turned into circles. Recent studies have shown that different molecular events related to chromosomal damage can result in the formation of ecDNA. These events include chromothripsis, religation of DNA, breakage–fusion–bridge (BFB) cycles, fork stalling, and template switching [15].

The term "chromothripsis" refers to chromosome-shattering events. In these events, one or a few chromosomes undergo tens to hundreds of closely grouped DNA breaks at the same time. This leads to DNA fragment rearrangements and a significant loss of DNA sequences. During this process, chromosome rearrangement can result in tiny circular DNA fragments, like ecDNA or "double minutes" [18]. Unlike multiple consecutive DNA rearrangements seen in punctuated equilibrium events, such as the breakage–fusion–bridge cycle, chromothripsis occurs as a single event. Researchers investigated the religation of DNA, which is part of ecDNA formation. They used the CRISPR-Cas9 system to create ecDNA from DNA fragments taken from chromosomes in human cell lines [19]. Additionally, a case study showed that the amplification of MYC on ecDNA was linked to the deletion of MYC in leukemia patients' chromosomes [20]. This suggests that ecDNA can originate from chromosomal DNA deletions. Sequence analysis in glioma cells revealed that ecDNA may form through a process similar to V(D)J recombination, involving DNA fragments from chromosomal deletions or rearrangements [15]. To create ecDNA without chromothripsis, deleted chromosomal DNA may ligate according to one model for another way ecDNA is produced (Figure 9.1) [9, 15]. However, further research is needed to uncover the molecular pathways that may play a role in this. DNA breakage starts the BFB cycle.

This is followed by telomere shortening and chromosomal end-to-end joining, which creates an anaphase bridge with two centromeres. Structural changes in chromosomes arise when the bridge breaks down and recombines [9, 15]. This cycle continues in daughter cells because they inherit chromosomes with an unusually high number of centromeres. ecDNA may form from the rearrangement of end products from multiple BFB cycle rounds. The BFB cycle mechanism leads to chromosomal organization and gene amplification after several cell cycles, contributing to cancer development [21]. Research shows that ecDNA can also come from replication fork stalling and template swapping. When a DNA lesion appears on the template strand, the DNA replication

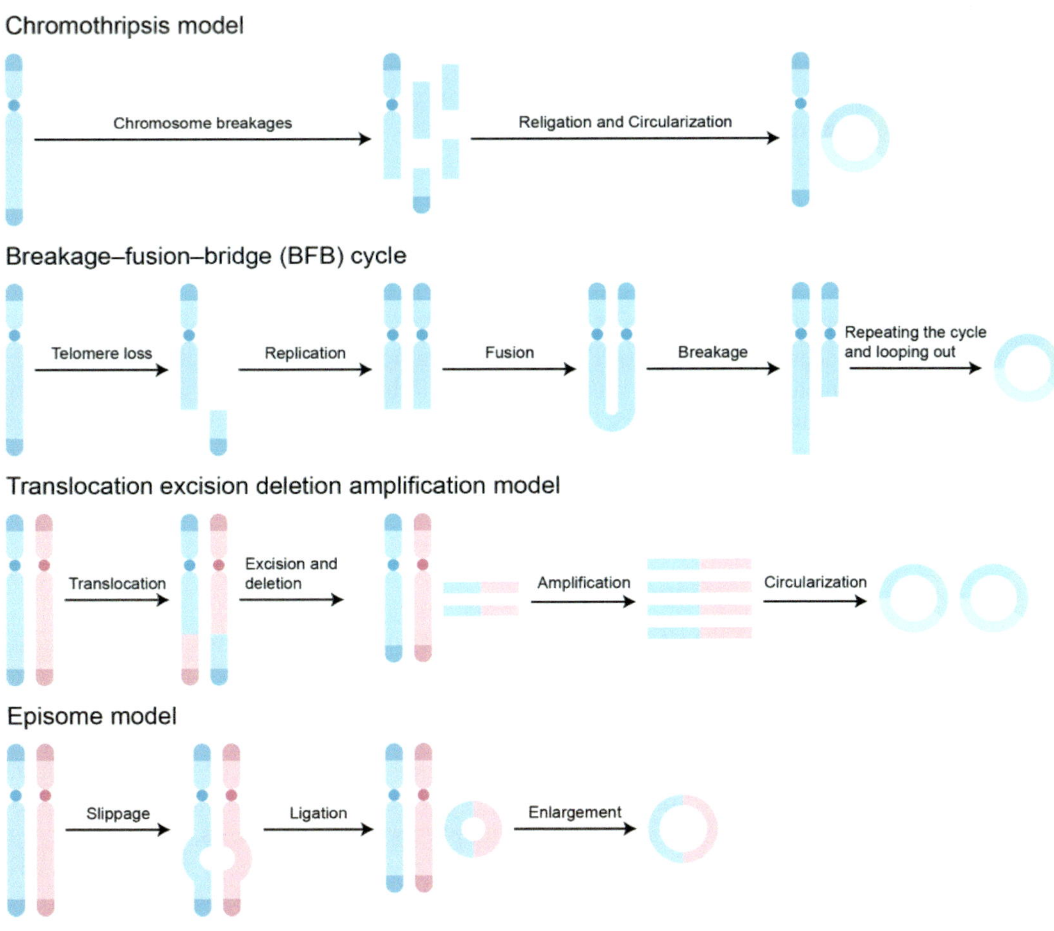

FIGURE 9.1 ecDNA formation.

fork stops. The lagging strand then separates and joins the DNA synthesis of a nearby replication fork. After several rounds of this separation and incorporation with the neighboring fork, the lagging strand returns to link with the original template strand. This results in a non-complementary sequence on the returning lagging strand, which appears as single-stranded DNA. Later, this single-stranded DNA can replicate and form ecDNA [15].

9.2.2 Dynamics of ecDNA

During cell division, ecDNA is dispersed randomly to daughter cells because they lack centromeres. Significant intratumoral heterogeneity results from this non-Mendelian inheritance, as distinct cancer cells within a single tumor may have different ecDNA copy numbers and, consequently, different levels of oncogene expression. Rapid variations in gene dosage are made possible by the high copy numbers and random segregation. Certain pressures, like medication treatment, can give cells with more copies of oncogene-bearing ecDNA a selective edge. Tumor evolution is accelerated by this quick adaptability [22].

One of the main methods by which cancer cells become resistant to different treatments, such as chemotherapy and targeted therapies, is through ecDNA. Increased expression of drug targets due to high copy numbers of oncogenes on ecDNA overwhelms the therapeutic agent. Oncogene amplification can be further stabilized by homogeneously staining regions (HSRs), which are formed when ecDNA dynamically reintegrates into chromosomes [23]. Crucially, ecDNA may resurface after treatment pressure is removed, which could result in relapse. Drug accessibility or the reaction to agents that damage DNA may also be impacted by the open chromatin structure of ecDNA. Because of its circular shape and high accessibility, ecDNA can change the transcriptional landscape by forming distinct cis-enhancer-promoter interactions and trans ecDNA particle cooperative regulatory interactions. Additionally, they can group in particular "hubs" inside the nucleus, which could help with gene expression [24, 25].

9.3 ECDNA IN CANCER: A MAJOR DRIVER OF DISEASE PROGRESSION AND THERAPEUTIC RESISTANCE

Extrachromosomal DNA (ecDNA) has become a key player in cancer, playing a major role in both treatment resistance and the advancement of the disease. ecDNA exists as tiny, circular DNA fragments that live independently within the cell's nucleus, in contrast to chromosomal DNA, which is neatly arranged into linear structures. It was once thought to be uncommon, but it is now known to be common in many aggressive cancers. Through a variety of mechanisms, including boosting oncogene expression by increasing copies or promoting chromosomal interactions, and serving as a source for DNA recombination by integrating into and being removed from chromosomes, ecDNA contributes to the development and evolution of cancer [15].

9.3.1 Driving Disease Progression

In eukaryotes, extrachromosomal DNA (ecDNA) is a special type of circular DNA that exists separately from regular chromosomes. In biology, these circular molecules are important, especially in cancer research [9]. The mystery of ecDNA is slowly being unraveled due to advances in sequencing technologies like CCDA-seq and AmpliconArchitect [26]. As a result, we are beginning to understand its purpose and how it is used. The most important studies focus on ecDNA's role in cancer. Numerous studies have shown how important ecDNA is in cancer development and how it speeds up cancer progression [27]. ecDNA often contains multiple copies of oncogenes, which are genes that promote cell growth and division. Since ecDNA can be present in high copy numbers and has an open chromatin structure, these oncogenes are usually expressed at high levels, leading to fast tumor growth and spread. ecDNA does not follow the typical rules of Mendelian inheritance. During cell division, it is randomly distributed to daughter cells [9]. This uneven distribution means some daughter cells can end up with many more copies of ecDNA, especially those with oncogenes. This "jackpot effect" speeds up tumor evolution by creating a diverse population of cancer cells. Some of these cells are better able to handle changes in their environment and respond to different pressures. Overall genomic instability in cancer cells is influenced by the dynamics and presence of ecDNA. They may develop as a result of chromothripsis (the shattering and rearrangement of chromosomes) or even reintegration into chromosomes, which can result in additional genomic changes and the development of novel gene regulatory interactions [28]. Tumors can quickly alter their genetic composition thanks to this genomic plasticity, which is essential for their progression. According to studies, ecDNA is more common in metastatic and treated tumors, which may indicate that it helps cancer cells spread to other parts of the body [29].

9.3.2 Contributing to Therapeutic Resistance

For cancer patients, extrachromosomal DNA (ecDNA) poses a significant challenge. Because ecDNA promotes massive oncogene transcription and rapid genome evolution, it makes tumors resistant to treatment, which lowers patient survival. ecDNAs can emerge during the initiation, growth, and spread of cancers. They have special biological characteristics that give malignant cells an advantage [30].

ecDNA can attach to or detach from chromosomal DNA. This leads to structural changes and genomic rearrangements in cancer cells. Recent progress in multi-omics analysis, which reveals the genomic and epigenetic traits of ecDNA, is expected to significantly enhance precision cancer treatment [15]. The non-Mendelian inheritance and high copy number of ecDNA allow cancer cells to quickly increase or decrease levels of specific oncogenes or drug-resistance genes in response to treatment pressure. If a drug targets a gene on a chromosome, cells may struggle to overcome that pressure [9]. However, with ecDNA, cells can swiftly adjust the copy number of the relevant ecDNA, thus changing the expression of resistance-related genes. The random segregation of ecDNA during cell division creates significant genetic diversity within the tumor. This means that even if a drug kills most cancer cells, a subgroup of cells with a favorable ecDNA profile, such as those with high numbers of resistance genes, can survive and multiply, leading to treatment failure and relapse. Cancer cells with ecDNA can also show a remarkable ability to switch their dependency from one oncogene to another under treatment pressure [9, 15]. For instance, if a targeted therapy blocks a specific oncogene on ecDNA, the tumor might respond by amplifying a different oncogene on another ecDNA, allowing it to keep growing and resist the initial drug. ecDNA can carry and amplify genes linked to drug resistance, including those for efflux pumps, such as ABC transporters like ABCB1/P-glycoprotein, which actively remove chemotherapy drugs from the cell and render the treatment ineffective [3].

9.4 LEVERAGING ECDNA KNOWLEDGE FOR PERSONALIZED MEDICINE

Recent advancements in cancer research have highlighted the significant role of extrachromosomal DNA (ecDNA) in tumor evolution, drug resistance, and patient outcomes [31]. Unlike linear chromosomal DNA, ecDNA exists as circular, autonomously replicating DNA molecules within the nucleus. Its unique characteristics make it a promising target for personalized medicine approaches. As adaptation experts, cancer cells frequently use permanent genetic alterations to withstand therapy [26]. Traditional drug discovery ignores the fast-acting genomic responses on extrachromosomal DNA (ecDNA), which are more difficult to detect and even more difficult to target, in favor of blocking transient, protein-level responses. ecDNA is a dynamic structure that can replicate quickly and segregate unevenly during cell division, in contrast to chromosomal mutations, deletions, or stable amplifications integrated within homogeneous staining regions (HSRs). It is thought that ecDNA contributes to tumor heterogeneity by generating genetic and molecular variations between cells that affect how well drugs are effective for treatment [9, 15].

9.4.1 Diagnostic and Prognostic Biomarkers

ecDNA characterization methods are a crucial research area on ecDNA, and cytogenetic techniques, such as 4′,6-diamidino-2-phenylindole (DAPI) staining and fluorescence in situ hybridization (FISH), have been employed to detect and quantify ecDNA elements6. Sequencing-based methods, including AmpliconArchitect, AmpliconReconstructor, Circle_Finder, and Circle-Map, deduce the ecDNA structures from whole-genome sequencing (WGS) data. Methods like Circle-Seq and CRISPR-CATCH also analyze the targeted ecDNA, and ecTag method can also be employed, which helps in the visualization of ecDNA [9, 15, 32].

A method called AmpliconArchitect helps researchers to recognize and study the complex structure of amplicons. This method is capable of identifying numerous human-viral heterozygous extra chromatin amplicon structures from the different tumor types. PacBio sequencing and metaphase fluorescence in situ hybridization (FISH) techniques have been combined for a detailed examination of these structures. UPCI: SCC090 cell line is used by the researchers for mapping the fine structure of an extrachromosomal heterozygous amplicon. Two main features of ecDNA are that the number of copies is significantly increased, as well as structural variation (SV) of a particular ring [26].

ecDNA is an important tool for tracking tumor growth and predicting outcomes. It plays a key role in changes to oncogene copy numbers and is associated with drug resistance. For example, researchers suggest that longer ecDNAs are found more often in tumor samples from lung cancer patients than in their matched controls. Recently, a study using next-generation sequencing (NGS) examined extrachromosomal DNA from the plasma of 6 lung adenocarcinomas (LADs) and 10 healthy people [33]. This analysis showed that nine ecDNAs were more common in cancer samples. Additionally, four ecDNAs were exclusively found in LADs and showed promise as biomarkers for early detection. Similarly, another study on LAD revealed that the expression patterns of several ecDNAs differed significantly between LAD and normal tissues. When distinguishing patients with LAD from healthy individuals, the serum levels of two ecDNAs displayed higher areas under the curve (AUCs) compared to serum carcinoembryonic antigen (CEA) and CYFRA21-1 (cytokeratin 19 fragments), which are regarded as some of the most reliable biomarkers for LAD [15].

9.4.2 Guiding Treatment Decisions (Therapeutic Targets)

Metaphase DNA fluorescence in situ hybridization (FISH) is still the best method for detecting ecDNA. Metaphase FISH offers single-molecule resolution and can be examined using computational tools like EcSeg [34]. The advent of sequencing provides an advantage for understanding the genomic rearrangements. Two newly developed methods for analyzing are Complete Reconstruction of Amplifications with Long reads (CoRAL) and Deconvolve Extrachromosomal Circular DNA Isoforms from Long-read data. Even the machine-learning techniques have been used to generate tools for ecDNA detection from whole-exome sequencing [35].

A previous study investigated that ecDNAs were prevalent in at least 17 types of cancer and were present at higher levels in WGS samples from cancer tissues and cell lines derived from patients with cancer compared with healthy controls [15]. ecDNA profiles also help to correlate with different types of cancer, including glioblastoma, neuroblastoma, ovarian cancer, leukemia, esophageal cancer, gastric cardia cancer, colorectal cancer, and lung cancer [36]. Tissue biopsy is a method considered to be accurate, but it is difficult to replicate, along with the disadvantage of invasive techniques [37]. In order to overcome this issue, liquid biopsy was introduced. Biomarkers used for liquid biopsy include circulating tumor cells (CTCs), circulating tumor DNA (ctDNA), and extracellular vehicles (EVs) [38]. Also, previous studies have shown that circular DNA can gain access through cell membranes much more easily than linear DNA, and studies have also shown that extrachromosomal circular DNA can be released into circulation from *in situ* tumors. ecDNa is considered a great technique to identify the tumor progression. Recently, next-generation sequencing (NGS) of extrachromosomal DNA from plasma samples of 6 lung adenocarcinomas (LADs) and 10 healthy controls suggested that the frequencies of nine ecDNAs were higher in cancer samples [15].

Targeting ecDNA is turning into a promising cancer treatment approach because it plays a critical role in the development of drug resistance and the progression of cancer [39]. By reducing the expression of drug resistance genes on ecDNA in human and hamster cell lines, hydroxyurea has shown promise in restoring sensitivity to vinblastine in cancer cells [40]. According to another study, hydroxyurea at low doses may reduce ecDNAs in some cancer patients, resulting in a longer period of stable disease. Despite the fact that hydroxyurea is not yet a clinical anti-cancer treatment, new drugs could be developed using this mechanism to reduce extrachromosomal amplicons that increase the sensitivity of cancer cells [15].

9.4.3 Monitoring Treatment Response and Relapse

Tracking changes in ecDNA profiles in ctDNA during and after treatment can provide real-time insights into tumor response. A decrease in oncogene-carrying ecDNA might indicate treatment effectiveness, while an increase could signal resistance or an upcoming relapse [41]. The return of specific ecDNA amplicons in ctDNA after successful treatment could act as an early sign of minimal residual disease or relapse, leading to earlier intervention [42]. Circulating tumor DNA (ctDNA) has become a valuable biomarker for monitoring treatment responses and predicting relapse across various cancers. Its ability to offer real-time insights into tumor behaviors allows for timely changes in treatment plans. ecDNA can increase oncogene levels, leading to higher expression and driving tumor growth and progression. The amplification of drug resistance genes on ecDNA can cause cancer cells to become resistant to chemotherapy and other treatments. ecDNA also adds genetic diversity within the cancer cell population, making it hard to target all cells with a single therapy [43].

9.4.4 Personalized Drug Development

Using patient-derived organoids or xenografts with known ecDNA profiles can help test new therapies pre-clinically and find personalized treatment options [44]. The unique circular shape and replication methods of ecDNA offer chances to develop specific inhibitors targeting ecDNA maintenance or its related proteins [15]. This approach could reduce unwanted effects on normal cells. As research in life sciences advances, extrachromosomal DNA (ecDNA) has emerged as an important biological phenomenon attracting more attention [45]. ecDNA plays a key role in managing gene expression, maintaining genetic stability in cells, and influencing disease development, including both normal physiological functions and pathological conditions. Recent studies into the role of ecDNA in aging and neurodegenerative diseases provide new insights into these complex illnesses [46].

The recent introduction of whole genome sequencing (WGS) into clinical practice has become a game-changer. A new concept for using WGS is newborn screening, which can become an essential tool for disease prevention and treatment optimization at an early age [47]. According to many, pharmacogenomics (PGx) is a key component of the personalized medicine concept of single-nucleotide polymorphisms (SNPs). SNPs make the difference in observed drug effects by modulating the activity of their protein product [48]. GenomAD, an online resource of genome sequencing data, contains allelic and genotype frequencies of drug-metabolizing enzymes for specific populations, helping the scientific community and physicians optimize the treatment [49]. Gene editing technologies, including clustered regularly interspaced short palindromic repeats and CRISPR-associated protein 9 (CRISPR-Cas9), will likely redefine the margins of the original personalized medicine concept. Cell therapy is already taking a significant place in the personalized medicine concept, particularly in patients with cancer [50].

There is tremendous potential in the new concept of "personalized regenerative medicine" focused on mesenchymal stem cells, known for their ability to treat tissue defects and regulate immune responses. There are important ethical issues related to personalized medicine. One is pricing and accessibility. It is expected that each "disease" will be sliced and stratified into several driving mechanisms, and each will require the development of its own treatment [51, 52].

9.5 CHALLENGES AND FUTURE DIRECTIONS

The field of ecDNA has made great progress over the past decade. Researchers have discovered the prevalence and significance of ecDNA amplifications and have identified a promising therapeutic target for cancers with ecDNA. Despite these advancements, several challenges remain to fully tap into this field's potential and improve outcomes for patients with ecDNA-positive cancers [53].

These challenges include the following:

(i) Developing new diagnostic tests for ecDNA in patient tumors that need fewer resources compared to whole genome sequencing and are more suitable for clinical use [54].
(ii) Conducting long-term studies to understand the role of ecDNA during all stages of tumor initiation and progression [15].
(iii) Expanding ecDNA research to cover more tumor types and experimental models. A noteworthy area of study is the impact of ecDNA on the tumor microenvironment.

The link between ecDNAs and relatively immune-cold tumors, along with their ability to encode immunomodulatory genes, suggests that ecDNAs might have an immunosuppressive role in cancer. Understanding how ecDNA may influence the innate and adaptive immune systems could lead to new treatments for ecDNA-driven malignancies. A valuable tool for these investigations will be the newly developed technique for producing ecDNAs in immunocompetent models [15, 35].

There are several important questions that need to be answered before ecDNA can be used in clinical settings. To create effective therapies targeting ecDNA, we must better understand the molecular mechanisms that govern ecDNA dynamics and how it affects the interactions between cancer cells and the immune system [15]. Additionally, researchers have suggested that the shapes of ecDNA influence its functions, which means more research in this area is needed. More studies should also be done to differentiate between ecDNA found in normal tissues and that in tumors, when possible [9]. We need to investigate the lifespan and amount of ecDNA to see if it can serve as a biomarker detectable in liquid biopsies. Bioinformatic analyses could help identify harmful characteristics of ecDNA and clarify the connections between its molecular properties and cancer [31]. In summary, our understanding of ecDNA and its role in tumor development and evolution has improved significantly in recent years. However, many questions remain about the functions and clinical significance of ecDNA. As we learn more about its role in cancer development, its characteristics may be used for diagnosing, predicting outcomes, and treating cancer patients.

REFERENCES

1. Zhu, X., et al., *Cancer evolution: A means by which tumors evade treatment*. Biomedicine & Pharmacotherapy, 2021. **133**: p. 111016.
2. Zhao, L., et al., *Amazing roles of extrachromosomal DNA in cancer progression*. Biochimica et Biophysica Acta (BBA)-Reviews on Cancer, 2023. **1878**(1): p. 188843.
3. Huang, Q., et al., *Insight on ecDNA-mediated tumorigenesis and drug resistance*. Heliyon, 2024. **10**(6): p. e27733.
4. Li, H., et al., *The application of nanotechnology in immunotherapy based combinations for cancer treatment*. Recent Patents on Anti-Cancer Drug Discovery, 2023. **18**(1): p. 53–65.
5. Li, J.-Y., et al., *Chemotherapeutic and targeted agents can modulate the tumor microenvironment and increase the efficacy of immune checkpoint blockades*. Molecular Cancer, 2021. **20**: p. 1–21.

6. Siegelmann-Danieli, N., et al., *Personalized medicine in oncology requires teamwork between medical oncologists and molecular pathologists*. Harefuah, 2022. **161**(4): p. 233–238.
7. Li, Q., et al., *Future challenges and prospects for personalized epigenetics*. In *Personalized Epigenetics*. 2024, Elsevier. p. 721–744.
8. Nguyen, D.D., et al., *The interplay of mutagenesis and ecDNA shapes urothelial cancer evolution*. Nature, 2024. **635**(8037): p. 219–228.
9. Liao, Z., et al., *Classification of extrachromosomal circular DNA with a focus on the role of extrachromosomal DNA (ecDNA) in tumor heterogeneity and progression*. Biochimica et Biophysica Acta (BBA)-Reviews on Cancer, 2020. **1874**(1): p. 188392.
10. Fox, B.I. and B.G. Felkey, *Precision medicine: Considerable potential combined with important challenges*. Hospital Pharmacy, 2016. **51**(11): p. 950–951.
11. Popova, L. and V.J. Carabetta, *The use of next-generation sequencing in personalized medicine*. In *High Throughput Gene Screening: Methods and Protocols*. 2024, Springer. p. 287–315.
12. Stefanicka-Wojtas, D. and D. Kurpas, *Personalised medicine—implementation to the healthcare system in Europe (focus group discussions)*. Journal of Personalized Medicine, 2023. **13**(3): p. 380.
13. Tsiakanikas, P., et al., *Beyond the chromosome: Recent developments in decoding the significance of extrachromosomal circular DNA (eccDNA) in human malignancies*. Life, 2024. **14**(8): p. 922.
14. Hung, K.L., et al., *ecDNA hubs drive cooperative intermolecular oncogene expression*. Nature, 2021. **600**(7890): p. 731–736.
15. Dong, Y., et al., *Extrachromosomal DNA (ecDNA) in cancer: Mechanisms, functions, and clinical implications*. Frontiers in Oncology, 2023. **13**: p. 1194405.
16. Xing, J., et al., *Progress on the role of extrachromosomal DNA in tumor pathogenesis and evolution*. Clinical Genetics, 2021. **99**(4): p. 503–512.
17. Dennin, R.H., *Overlooked: Extrachromosomal DNA and their possible impact on whole genome sequencing*. The Malaysian Journal of Medical Sciences: MJMS, 2018. **25**(2): p. 20.
18. Krupina, K., A. Goginashvili, and D.W. Cleveland, *Scrambling the genome in cancer: Causes and consequences of complex chromosome rearrangements*. Nature Reviews Genetics, 2024. **25**(3): p. 196–210.
19. Seeger, C., *A CRISPR-based system to investigate HBV cccDNA biology*. Journal of Virology, 2023. **97**(10): p. e0118523.
20. Zhang, H., et al., *Characterization and integrated analysis of extrachromosomal DNA amplification in hematological malignancies*. Neoplasia, 2024. **56**: p. 101025.
21. Selvarajah, S., et al., *The breakage–fusion–bridge (BFB) cycle as a mechanism for generating genetic heterogeneity in osteosarcoma*. Chromosoma, 2006. **115**: p. 459–467.
22. Mischel, P.S., *Abstract SY03-02: Extrachromosomal DNA (ecDNA): Cancer's dynamic circular genome*. Cancer Research, 2022. **82**(12_Supplement): p. SY03-02.
23. Ashique, S., et al., *Impact of ecDNA: A mechanism that directs tumorigenesis in cancer drug resistance-a review*. Chemico-Biological Interactions, 2022. **363**: p. 110000.
24. Lange, J.T., et al., *The evolutionary dynamics of extrachromosomal DNA in human cancers*. Nature Genetics, 2022. **54**(10): p. 1527–1533.
25. Scanu, E., B. Werner, and W. Huang, *Population dynamics of multiple ecDNA types*. arXiv preprint arXiv:2411.14588, 2024.
26. Wu, S., et al., *Extrachromosomal DNA (ecDNA): Unveiling its role in cancer progression and implications for early detection*. Heliyon, 2023. **9**(11).
27. Arshadi, A., *Amplificazione genomica nel cancro: Circolarizzazione degli ampliconi tramite CRISPR-C, conseguenze trascrizionali e implicazioni terapeutiche,* 2024.
28. Verhaak, R.G., V. Bafna, and P.S. Mischel, *Extrachromosomal oncogene amplification in tumour pathogenesis and evolution*. Nature Reviews Cancer, 2019. **19**(5): p. 283–288.
29. Turner, K.M., et al., *Extrachromosomal oncogene amplification drives tumour evolution and genetic heterogeneity*. Nature, 2017. **543**(7643): p. 122–125.
30. Tang, J., et al., *Enhancing transcription–replication conflict targets ecDNA-positive cancers*. Nature, 2024. **635**(8037): p. 210–218.
31. Li, Z., et al., *Pioneering insights of extrachromosomal DNA (ecDNA) generation, action and its implications for cancer therapy*. International Journal of Biological Sciences, 2022. **18**(10): p. 4006.
32. Wang, S., et al., *Machine learning-based extrachromosomal DNA identification in large-scale cohorts reveals its clinical implications in cancer*. Nature Communications, 2024. **15**(1): p. 1515.
33. Wu, X., et al., *Identification and characterization of extrachromosomal circular DNA in plasma of lung adenocarcinoma patients*. International Journal of General Medicine, 2022: p. 4781–4791.

34. Arshadi, A., et al., *Advancements in focal amplification detection in tumor/liquid biopsies and emerging clinical applications*. Genes, 2023. **14**(6): p. 1304.
35. Weiser, N.E., et al., *A guide to extrachromosomal DNA: Cancer's dynamic circular genome*. Cancer Discovery, 2025. **15**(6): p. 1105–1114.
36. Mao, X., et al., *Insights into extrachromosomal DNA in cancer: Biogenesis, methodologies, functions, and therapeutic potential*. Advanced Biology, 2025. **9**(3): p. 2400433.
37. Hashimoto, K., et al., *Limitations and usefulness of biopsy techniques for the diagnosis of metastatic bone and soft tissue tumors*. Annals of Medicine and Surgery, 2021. **68**: p. 102581.
38. Wang, X., et al., *Research progress of CTC, ctDNA, and EVs in cancer liquid biopsy*. Frontiers in Oncology, 2024. **14**: p. 1303335.
39. Luo, J., et al., *Extrachromosomal circular DNA in cancer drug resistance and its potential clinical implications*. Frontiers in Oncology, 2023. **12**: p. 1092705.
40. Komal, L. Singh, and S.S. Ganti, Targeting Extrachromosomal DNA (ecDNA) in Cancer: A New Era of CHK1 Inhibition and Personalized Treatments. 2025, Bentham Science Publishers.
41. Chi, Y., et al., *Dynamic analysis of circulating tumor DNA to predict the prognosis and monitor the treatment response of patients with metastatic triple-negative breast cancer: A prospective study*. Elife, 2023. **12**: p. e90198.
42. Gale, D., et al., *Residual ctDNA after treatment predicts early relapse in patients with early-stage non-small cell lung cancer*. Annals of Oncology, 2022. **33**(5): p. 500–510.
43. Kim, H., et al., *Mapping extrachromosomal DNA amplifications during cancer progression*. Nature Genetics, 2024. **56**(11): p. 2447–2454.
44. Berezovsky, A., Platelet Derived Growth Factor Receptor Alpha (PDGFRA) Oncogene Dependency in Glioblastoma. 2023, Wayne State University.
45. Yang, Q.-L., et al., *Modern biology of extrachromosomal DNA: A decade-long voyage of discovery*. Cell Research, 2025: p. 1–12.
46. Ma, Y.-N., et al., *Extrachromosomal DNA: Molecular perspectives in aging and neurodegenerative diseases*. Intractable & Rare Diseases Research, 2024. **13**(4): p. 251–254.
47. Kingsmore, S.F., et al., *A genome sequencing system for universal newborn screening, diagnosis, and precision medicine for severe genetic diseases*. The American Journal of Human Genetics, 2022. **109**(9): p. 1605–1619.
48. Yanase, K., et al., *Functional SNPs of the breast cancer resistance protein-therapeutic effects and inhibitor development*. Cancer Letters, 2006. **234**(1): p. 73–80.
49. Khan, A., et al., *Genomic medicine and personalized treatment: A narrative review*. Annals of Medicine and Surgery, 2025: p. 10.1097.
50. Primorac, D. and A. Ciechanover, *Personalized medicine: The future is here*. Croatian Medical Journal, 2024. **65**(3): p. 169.
51. Primorac, D., et al., *Compassionate mesenchymal stem cell treatment in a severe COVID-19 patient: A case report*. Croatian Medical Journal, 2021. **62**(3): p. 288–296.
52. Dimova, A., et al., *Novel approach in rectovaginal fistula treatment: Combination of modified martius flap and autologous micro-fragmented adipose tissue*. Biomedicines, 2023. **11**(9): p. 2509.
53. Rehman, S.U., *The multifaceted role of extrachromosomal DNA in cancer: Functions and clinical implications*. Bangladesh Journal of Medicine, 2025. **36**(2): p. 82–91.
54. Noorani, I., P.S. Mischel, and C. Swanton, *Leveraging extrachromosomal DNA to fine-tune trials of targeted therapy for glioblastoma: Opportunities and challenges*. Nature Reviews Clinical Oncology, 2022. **19**(11): p. 733–743.

10 ecDNA Formation
Tumor Immune Escape

Mohammad Habeeb,[1] Hariharan Thirumalai Vengateswaran,[1] Darshan S.,[2] and Ciniraj Raveendran[3]

1 Department of Pharmaceutics, Crescent School of Pharmacy
B.S. Abdur Rahman Crescent Institute of Science and Technology
Chennai, India
2 Rajiv Gandhi University of Health Sciences
Bangalore, Karnataka
3 Department of Radiation Oncology
Government Medical College Thiruvananthapuram
India

10.1 INTRODUCTION

10.1.1 Definition of Extrachromosomal DNA (ecDNA)

Extrachromosomal DNA (ecDNA) is defined as DNA outside a cell's chromosomes. A 1965 microscopic analysis confirmed that cancer cells contained small DNA fragments outside of the chromosomes. Because these fragments frequently occur in pairs, they were given the term "Double Minutes (DMs)." Further study has made it possible to classify extrachromosomal circular DNA with greater accuracy. Since ecDNA is more visible than chromosomal DNA and has been found to be a significant carrier of amplified oncogenes, it causes an increase in transcriptional activity. Some tumor cells possess large copy numbers of oncogenes that can significantly affect tumor heterogeneity and treatment resistance because ecDNA is inherited in a non-Mendelian manner during mitosis due to its lack of a centromere. Recent research employing cytogenetics and whole-genome sequencing (WGS) techniques revealed that ecDNA was infrequently detected in healthy tissues but present in a variety of cancer types [1].

10.1.2 Overview of ecDNA's Role in Cancer

The existence of ecDNA had a significant association with the degree of tumor immune infiltration in cancer cells, according to the analysis of WGS and gene expression information from 1684 cancer victims. It is well known that ecDNA plays a vital part in cell malignant transformation, tumor heterogeneity, and drug resistance of tumor cells. One adjuvant treatment option against cancer is to block the production of ecDNA or clear the existing ecDNA. The previous description of tumor cell resistance includes some mechanisms for reducing ecDNA production. Additionally, the removal of cells containing ecDNA can be used as a potential therapeutic intervention, particularly through the immune system. The fact that tumors carrying ecDNA had lower expression of the two major histocompatibility complexes (MHC), MHC-I and MHC-II, is one possible mechanism explaining this relationship. As a result, numerous studies have linked ecDNA to the development of tumors under restrictive environmental conditions and metastatic circumstances [2].

DOI: 10.1201/9781003655299-10

10.1.3 Brief Introduction to Tumor Immune Escape

Tumor immune escape is possible by ecDNA through several workflows. Regardless of having identical tumor mutation burden and neoantigen load to tumors lacking ecDNA, cancers with ecDNA exhibit considerably lower immune cell infiltration and cytotoxic T-cell activity. A less active immune response is indicated by the tumor microenvironment being less immune-enriched and more immune-depleted in ecDNA-positive tumors. Crucially, tumors with ecDNA have down-regulated expression of MHC class I and class II antigen presentation genes, delaying the tumor cells' ability to present antigens and elude immune detection. Since enhanced immune checkpoint signaling does not cause this immunological escape, immune checkpoint inhibitors may not be as effective in these cancers [3]. Thus, by inhibiting antigen-presenting pathways, ecDNA fabrication is a tumor-intrinsic process that promotes immune evasion.

10.2 MECHANISMS OF ECDNA FORMATION

10.2.1 Chromothripsis

Chromothripsis is the term for chromosome fragmenting events in which one or a few chromosomes experience 10 to 100 localized DNA breaks at the same time, leading to DNA fragment rearrangements and an extensive loss of DNA sequences. Chromosome reorganization during this process may result in the formation of extremely small circular DNA fragments, such as ecDNA, or "double minutes." In contrast to several consecutive DNA reorganizations in punctuated equilibrium occurrences like the breakage–fusion–bridge cycle, chromothripsis is a single, one-off occurrence. The majority of cells are harmed by the sudden changes in genetic composition that occur during chromosomal fragmentation, which favor the survival of cells with the changed phenotype. As a result, these cells become dominant in cell clones and aid in the development of malignant transformation. Acute myeloid leukemia and medulloblastoma are two examples of research that have shown a correlation between chromosomal breakage and ecDNA formation. In both cases, dominant-cell selection following chromosomal fragmentation may be linked to ecDNA formation [4]. Although there is proof that chromosomal fragmentation in some cancer types produces ecDNA, there is an absence of research at the pan-cancer level that shows how closely chromosome fragmentation and ecDNA genesis are related.

10.2.2 Breakage–Fusion–Bridge Cycles

The BFB cycle in maize during meiosis, mitosis, and endosperm production was discovered in 1939. Sister chromatids fuse together as a result of recently broken chromosomal ends during meiotic mitosis, starting the breakage–fusion–bridge cycle, causing the centromeres of the dicentric chromatid to separate after a bridge arrangement. The breakage–fusion–bridge cycle causes large deletions and focal amplifications that resemble ladders of DNA, and it may also result in the formation of double minutes (DM), a term that was frequently used in early investigations to describe extrachromosomal frameworks. This is due to the fragmentation that could occur anywhere between the two centromeres, and the cycle keeps going in successive mitoses during development [5]. It has been recognized that the BFB cycle contributes to the development of tumorigenic changes in structure in certain malignancies, resulting in the synthesis of gigantic aberrant chromosomes (cancer-related neochromosomes) that include highly amplified oncogenes (e.g., CDK4, HMGA2, & MDM2). Additionally, because the BFB cycle happens after chromothripsis events, these neochromosomes can occasionally have pieces from several chromosomes stitched together. Despite being present in only around 3% of malignancies, BFB-mediated neochromosomes are highly prevalent in specific, uncommon tumors [6].

10.2.3 Replication Stress and DNA Damage

Endogenous DNA damage is a result of replication stress because halted replication forks are brittle and prone to breaking, forming DNA lesions [7]. Replication stress develops when restrictions prevent DNA replication forks from moving consistently, which causes fork stalling or collapse [8]. DNA damage, especially double-strand breaks, can develop from this, which increases chromosomal instability. ecDNA can develop under such stress as a result of chromosome breakage or inadequate DNA repair. DNA damage and continuous replication stress are associated with the production of ecDNA, which is commonly found in cancer cells and frequently harbors oncogenes [9]. To combat these risks, cells initiate checkpoint pathways such as ATR-CHK1, which stabilize replication forks and encourage repair. Persistent replication stress, however, might overpower these defenses and permit the accumulation of ecDNA. Therefore, replication stress and DNA damage encourage the production and maintenance of ecDNA in addition to destabilizing the genome [10].

10.2.4 Telomere Dysfunction

Breakage–fusion–bridge (B/F/B) cycles, in which chromosomes without telomeres constantly fuse and break during cell division, telomere rupture, lead to chromosome instability. Massive inverted repeats and widespread DNA amplification are the results of these cycles. Additionally, B/F/B cycle-generated amplified areas remain highly unstable and can result in fragments of ecDNA. These ecDNA fragments can contribute to gene amplification in malignancy by either remaining as double minute chromosomes or reintegrating into other chromosomal sites. Thus, telomere degeneration leads to the development of tumors by destabilizing chromosomes and encouraging the generation of ecDNA [11].

10.3 ECDNA AND ONCOGENE AMPLIFICATION

10.3.1 Common Oncogenes Found on ecDNA

Massive circular ecDNA has been shown to include multiple oncogenes that were previously believed to be found mainly on linear chromosomes. According to the Mittelman database of chromosomal abnormalities, ecDNA was historically considered to be a rare feature of malignancies (1.4% of tumors) with unknown relevance. However, recent research indicates that highly amplified, oncogene-containing ecDNAs are prevalent in cancer (25 among 29 kinds). Even after controlling for copy number, there is still a definite increase in oncogene expression. By integrating regulatory elements that are too distant to interact on a linear chromosome closer together on a circle, the circular layout generates new cis-regulatory interactions. Particularly, major oncogenes like CDK4, EGFR, MYC, and MDM2 are commonly found in ecDNA **(Table 10.1)**, and they are linked to aggressive tumor activity and resistance to treatment [12].

TABLE 10.1
Common Oncogenes Linked with Tumor Types Based on ecDNA

Oncogene	Tumor Types	Primary Oncogenic Function	References
CDK4	Glioblastoma, Sarcoma	Cell cycle regulation	[13]
MDM2	Liposarcoma, Glioma	p53 inhibition	[14]
AKT1	Breast, Prostate	PI3K/AKT pathway signaling	[14]
E2F3	Bladder, Retinoblastoma	Transcription factor	[14]
NEDD9	Melanoma, Lung	Cell migration, invasion	[14]
EGFR	Glioblastoma, Lung	Receptor tyrosine kinase	[14]
MYCL	Small-cell lung cancer	MYC family transcription	[14]
PDGFRA	Glioma, GIST	Growth factor signaling	[14]
SOX2	Lung, Esophageal	Stemness, lineage control	[14]
TERT	Multiple	Telomerase activation	[14]
MYC	Neuroblastoma, Breast	Global transcriptional control	[14]
ERBB2	Breast, Gastric	HER2 receptor signaling	[14]

10.3.2 Advantages of ecDNA-Based Amplification

Cells that have ecDNA amplicons carrying oncogenes have a competitive advantage through several types of techniques. The 1979 patients designed amplicons showed a high enrichment for oncogenes, and the random distribution of breakpoints close to the oncogene locus suggests that ecDNA originates at random loci. The loci that contain regulatory elements and oncogenes have a selective advantage. Additional selective advantages are obtained by increasing ecDNA expression independent of gene dosage and by unequally segregating the circular amplicons, which increases the number of daughter cells. As a result, cells exhibit enhanced oncogene expression and many copies. Circle reintegration could provide further selective advantages [15].

10.3.3 Impact on Tumor Heterogeneity and Evolution

Joshua T. Lange et al. discovered that oncogenes on circular DNA segments that randomly segregate during cell division are amplified by ecDNA, which causes a high degree of cell-to-cell variability in the amount of ecDNA copies inside tumors. Compared to chromosomal amplifications, considerable intratumoral genetic variability is synthesized by this random inheritance. By enabling certain cells to rapidly acquire or lose oncogene copies, such heterogeneity facilitates rapid tumor adaptability to environmental stimuli, including drugs. Tumors that contain ecDNA can therefore evolve and acquire resistance considerably more quickly than those that only rely on chromosomal alterations. In general, ecDNA promotes aggressive tumor growth and is associated with poor clinical results.

10.4 ECDNA AND TUMOR IMMUNE ESCAPE

Oncogene overexpression is promoted by (ecDNA), which helps malignancies escape immune surveillance (Figure 10.1). This stimulates the proliferation of cancer and makes immune-based treatments less effective.

10.4.1 Overexpression of Immune Checkpoint Inhibitors

The inadequate response to immune checkpoint inhibitors (ICIs) may be caused by the observation that tumors containing ecDNA frequently show decreased expression of immune checkpoint genes (e.g., PD-L1, CTLA4) rather than overexpression [17, 16].

10.4.1.1 PD-L1 Amplification on ecDNA

Some malignancies can amplify PD-L1 (CD274) on ecDNA. A head and neck cancer sample proved to contain a hybrid ecDNA amplicon that included and amplified the immune-regulating ligand PD-L1. Lung, kidney, and bladder tumor subsets have also exhibited PD-L1 amplification via ecDNA. This pathway has been associated with poor clinical outcomes and contributes to oncogene amplification [18]. PD-L1 amplification on ecDNA is comparatively uncommon and context-dependent, as shown by a finding that ecDNA-positive tumors generally have lower PD-L1 expression and are frequently non-responsive to checkpoint inhibitor therapy. The genetic adaptability of tumors and their capacity for immune evasion are emphasized by the presence of PD-L1 on ecDNA.

10.4.1.2 Other Immune Checkpoint Molecules on ecDNA

Thirty-four percent of ecDNA-containing tumors contained known immunomodulatory genes amplified on ecDNA, the majority of which were co-amplified with adjacent oncogenes. Leukocyte-mediated cytotoxicity, the negative regulation of lymphocyte activation, and the negative regulation of immunological effector processes were among the processes in which these immunomodulatory

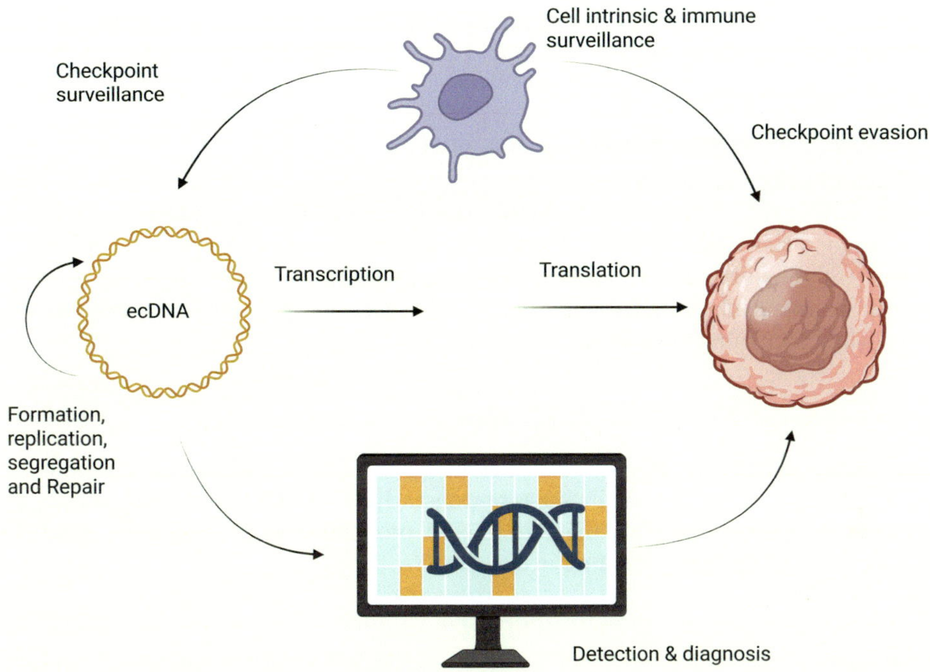

FIGURE 10.1 ecDNA in tumor immune escape mechanisms. (From Ref. [16]).

genes were implicated. Oncogenes were absent from 41.5% of tumors with immunomodulatory genes expressed on ecDNA in certain cases, indicating a functional role for these elements. T-cell infiltration in tumors is decreased when ecDNAs containing immunomodulatory genes are present. As a result, ecDNA may contribute to immune evasion in cancer by amplifying a variety of immunomodulatory genes but not necessarily typical checkpoint molecules like PD-L1 and CTLA4 [19]. B7-H3 (CD276), B7-H4 (B7x), VISTA, B7-H2 (ICOS-L), B7-H6, HHLA2, TIM-3, Siglec-15, and catabolic enzymes like IDO1 are additional immune checkpoint molecules in addition to PD-1, PD-L1, and CTLA-4 [20].

10.4.2 Alterations in Antigen Presentation

Both MHC class I and class II antigen presentation genes have significantly reduced expression in tumors containing ecDNA, which impairs antigen presentation and lowers immune cell infiltration [21].

10.4.2.1 MHC Class I Downregulation

In 40–90% of human malignancies, MHC class I downregulation is a common immune escape strategy in cancer and is frequently associated with a worse prognosis [22]. The expression of the MHC class I antigen presentation gene is significantly suppressed in malignancies containing ecDNA, which impairs CD8+ T-cell identification and cytotoxicity. This downregulation is caused via transcriptional and post-transcriptional regulation, along with epigenetic silencing mechanisms such as promoter hypermethylation [23]. Many malignancies, especially melanoma, breast, colorectal, and cervical cancers, have been shown to exhibit decreased or absent MHC class I surface expression, which has been associated with a higher risk of metastasis and a shorter survival time

[24]. Both reversible epigenetic ("soft lesions") and genetic ("hard lesions") alterations can cause this downregulation, with both of them potentially leading to therapeutic intervention. Preclinical models showed progress in restoring MHC class I expression with epigenetic drugs, which could improve the effectiveness of immunotherapies.

10.4.2.2 Defects in Antigen Processing Machinery

Significant abnormalities in the antigen processing machinery are observed in tumors containing ecDNA, especially in the expression of the MHC class I and class II antigen presentation genes [25]. Decreased immune cell infiltration and cytotoxic T-cell proliferation in the tumor microenvironment are due to this downregulation, which also affects the tumor cells' capacity to carry antigens to immune cells. Consequently, ecDNA-containing cancers have a higher chance of evading immune surveillance while remaining unchecked.

10.4.3 Cytokine and Chemokine Modulation

ecDNA can stimulate immune cells to release cytokines like TNF-α and IL-6, which both increase tumorigenic activity and immunosuppressive responses [26]. Furthermore, eccDNA can bind to innate immune receptors to regulate the synthesis of proinflammatory cytokines, which further affects cytokine and chemokine signaling in cancers [27].

10.4.3.1 Overexpression of Immunosuppressive Factors

Immunomodulatory and inflammatory genes, such as those that impact immunological effector functions and lymphocyte-mediated immunity, are commonly amplified by ecDNA. Reduced T-cell infiltration is exhibited in tumors carrying ecDNA containing these genes, which helps build an immunosuppressive tumor microenvironment. These increased genes are implicated in immunological escape and immune response suppression, based on gene ontology analysis. Since certain ecDNAs exclusively carry immunomodulatory genes, this immunosuppressive impact is not restricted to co-amplification with oncogenes. In cancer patients, the presence of ecDNA is related to undesirable outcomes, metastasis, and resistance to treatment. Therefore, immunosuppressive factor amplification via ecDNA is essential for tumor growth and immune evasion [28].

10.4.3.2 Disruption of Immune Cell Recruitment

Tumor ecDNA is associated with reduced immune cell infiltration and recruitment. ecDNA-containing tumors show reduced expression of MHC class I and II antigen presentation genes along with decreased cytotoxic T-cell function. Tumor cell immune evasion is more simpler by this downregulation of antigen presentation machinery, which compromises immune detection. In tumors that express ecDNA, the tumor microenvironment is less fibrotic and more immune-depleted. The downregulation of immune checkpoint gene expression (PD-L1, CTLA4, etc.) indicates that immune evasion fails to occur through the checkpoint pathway. The production of ecDNA is a tumor-intrinsic immune escape mechanism that mainly inhibits the presentation of antigens and the recruitment of immune cells [29].

10.4.4 Resistance to Immune-Mediated Cell Death

Suppressing MHC class I and II antigen presentation genes, tumors containing ecDNA show resistance to immune-mediated cell death by making it more difficult for cytotoxic T-cells to recognize them. An immuno-depleted tumor microenvironment and decreased immune cell infiltration result. Consequently, cancers that express ecDNA may evade immune monitoring and survive immunotherapy [30].

10.4.4.1 Amplification of Anti-Apoptotic Genes

Anti-apoptotic gene amplification on ecDNA has been experimentally demonstrated in cancer models [31]. Mice with engineered ecDNAs containing Mdm2 showed high levels of overexpression of this anti-apoptotic gene, which promoted aggressive tumor development, cell transformation, and immortalization. The transcriptional activity of these ecDNAs was shown in vivo by their open chromatin and significant upregulation of every gene within the amplicon. The functional impact of anti-apoptotic gene amplification on ecDNA was demonstrated by treatment with an MDM2 antagonist, which suppressed proliferation and reactivated p53 signaling. Tumor growth, treatment resistance, and a poor prognosis in ecDNA-driven malignancies are all a result of this mechanism. Therefore, ecDNA-mediated anti-apoptotic gene amplification plays an essential part in the aggressiveness and survival of cancer [32].

10.4.4.2 Alterations in Death Receptor Signaling

Cancer cell death receptor signaling can be indirectly altered by oncogene amplification on ecDNA. ecDNA helps tumor cells in escaping apoptosis, including death receptor-mediated cell death pathways, by promoting the high-level expression of genes that promote growth and survival. This causes tumors to be more resistant to treatments like immune-based therapies and some chemotherapeutics that depend on activating death receptors. Cancer cell adaptation and survival in the face of death receptor signaling challenges are further supported by the rapid growth and dynamic regulation provided by ecDNA [33].

10.5 ECDNA-MEDIATED IMMUNE EVASION STRATEGIES

10.5.1 Rapid Adaptation to Immune Pressure

ecDNA produces significant intratumoral heterogeneity and flexible oncogene expression, and cancers are able to adapt to immune pressure [34]. Differences in ecDNA copy number effectively correlate with gene expression and growth, according to a recent study using single-cell multi-omics. This difference rapidly recovers following disruption, indicating active, population-level adaptability [35]. MYC-carrying ecDNA in pancreatic cancer causes tumor cells to modify their growth requirements and transform into more aggressive forms when under stress, exhibiting "bet-hedging" against environmental and immunological challenges [36].

10.5.2 Heterogeneity in Immune-Related Gene Expression

ecDNA-positive tumors are characterized by heterogeneity in immune-related gene expression, with persistent downregulation of MHC class I and II antigen presentation genes leading to reduced immune cell infiltration. According to a 2024 review, ecDNA-mediated immune escape is mostly caused by suboptimal antigen presentation as compared to increased immune checkpoint gene expression. Overall, these findings highlight that the main mechanisms for immune escape and treatment resistance in cancer are ecDNA-induced heterogeneity in immune gene expression [16].

10.5.3 Enhanced Plasticity in Response to Immunotherapy

MYC ecDNA in pancreatic cancer permits cells to shift development dependency and morphologically respond under selective stress, with ecDNA copy number increasing or decreasing as desired [37]. This plasticity helps immune evasion, as ecDNA-driven heterogeneity develops subpopulations capable of surviving immune attack or therapy. Therefore, ecDNA improves tumor plasticity in reaction to immunotherapy through permitting rapid, reversible changes in oncogene dosage and gene expression [38].

10.6 CLINICAL IMPLICATIONS

10.6.1 Impact on Immunotherapy Efficacy

ecDNA amplification in cancers increases genetic heterogeneity and immune evasion. It is linked with decreased immunotherapy efficacy, especially with immune checkpoint inhibitors. Clinical research demonstrates that ecDNA-positive tumors are less susceptible to anti-PD-1 treatments because they frequently exhibit decreased immune cell infiltration and downregulated antigen presentation. For gastrointestinal malignancies in specific, ecDNA amplification was found to be a risk factor and a possible biomarker to inform immunotherapy choices in a recent pan-cancer investigation of more than 13,000 patients [39].

10.6.2 Potential Biomarkers for Treatment Response

PD-L1 expression, tumor mutation burden (TMB), microsatellite instability (MSI), mismatch repair deficiency (MMR), HER2 amplification, EGFR mutations, KRAS mutations, BRAF mutations, and ecDNA amplification are important biomarkers for predicting whether cancer treatments will be effective [40]. Furthermore, several types of cancers are diagnosed, and their response to treatment is monitored using small eccDNAs from genes like ALK and ETV6 as well as conventional protein biomarkers like CEA, CA19-9, and PSA [41].

10.6.3 Challenges in Targeting ecDNA-Driven Immune Escape

There are several major challenges to addressing ecDNA-driven immune evasion. First, ecDNA's random segregation and rapid number of copies fluctuation promote significant tumor heterogeneity, allowing tumor cells to rapidly modify and resist therapies, including immunotherapies. Second, immunology-based therapies are less successful because ecDNA increases oncogenes and immune evasion genes, which decrease antigen presentation and immune cell infiltration. Third, targeted techniques are made more difficult by the fact that existing treatments that cause DNA damage (such as radiation and chemotherapy) are not specific to ecDNA and may also damage healthy cells. Lastly, although strategies like BET inhibitors and replication stress targeting are being investigated, their clinical effectiveness and safety in removing ecDNA specifically without causing off-target consequences are still undetermined [42].

10.7 THERAPEUTIC APPROACHES

10.7.1 Combination Therapies Targeting ecDNA and Immune Checkpoints

To overcome resistance in ecDNA-positive tumors, combination treatments that target both immunological checkpoints and ecDNA-driven mechanisms have shown promise. According to recent research, ecDNA-mediated oncogene amplification can be suppressed and tumor regression improved in preclinical models by combining CHK1 inhibitors, which take advantage of the transcription-replication conflict and replication stress specific to ecDNA, with targeted therapies such as FGFR inhibitors [43].

10.7.2 Strategies to Enhance Antigen Presentation

The epigenetic drugs are a key component of strategies to improve antigen presentation in cancers. Histone deacetylase inhibitors (HDACis) and DNA methyltransferase inhibitors (DNMTis) can upregulate antigen presentation machinery and restore MHC class I expression, increasing the visibility of tumor cells to T lymphocytes. To increase tumor immunogenicity further, these substances can reactivate cancer/testis antigens (CTAs) and other cancer-associated antigens. Combining immunotherapy and epigenetic treatment is a potential strategy for preventing immune evasion by improving antigen presentation and anti-tumor immunity [44, 45].

10.7.3 Novel Approaches to Target ecDNA-Specific Vulnerabilities

Strategies emerging from recent research, including CHK1 Inhibition, Disrupting Transcription-Replication Conflict, Gene Targets for ecDNA Maintenance, Spatial as well as Single-Cell Technologies, and Drug Library Screening, offer novel approaches to target ecDNA-specific vulnerabilities efficiently. These strategies aim to overcome resistance mechanisms in aggressive tumors and provide more potent, cancer-specific treatments while taking advantage of the distinct biology of ecDNA.

10.8 FUTURE DIRECTIONS

10.8.1 Emerging Technologies for ecDNA Detection and Analysis

Numerous tumor tissues have been shown to contain ecDNA, and mounting data indicate that it is crucial for the development and heterogeneity of tumor, and, through reintegration, the genes and regulatory functions carried by ecDNA can cause mutations, increase tumor heterogeneity through non-Mendelian inheritance, and upregulate gene expression. Since ecDNA is linked to both patient prognosis and drug resistance in a number of cancer types, techniques for detecting ecDNA for liquid biopsies may be developed for use in prognostic and diagnostic applications in addition to guiding treatment plans aimed at overcoming drug resistance [46].

The application of ecDNA in clinical practice is contingent upon the resolution of numerous crucial concerns. A better understanding of the molecular mechanism behind ecDNA dynamics and how ecDNA influences the interactions between cancer cells and the body's immune system is necessary to create effective ecDNA-targeted therapies. Furthermore, it has been proposed that ecDNA's topological configurations influence its function.

10.8.2 Integrating ecDNA Status in Personalized Immunotherapy

Integrating ecDNA status into personalized immunotherapy represents a promising frontier in oncology, offering potential enhancement of treatment efficacy and overcoming resistance mechanisms in various fields mentioned next:

- ecDNA as biomarker for immune therapy response
- Targeting ecDNA to enhance immunogenicity
- Development of eCDNA direct therapeutics
- Integration into clinical practice such as whole genome sequencing [47, 48]

A distinct population of patients with ecDNA-positive tumors exhibits characteristics such as MSS, Low TIS, and PD-L1 expression that indicate non-response to checkpoint inhibitor therapy. Determining the tumor's ecDNA status could therefore be useful as an extra patient selection method for checkpoint of inhibitor treatment. Since ecDNA is not only found in stomach tumors, this chapter emphasizes the need for creating a clinical diagnostic test for ecDNA status as well as the necessity for more investigation into the biology of ecDNA, how it affects immunotherapy response, and possible ecDNA-directed treatment [49].

10.8.3 Potential for ecDNA-Targeted Therapies to Overcome Immune Escape

ecDNAs are free-floating circular DNA fragments that are thought to increase the aggressiveness of malignancies and aid in their resistance to various treatments. Researchers have discovered a method that may be able to eradicate tumors driven by these DNA fragments.

In the study centered on treatment, Dr. Mischel and his associates demonstrated that ecDNAs can also cause cancer cells to have significant trouble copying their DNA in order to split into two new cells. Additionally, they discovered that BBI-2779, an investigational medication, could take advantage of certain vulnerabilities. When used with a medication that targets the genetic mutation that causes cancer, BBI-2779 rapidly reduced tumor size and prevented it from spreading for comparatively lengthy periods of time in mice whose tumors had ecDNAs with the mutation.

A closely comparable medication, BBI-355, is currently being investigated in a limited clinical trial in people. Depending on the genetic alterations in the participants' tumors, it can be used alone or in conjunction with one of many targeted therapies.

It is well known that only cancer cells have ecDNAs, and they seem to have a significant and varied impact on cancer cells, as noted by Ian Fingerman, Ph.D., of NCI's Division of Cancer Biology [50].

10.9 CONCLUSION

10.9.1 Summary of ecDNA's Role in Tumor Immune Escape

ecDNA is a circular genetic element that is particular to tumors. ecDNA has been linked to poor prognoses for cancer patients, has been shown to exhibit open chromatin structure, and facilitates oncogene amplification and unequal segregation of genetic material. Although the ability to evade the immune system is a common characteristic of cancer growth, little is known about the intrinsic components of the tumor that influence immune evasion.

In this chapter, we demonstrate that ecDNA is linked to indicators of tumor immune evasion and that ecDNA acquisition may be one of the strategies tumor cells use to evade immunological monitoring. ecDNA-containing tumors often have similar TMB and neoantigen loads, but they also have decreased cytotoxic T-cell activity and immune cell infiltration. Tumor microenvironment containing ecDNA exhibits a decrease in immune-enriched fibrotic types and an increase in immune-depleted fibrotic types. Tumors containing ecDNA exhibit reduced expression of both MHC class I and class II antigen presentation genes, which may be the fundamental mechanism for immune evasion linked to ecDNA. This chapter shows that the production of ecDNA is a way for cancer cells to evade the immune system [51].

10.9.2 Importance of Understanding ecDNA Biology for Improving Cancer Immunotherapy

Current knowledge of the role of circular DNA outside of the chromosome in the development of cancer. In cancer research and treatment, a number of recent studies have shown the growing significance of extrachromosomal circular DNA on carcinogenic characteristics and regulatory capacity.

There is increasing evidence that circular DNA plays a functional role in fostering genetic variability and hastening the development of human cancer diseases. Amplification of circular DNA elements carrying oncogenes (such EGFR and c-Myc) may also give cancer cells benefits that improve their ability to adapt to changing environmental stress [51].

Recent studies have shown that ecDNA can contribute to immune escape by:

- Reducing the expression of MHC class I molecules, impairing antigen presentation to cytotoxic T-cells
- Promoting tumor heterogeneity, allowing subclonal populations to escape immune surveillance

A deep understanding of these mechanisms enables the identification of biomarkers predictive of immunotherapy resistance.

10.9.3 Future Perspectives on Targeting ecDNA to Enhance Anti-Tumor Immunity

This chapter demonstrates the major future direction for the ecDNA in cancer management. However, because artefacts may easily result from the amplification of small amounts of material, single-cell ecDNA sequencing techniques will need to be thoroughly benchmarked, and several future directions are mentioned next:

- Development of ecDNA-specific inhibitors
- CRISPR-based gene-editing approaches
- Exploiting ecDNA-induced cGAS-STING activation
- Combination with immune checkpoint inhibitor
- Biomarker development for ecDNA profiling
- Targeting ecDNA packaging and segregation mechanisms
- Nanoparticle-based delivery of ecDNA-targeting agents
- Immunogenic modulation of tumor microenvironment
- Preclinical ecDNA tumor models

REFERENCES

[1] H. Wu, S. Liu, D. Wu, H. Zhou, G. Wu, Tumor extrachromosomal DNA: Biogenesis and recent advances in the field, Biomed Pharmacother 174 (2024) 116588. https://doi.org/10.1016/J.BIOPHA.2024.116588.

[2] Y. Dong, Q. He, X. Chen, F. Yang, L. He, Y. Zheng, Extrachromosomal DNA (ecDNA) in cancer: Mechanisms, functions, and clinical implications, Front Oncol 13 (2023) 1194405. https://doi.org/10.3389/FONC.2023.1194405/BIBTEX.

[3] T. Wu, C. Wu, X. Zhao, G. Wang, W. Ning, Z. Tao, F. Chen, X.S. Liu, Extrachromosomal DNA formation enables tumor immune escape potentially through regulating antigen presentation gene expression, Sci Rep 12 (2022) 1–9. https://doi.org/10.1038/S41598-022-07530-8;SUBJMETA=114,250,631,67;KWRD=CANCER,COMPUTATIONAL+BIOLOGY+AND+BIOINFORMATICS,IMMUNOLOGY.

[4] T. Rausch, D.T.W. Jones, M. Zapatka, A.M. Stütz, T. Zichner, J. Weischenfeldt, N. Jäger, M. Remke, D. Shih, P.A. Northcott, E. Pfaff, J. Tica, Q. Wang, L. Massimi, H. Witt, S. Bender, S. Pleier, H. Cin, C. Hawkins, C. Beck, A. Von Deimling, V. Hans, B. Brors, R. Eils, W. Scheurlen, J. Blake, V. Benes, A.E. Kulozik, O. Witt, D. Martin, C. Zhang, R. Porat, D.M. Merino, J. Wasserman, N. Jabado, A. Fontebasso, L. Bullinger, F.G. Rücker, K. Döhner, H. Döhner, J. Koster, J.J. Molenaar, R. Versteeg, M. Kool, U. Tabori, D. Malkin, A. Korshunov, M.D. Taylor, P. Lichter, S.M. Pfister, J.O. Korbel, Genome sequencing of pediatric medulloblastoma links catastrophic DNA rearrangements with TP53 mutations, Cell 148 (2012) 59–71. https://doi.org/10.1016/J.CELL.2011.12.013/ATTACHMENT/D57E6D3A-3280-407E-B57E-CCFBD5D70755/MMC8.PDF.

[5] M. Wu, K. Rai, Demystifying extrachromosomal DNA circles: Categories, biogenesis, and cancer therapeutics, Comput Struct Biotechnol J 20 (2022) 6011–6022. https://doi.org/10.1016/J.CSBJ.2022.10.033/ASSET/B7589342-A6B8-47E9-B625-171BD74894AF/MAIN.ASSETS/GR3.JPG.

[6] S.K. Bohlander, P.M. Kakadiya, A. Coysh, Chromosome rearrangements and translocations, Encycl Cancer (2019) 389–404. https://doi.org/10.1016/B978-0-12-801238-3.65105-X.

[7] J. Tang, N.E. Weiser, G. Wang, S. Chowdhry, E.J. Curtis, Y. Zhao, I.T. Lo Wong, G.K. Marinov, R. Li, P. Hanoian, E. Tse, S.G. Mojica, R. Hansen, J. Plum, A. Steffy, S. Milutinovic, S.T. Meyer, J. Luebeck, Y. Wang, S. Zhang, N. Altemose, C. Curtis, W.J. Greenleaf, V. Bafna, S.J. Benkovic, A.B. Pinkerton, S. Kasibhatla, C.A. Hassig, P.S. Mischel, H.Y. Chang, Enhancing transcription–replication conflict targets ecDNA-positive cancers, Nature 635 (2024) 210–218. https://doi.org/10.1038/s41586-024-07802-5.

[8] T. Wilhelm, M. Said, V. Naim, Dna replication stress and chromosomal instability: Dangerous liaisons, Genes (Basel) 11 (2020) 1–35. https://doi.org/10.3390/GENES11060642.

[9] X. Kang, X. Li, J. Zhou, Y. Zhang, L. Qiu, C. Tian, Z. Deng, X. Liang, Z. Zhang, S. Du, S. Hu, N. Wang, Z. Yue, Y. Xu, Y. Gao, J. Dai, Z. Wang, C. Yu, J. Chen, Y. Wu, L. Chen, Y. Yao, S. Yao, X. Yang, L. Yan, Q. Wen, O.M. Depies, K. Chan, X. Liang, G. Li, Z. Zi, X. Liu, H. Gan, Extrachromosomal DNA replication and maintenance couple with DNA damage pathway in tumors, Cell 188 (2025). https://doi.org/10.1016/J.CELL.2025.04.012/ASSET/B40FDD63-4AFA-409E-87CC-B04926F05BCB/MAIN.ASSETS/FX1.JPG.

[10] K. Yoshida, M. Fujita, DNA damage responses that enhance resilience to replication stress, Cell Mol Life Sci 78 (2021) 6763–6773. https://doi.org/10.1007/S00018-021-03926-3.

[11] J.P. Murnane, Telomere dysfunction and chromosome instability, Mutat Res—Fundam Mol Mech Mutagen 730 (2012) 28–36. https://doi.org/10.1016/J.MRFMMM.2011.04.008.

[12] L.T. Pecorino, R.G.W. Verhaak, A. Henssen, P.S. Mischel, Extrachromosomal DNA (ecDNA): An origin of tumor heterogeneity, genomic remodeling, and drug resistance, Biochem Soc Trans 50 (2022) 1911–1920. https://doi.org/10.1042/BST20221045.

[13] M. Karami Fath, N. Karimfar, A. Fazlollahpour Naghibi, S. Shafa, M. Ghasemi Shiran, M. Ataei, H. Dehghanzadeh, M. Nabi Afjadi, T. Ghadiri, Z. Payandeh, V. Tarhriz, Revisiting characteristics of oncogenic extrachromosomal DNA as mobile enhancers on neuroblastoma and glioma cancers, Cancer Cell Int 22 (2022) 1–14. https://doi.org/10.1186/S12935-022-02617-8.

[14] K. Adelman, B.J.E. Martin, ecDNA party bus: Bringing the enhancer to you, Mol Cell 81 (2021) 1866–1867. https://doi.org/10.1016/J.MOLCEL.2021.04.017.

[15] C. Bailey, M.J. Shoura, P.S. Mischel, C. Swanton, Extrachromosomal DNA—relieving heredity constraints, accelerating tumour evolution, Ann Oncol 31 (2020) 884–893. https://doi.org/10.1016/J.ANNONC.2020.03.303/ASSET/832E2AA1-829C-494D-8618-A1F7204726FC/MAIN.ASSETS/GR3.JPG.

[16] Q.L. Yang, Y. Xie, K. Qiao, J.Y.S. Lim, S. Wu, Modern biology of extrachromosomal DNA: A decade-long voyage of discovery, Cell Res 35 (2025) 11–22. https://doi.org/10.1038/s41422-024-01054-8.

[17] X. Mao, G. Rao, G. Li, S. Chen, Insights into extrachromosomal DNA in cancer: Biogenesis, methodologies, functions, and therapeutic potential, Adv Biol 9 (2025) 2400433. https://doi.org/10.1002/ADBI.202400433;JOURNAL:JOURNAL:23667478;WGROUP:STRING:PUBLICATION.

[18] J. Pang, N. Nguyen, J. Luebeck, L. Ball, A. Finegersh, S. Ren, T. Nakagawa, M. Flagg, S. Sadat, P.S. Mischel, G. Xu, K. Fisch, T. Guo, G. Cahill, B. Panuganti, V. Bafna, J. Califano, Extrachromosomal DNA in HPV mediated oropharyngeal cancer drives diverse oncogene transcription, Clin Cancer Res 27 (2021) 6772. https://doi.org/10.1158/1078-0432.CCR-21-2484.

[19] C. Bailey, O. Pich, K. Thol, T.B.K. Watkins, J. Luebeck, A. Rowan, G. Stavrou, N.E. Weiser, B. Dameracharla, R. Bentham, W.-T. Lu, J. Kittel, S.Y.C. Yang, B.E. Howitt, N. Sharma, M. Litovchenko, R. Salgado, K.L. Hung, A.J. Cornish, D.A. Moore, R.S. Houlston, V. Bafna, H.Y. Chang, S. Nik-Zainal, N. Kanu, N. McGranahan, Genomics England Consortium, A.M. Flanagan, P.S. Mischel, M. Jamal-Hanjani, C. Swanton, Origins and impact of extrachromosomal DNA, Nature 635 (2024) 193–200. https://doi.org/10.1038/s41586-024-08107-3.

[20] K. Mortezaee, J. Majidpoor, Alternative immune checkpoints in immunoregulatory profile of cancer stem cells, Heliyon 9 (2023). https://doi.org/10.1016/j.heliyon.2023.e23171.

[21] T. Wu, C. Wu, X. Zhao, G. Wang, W. Ning, Z. Tao, F. Chen, X.-S. Liu, Extrachromosomal DNA is associated with decreased immune cell infiltration and antigen presentation, represents a potential cancer immune evasion mechanism, BioRxiv (2022) 2022.02.04.479205. https://doi.org/10.1101/2022.02.04.479205.

[22] A.M. Cornel, I.L. Mimpen, S. Nierkens, MHC class I downregulation in cancer: Underlying mechanisms and potential targets for cancer immunotherapy, Cancers (Basel) 12 (2020) 1–33. https://doi.org/10.3390/CANCERS12071760.

[23] K. Dhatchinamoorthy, J.D. Colbert, K.L. Rock, Cancer immune evasion through loss of MHC class I antigen presentation, Front Immunol 12 (2021) 636568. https://doi.org/10.3389/FIMMU.2021.636568/XML/NLM.

[24] B.C. Taylor, J.M. Balko, Mechanisms of MHC-I downregulation and role in immunotherapy response, Front Immunol 13 (2022) 844866. https://doi.org/10.3389/FIMMU.2022.844866/XML/NLM.

[25] Y. Zhao, L. Yu, S. Zhang, X. Su, X. Zhou, Extrachromosomal circular DNA: Current status and future prospects, Elife 11 (2022) e81412. https://doi.org/10.7554/ELIFE.81412.

[26] S. Zheng, Y. Li, L. Wang, Q. Wei, M. Wei, T. Yu, L. Zhao, Extrachromosomal circular DNA and their roles in cancer progression, Genes Dis 12 (2023) 101202. https://doi.org/10.1016/J.GENDIS.2023.101202.

[27] L. Yang, R. Jia, T. Ge, S. Ge, A. Zhuang, P. Chai, X. Fan, Extrachromosomal circular DNA: Biogenesis, structure, functions and diseases, Signal Transduct Target Ther 7 (2022) 1–21. https://doi.org/10.1038/s41392-022-01176-8.

[28] C. Bailey, O. Pich, K. Thol, T.B.K. Watkins, J. Luebeck, A. Rowan, G. Stavrou, N.E. Weiser, B. Dameracharla, R. Bentham, W.-T. Lu, J. Kittel, S.Y.C. Yang, B.E. Howitt, N. Sharma, M. Litovchenko, R. Salgado, K.L. Hung, A.J. Cornish, D.A. Moore, R.S. Houlston, V. Bafna, H.Y. Chang, S. Nik-Zainal, N. Kanu, N. McGranahan, Genomics England Consortium, A.M. Flanagan, P.S. Mischel, M. Jamal-Hanjani, C. Swanton, Origins and impact of extrachromosomal DNA, Nature 635 (2024) 193–200. https://doi.org/10.1038/s41586-024-08107-3.

[29] W. Lv, Y. Zeng, C. Li, Y. Liang, H. Tao, Y. Zhu, X. Sui, Y. Li, S. Jiang, Q. Gao, E. Rodriguez-Fos, G. Prasad, Y. Wang, R. Zhou, Z. Xu, X. Pan, L. Chen, X. Xiang, H. Teng, C. Sun, T. Qin, W. Dong, Y. Li, X. Lan, X. Li, L. Lin, L. Bolund, H. Yang, R.G.W. Verhaak, B.M. Faltas, J.B. Hansen, S. Wu, P.S. Mischel, A.G. Henssen, V. Bafna, J. Luebeck, B. Regenberg, Y. Luo, C. Lin, P. Han, Spatial–temporal diversity of extrachromosomal DNA shapes urothelial carcinoma evolution and the tumor immune microenvironment, Cancer Discov (2025) OF1–OF22. https://doi.org/10.1158/2159-8290.CD-24-1532.

[30] Pan-cancer analysis of whole genomes identifies driver rearrangements promoted by LINE-1 retrotransposition, (n.d.). https://compgen.hms.harvard.edu/publications/pan-cancer-analysis-whole-genomes-identifies-driver-rearrangements-promoted-line (accessed May 18, 2025).

[31] S. Wu, V. Bafna, H.Y. Chang, P.S. Mischel, Extrachromosomal DNA: An emerging hallmark in human cancer, Annu Rev Pathol Mech Dis 17 (2021) 367–386. https://doi.org/10.1146/ANNUREV-PATHMECHDIS-051821-114223.

[32] D. Pradella, M. Zhang, R. Gao, M.A. Yao, K.M. Gluchowska, Y. Cendon-Florez, T. Mishra, G. La Rocca, M. Weigl, Z. Jiao, H.H.M. Nguyen, M. Lisi, M.M. Ozimek, C. Mastroleo, K. Chen, F. Grimm, J. Luebeck, S. Zhang, A.A. Zolli, E.G. Sun, B. Dameracharla, Z. Zhao, Y. Pritykin, C. Sigel, H.Y. Chang, P.S. Mischel, V. Bafna, C.R. Antonescu, A. Ventura, Engineered extrachromosomal oncogene amplifications promote tumorigenesis, Nature 637 (2024) 955–964. https://doi.org/10.1038/s41586-024-08318-8.

[33] J.T. Lange, J.C. Rose, C.Y. Chen, Y. Pichugin, L. Xie, J. Tang, K.L. Hung, K.E. Yost, Q. Shi, M.L. Erb, U. Rajkumar, S. Wu, S. Taschner-Mandl, M. Bernkopf, C. Swanton, Z. Liu, W. Huang, H.Y. Chang, V. Bafna, A.G. Henssen, B. Werner, P.S. Mischel, The evolutionary dynamics of extrachromosomal DNA in human cancers, Nat Genet 54 (2022) 1527–1533. https://doi.org/10.1038/S41588-022-01177-X;SUBJMETA=114,208,631,67,80;KWRD=CANCER,CELL+BIOLOGY,COMPUTATIONAL+BIOLOGY+AND+BIOINFORMATICS,GENETICS.

[34] X. Kang, X. Li, J. Zhou, Y. Zhang, L. Qiu, C. Tian, Z. Deng, X. Liang, Z. Zhang, S. Du, S. Hu, N. Wang, Z. Yue, Y. Xu, Y. Gao, J. Dai, Z. Wang, C. Yu, J. Chen, Y. Wu, L. Chen, Y. Yao, S. Yao, X. Yang, L. Yan, Q. Wen, O.M. Depies, K. Chan, X. Liang, G. Li, Z. Zi, X. Liu, H. Gan, Extrachromosomal DNA replication and maintenance couple with DNA damage pathway in tumors, Cell 188 (2025). https://doi.org/10.1016/J.CELL.2025.04.012/ASSET/B40FDD63-4AFA-409E-87CC-B04926F05BCB/MAIN.ASSETS/FX1.JPG.

[35] Y. Wang, O. Cope, J. Chen, A. Mehta, D. Fleifel, C.G. Ford, P. Benhamie, S. Haase, S. Gulec, T. Elston, P.M. Spanheimer, C.A. Tomblin, A.M. Rojas, T. Tate, J. Purvis, J. Wang, J.M. Dahl, S. Wolff, J.G. Cook, E. Brunk, Extrachromosomal DNA gives cancer a new evolutionary pathway, BioRxiv (2025) 2025.04.26.650733. https://doi.org/10.1101/2025.04.26.650733.

[36] E. Fiorini, A. Malinova, D. Schreyer, D. Pasini, M. Bevere, G. Alessio, D. Rosa, S. D'Agosto, L. Azzolin, S. Milite, S. Andreani, F. Lupo, L. Veghini, S. Grimaldi, S. Pedron, M. Castellucci, C. Nourse, R. Salvia, G. Malleo, A. Ruzzenente, A. Guglielmi, M. Milella, R.T. Lawlor, C. Luchini, A. Agostini, C. Carbone, C. Pilarsky, A. Sottoriva, A. Scarpa, D.A. Tuveson, P. Bailey, V. Corbo, MYC ecDNA promotes intratumour heterogeneity and plasticity in PDAC, Nature (2025). https://doi.org/10.1038/S41586-025-08721-9.

[37] K. Turner, H. Hemmati, C. DuPai, J. Bibay, C. Klebs, J. Chen, D. Liao, C. Hassig, S. Kasibhatla, Abstract B092: Intrinsic genomic plasticity of extrachromosomal DNA (ecDNA) enables oncogene amplified tumor cells to develop rapid acquired resistance to targeted therapy, Mol Cancer Ther 22 (2023) B092. https://doi.org/10.1158/1535-7163.TARG-23-B092.

[38] Extrachromosomal DNA carrying MYC drives pancreatic cancer aggression, (n.d.). https://www.genengnews.com/topics/cancer/extrachromosomal-dna-carrying-myc-drives-pancreatic-cancer-aggression/ (accessed May 19, 2025).

[39] S. Wang, C.Y. Wu, M.M. He, J.X. Yong, Y.X. Chen, L.M. Qian, J.L. Zhang, Z.L. Zeng, R.H. Xu, F. Wang, Q. Zhao, Machine learning-based extrachromosomal DNA identification in large-scale cohorts reveals its clinical implications in cancer, Nat Commun 15 (2024) 1–17. https://doi.org/10.1038/s41467-024-45479-6.

[40] A. Passaro, M. Al Bakir, E.G. Hamilton, M. Diehn, F. André, S. Roy-Chowdhuri, G. Mountzios, I.I. Wistuba, C. Swanton, S. Peters, Cancer biomarkers—emerging trends and clinical implications for personalized treatment, Cell 187 (2024) 1617. https://doi.org/10.1016/J.CELL.2024.02.041.

[41] X. Luo, L. Zhang, J. Cui, Q. An, H. Li, Z. Zhang, G. Sun, W. Huang, Y. Li, C. Li, W. Jia, L. Zou, G. Zhao, F. Xiao, Small extrachromosomal circular DNAs as biomarkers for multi-cancer diagnosis and monitoring, Clin Transl Med 13 (2023). https://doi.org/10.1002/CTM2.1393.

[42] S.U. Rehman, The multifaceted role of extrachromosomal DNA in cancer: Functions and clinical implications, Bangladesh J Med 36 (2025) 82–91. https://doi.org/10.3329/BJM.V36I2.79673.

[43] J. Tang, N.E. Weiser, G. Wang, S. Chowdhry, E.J. Curtis, Y. Zhao, I.T. Lo Wong, G.K. Marinov, R. Li, P. Hanoian, E. Tse, S.G. Mojica, R. Hansen, J. Plum, A. Steffy, S. Milutinovic, S.T. Meyer, J. Luebeck, Y. Wang, S. Zhang, N. Altemose, C. Curtis, W.J. Greenleaf, V. Bafna, S.J. Benkovic, A.B. Pinkerton, S. Kasibhatla, C.A. Hassig, P.S. Mischel, H.Y. Chang, Enhancing transcription–replication conflict targets ecDNA-positive cancers, Nature 635 (2024) 210–218. https://doi.org/10.1038/s41586-024-07802-5.

[44] H. Luo, J. Shan, H. Zhang, G. Song, Q. Li, C.X. Xu, Targeting the epigenetic processes to enhance antitumor immunity in small cell lung cancer, Semin Cancer Biol 86 (2022) 960–970. https://doi.org/10.1016/J.SEMCANCER.2022.02.018.

[45] Z. Liu, Y. Ren, S. Weng, H. Xu, L. Li, X. Han, A new trend in cancer treatment: The combination of epigenetics and immunotherapy, Front Immunol 13 (2022) 809761. https://doi.org/10.3389/FIMMU.2022.809761.

[46] C. Lin, Y. Chen, F. Zhang, B. Liu, C. Xie, Y. Song, Encoding gene RAB3B exists in linear chromosomal and circular extrachromosomal DNA and contributes to cisplatin resistance of hypopharyngeal squamous cell carcinoma via inducing autophagy, Cell Death Dis 13 (2022) 1–10. https://doi.org/10.1038/s41419-022-04627-w.

[47] Y. Dong, Q. He, X. Chen, F. Yang, L. He, Y. Zheng, Extrachromosomal DNA (ecDNA) in cancer: Mechanisms, functions, and clinical implications, Front Oncol 13 (2023) 1194405. https://doi.org/10.3389/FONC.2023.1194405.

[48] K.M. Turner, V. Deshpande, D. Beyter, T. Koga, J. Rusert, C. Lee, B. Li, K. Arden, B. Ren, D.A. Nathanson, H.I. Kornblum, M.D. Taylor, S. Kaushal, W.K. Cavenee, R. Wechsler-Reya, F.B. Furnari, S.R. Vandenberg, P.N. Rao, G.M. Wahl, V. Bafna, P.S. Mischel, Extrachromosomal oncogene amplification drives tumour evolution and genetic heterogeneity, Nature 543 (2017) 122–125. https://doi.org/10.1038/NATURE21356;SUBJMETA=631,67,68;KWRD=CANCER,CANCER+GENETICS.

[49] J.H. Christiansen, N. Nguyen, R. Verhaak, H. Kim, V. Bafna, P. Mischel, C. Hassig, Extrachromosomal DNA (ecDNA) carrying amplified oncogenes as a biomarker for insensitivity to pembrolizumab treatment in gastric cancer patients., J Clin Oncol 38 (2020) 3123. https://doi.org/10.1200/JCO.2020.38.15_SUPPL.3123.

[50] J. Tang, N.E. Weiser, G. Wang, S. Chowdhry, E.J. Curtis, Y. Zhao, I.T. Lo Wong, G.K. Marinov, R. Li, P. Hanoian, E. Tse, S.G. Mojica, R. Hansen, J. Plum, A. Steffy, S. Milutinovic, S.T. Meyer, J. Luebeck, Y. Wang, S. Zhang, N. Altemose, C. Curtis, W.J. Greenleaf, V. Bafna, S.J. Benkovic, A.B. Pinkerton, S. Kasibhatla, C.A. Hassig, P.S. Mischel, H.Y. Chang, Enhancing transcription-replication conflict targets ecDNA-positive cancers, Nature 635 (2024) 210–218. https://doi.org/10.1038/S41586-024-07802-5.

[51] Y. Yan, G. Guo, J. Huang, M. Gao, Q. Zhu, S. Zeng, Z. Gong, Z. Xu, Current understanding of extrachromosomal circular DNA in cancer pathogenesis and therapeutic resistance, J Hematol Oncol 13 (2020) 1–16. https://doi.org/10.1186/S13045-020-00960-9/TABLES/3.

11 The Role of ecDNA in Tumor Heterogeneity and Evolution

Syed Mustafizur Rahaman,[1] *Misbaul Hoque,*[2]
Saba Wahid A.M. Khan,[3] *Radheshyam Pal,*[1] *Shreyasi Katari,*[4]
Mirazuddin Mollick,[5] *and Md. Yousuf Ansari*[5]

1 Bengal College of Pharmaceutical Sciences and Research
Durgapur, India
2 Gitanjali College of Pharmacy
Lohapur, Birbhum, India
3 H.K. College of Pharmacy
Department of Pharmaceutics
Jogeshwari West, Mumbai, India
4 Derozio Pharma institute Moyna
Purba Medinipur, India
5 College of Pharmacy, Maharishi Markandeshwar
(Deemed to be University)
Mullana, India

11.1 INTRODUCTION

11.1.1 Overview of Tumor Heterogeneity and Its Significance in Cancer Progression

Tumor heterogeneity describes the presence of multiple subpopulations of cancer cells within one tumor, or among multiple tumors in a single patient. These tumor cells can differ greatly with respect to their genotype (e.g., mutations, copy number variations), epigenotype (e.g., DNA methylation, histone modifications), phenotype (e.g., growth rates, metabolic activity, cell surface markers), and osteo-biological properties (e.g., metastatic potential, drug resistance, immune recognition) [1]. Tumor heterogeneity significantly influences the trajectory of cancer development, progression, and treatment response. A single tumor can exhibit significant variability among cancer cells regarding their genetic composition, behavior, and response to treatments. This variation is brought about by genetic changes, epigenetic modifications, environmental factors present in the tumor microenvironment, and selective pressures from therapies [2]. A major consequence of tumor heterogeneity is its influence on resistance to treatment. Therapies like chemotherapy, targeted therapy, or immunotherapy can work well on specific subsets of cancer cells, but some may naturally have or develop resistance mechanisms. These resistant cells endure the initial therapy, keep multiplying, and ultimately take ovcr the tumor mass, resulting in treatment failure and disease recurrence. This dynamic makes long-term cancer management much more complex and requires strategies that can tackle the entire spectrum of tumor cell variants [3]. Another effect of tumor heterogeneity is its impact on metastasis, the mechanism through which cancer disseminates from the primary location to different areas of the body. In a diverse tumor, certain subclones may have particular characteristics that enable them to penetrate adjacent tissue, access the bloodstream or lymphatic system, and establish colonies in remote organs. These subpopulations tend to be more aggressive and might evade detection or effective targeting in the early stages of treatment, leading to worse outcomes [4]. The advancement of the disease is significantly affected

DOI: 10.1201/9781003655299-11

by the existence of various subclones. As the tumor develops, some clones might acquire competitive benefits that enhance quicker growth, invasion, and resilience to immune responses. This clonal evolution not only contributes to the complexity of the disease but also leads to a dynamic landscape that evolves over time, frequently becoming harder to treat [5]. Tumor heterogeneity can also undermine diagnostic accuracy. When tissue samples are collected for biopsy and examination, they might not encompass the entire range of cellular variation present in the tumor. Consequently, essential subclones with clinical relevance—like those linked to drug resistance or metastasis—may be missed. This partial depiction of the tumor can confuse clinicians and result in less effective treatment choices [6]. In personalized medicine, it is crucial to recognize and consider tumor heterogeneity. Precision oncology seeks to customize therapies based on the distinctive genetic and molecular characteristics of a person's cancer. If only part of the tumor's diversity is taken into account, treatments might overlook subclones that fuel recurrence or progression. Thus, a thorough grasp of the diverse characteristics of tumours is essential for creating multi-targeted, adaptable, and more lasting treatment strategies that can efficiently address the intricacies of cancer [7].

11.1.2 Definition and Characteristics of Extrachromosomal DNA (ecDNA)

Extrachromosomal DNA (ecDNA) denotes circular DNA strands located in the cell nucleus that exist independently from the linear chromosomes. Although they are found in normal cells, they are especially common and heightened in cancer cells, significantly contributing to gene amplification and rapid evolution [8]. Extrachromosomal DNA (ecDNA) significantly contributes to cancer biology, especially in enhancing tumor aggressiveness, adaptability, and resistance to treatment. A key characteristic is its circular formation, which sets it apart from the linear chromosomes present in the nucleus. This circular structure not only provides structural integrity but also facilitates effective replication, aiding in its durability and swift expansion within cancerous cells [9]. An important feature of ecDNA is its episomal characteristic. In contrast to chromosomal DNA that is embedded in the genome and adheres to the standard cell cycle for replication, ecDNA replicates autonomously. It does not depend on chromosomal mechanisms for its replication and can thus experience fluctuations in copy number more flexibly. This independence offers a considerable evolutionary benefit, allowing cells to swiftly modify the levels of genes present on ecDNA in response to environmental stresses [10]. A major consequence of ecDNA in cancer is its ability to exist in very high copy numbers. Oncogenes found on these circular DNA structures can be greatly amplified—occasionally attaining hundreds or thousands of copies in one cell. This degree of amplification significantly boosts the activity of these cancer-promoting genes, fostering aggressive tumor behavior and speeding up growth. The enhancement of oncogenes is a key functional role of ecDNA in cancer. Genes like EGFR, MYC, and MDM2, commonly linked to cell growth and survival, are often located on ecDNA. The elevated expression of these genes leads to swift tumor advancement, heightened malignancy, and resistance to conventional treatments. Since these genes can be increased significantly beyond chromosomal region boundaries, ecDNA acts as a potent mechanism for oncogenic amplification [11]. The actions of ecDNA in cell division significantly influence tumor development. In contrast to chromosomes, which are accurately divided because of centromeres, ecDNA frequently does not have these elements, resulting in haphazard and unequal allocation between daughter cells. This uneven distribution creates significant variability within the tumor, enabling subpopulations of cancer cells to form unique genetic profiles. This mechanism enables quick adjustment and the formation of various cell clones, each possessing distinct survival benefits in varying circumstances, such as drug exposure [12]. Regarding cellular localization, ecDNA is typically located in the nucleus, but its physical properties during cell division vary from those of chromosomal DNA. Due to its lack of adherence to the rigid principles of mitotic segregation, it plays a role in the genetic instability and diversity that characterize aggressive cancers [13]. A notably concerning outcome of ecDNA in tumors is its involvement in facilitating drug resistance. Cancer cells can utilize ecDNA to enhance genes that counteract the effects of treatment agents. This swift upregulation enables cells to endure in the presence of drugs that would typically be effective, resulting in treatment failure. Due to its capacity to rapidly and autonomously increase gene copy number, ecDNA allows tumors to adapt swiftly, presenting a significant obstacle to enduring cancer management [14]. In total, the distinct characteristics of ecDNA—such as its configuration, replication patterns, and involvement in oncogene amplification—render it a significant contributor to cancer advancement and a challenging obstacle to effective treatment. Grasping its biology is crucial for the creation of precise therapies that can mitigate its impacts [15].

11.2 ECDNA AND ONCOGENE AMPLIFICATION

11.2.1 Role of ecDNA in Harboring and Amplifying Oncogenes (e.g., MYC, EGFR, MDM2)

ecDNA can harbor and amplify oncogenes more rapidly than chromosomal amplification and rapidly improves tumor cell growth. The circular structure of ecDNA makes it more approachable to the transcriptional machinery. Extrachromosomal DNA contributes to the progression of cancer and drug resistance. The prevalence of ecDNA rose after treatments like chemotherapy and correlated with metastasis and worse patient outcomes. In a normal human cell, the nucleus contains 23 pairs of chromosomes that enclose the DNA [16]. In cancer patients, chromosomal rearrangements can lead to the formation of extrachromosomal DNA (ecDNA). During this process, fragments of various chromosomes break apart and circularize to form ecDNA, which often clusters into hubs. These hubs are enriched with genes, particularly oncogenes, that drive tumor progression. The ecDNA carrying oncogenes can be actively transcribed into mRNA, leading to high levels of oncogene expression. This phenomenon, known as oncogene amplification, promotes rapid cancer growth and contributes to therapy resistance [17].

MYC, EGFR, MDM2, and other various factors are present in ecDNA that mainly play a role in harboring and amplifying oncogenes. MYC is a key regulator of cell growth and metabolism. Extrachromosomal DNA (ecDNA) can harbor the MYC gene, leading to the production of high copy numbers of this oncogene. Such amplification enhances MYC expression, which promotes uncontrolled tumor cell proliferation and survival. EGFR, also called epidermal growth factor receptor, is a type of factor amplified in ecDNA. EGFR transfers the information to mRNA, and then cancer spreads easily. This is called amplification [18]. EGFR promotes uncontrolled cell proliferation and survival signaling. MDM2 is also a type of oncogene that controls the P53; MDM2 helps to impair the tumor suppressor activity. MDM2 contributes the tumor formation. Amplification of MDM2 and other oncogenes has increased drug resistance in cancer therapies [19].

11.2.2 Impact on Gene Expression Regulation

The ecDNA has the property to alter the gene expression regulation in the cancer cells. Extrachromosomal DNA (ecDNA) acts as a switch that enables non-linear, and hyperactive transcriptional responses in gene expression regulation. This unique mechanism alters normal transcriptional control and contributes to tumor progression. Among its impacts, ecDNA drives abnormally high levels of oncogene expression, which plays a central role in cancer development and maintenance [20]. The oncogenes like MYC, EGFR, and MDM2 are the main factors regulating gene expression. The gene expression alteration can raise the chromatin accessibility by activating the histone marks and by repressive marks. Nuclear localization is another impact of gene expression regulation in the transcriptional hubs. They concentrate the transcription factors and RNA polymerase. Some other impacts are dynamic and unequal distribution and functional consequences like oncogene addiction, therapy evasion, and tumor aggressiveness [21].

11.2.3 Enhancer Hijacking and Epigenetic Modulation

Enhancer hijacking is a process where stronger progressive elements change their locations onto ecDNA. This process slowly activates oncogene expression. The ecDNAs are circular in nature.

So the inner part contains oncogenes, and the distance enhancers are arranged closely. There are also some abnormal arrangements in the Cis-regulatory elements enhancer present in the linear genome oncogenes [22]. MYC ecDNA is the overexpression in neuroblastoma and glioblastoma due to its chromosomal structure nature. The hijacking mechanism can depend on the formation of the ecDNA, such as chromothripsis and DNA looping. The linear genomes are not able to participate in the hijacking part, because they do not promote the de novo enhancer oncogene interactions [23].

To boost the facilitating genes expression ecDNA can perform other properties like epigenetic landscape. By following some extra features, the ecDNA can exhibit properties like hypomethylation, active chromatin marks, and absence of repressive marks. By enhancing hijacking and the epigenetic modulation of the oncogene addiction, tumor heterogeneity and drug resistance are raising their properties of work nature [24].

11.3 CONTRIBUTION OF ECDNA TO TUMOR HETEROGENEITY

11.3.1 Prevalence of ecDNA in Human Cancers

Recent large-scale genomic analyses have demonstrated that extrachromosomal DNA (ecDNA) is a common feature across a wide range of human solid tumors, rather than being a rare or incidental finding. A pivotal study by Turner and colleagues in 2017 analyzed over 2,500 tumor samples and found that nearly 40% contained ecDNA segments enriched with oncogene amplifications. Notably, in aggressive malignancies like glioblastoma, ecDNA was detected in up to 90% of cases, with genes such as EGFR and MYC preferentially amplified on ecDNA rather than on chromosomal loci [25]. Similarly, in high-risk neuroblastomas, amplification of MYCN was frequently observed on ecDNA structures. Additional cancers, including those of the lung, esophagus, and ovary, frequently showed ecDNA-based amplification of oncogenes like MET, ERBB2, and MDM2. Unlike traditional amplifications confined to homogeneously staining regions (HSRs), these ecDNA-associated amplifications represent a distinct genomic architecture. Further supporting its clinical relevance, Kim et al. (2020) demonstrated that ecDNA presence correlates with increased focal amplification, greater mutational complexity, and poorer patient prognosis.

11.3.2 Key Differences between ecDNA and Chromosomal Amplifications

Although both ecDNA and chromosomal DNA can harbor amplified oncogenes, they differ significantly in their biological behavior and impact. Chromosomal amplifications typically follow stable, Mendelian patterns of inheritance and are subject to cell cycle-regulated replication. In contrast, ecDNA operates outside these constraints. It replicates independently and is segregated randomly during cell division, allowing for unequal distribution among daughter cells. This non-Mendelian inheritance provides ecDNA with the capacity to rapidly amplify oncogenic signals and adapt dynamically within the tumor microenvironment, giving cancer cells a potential evolutionary edge. Additionally, chromatin architecture on ecDNA is more accessible, enriched with active histone modifications like H3K27ac, enabling hyper-transcription of embedded oncogenes [26]. Studies leveraging ATAC-seq and ChIP-seq have demonstrated that oncogenes on ecDNA exhibit transcriptional activity up to 50-fold higher than their chromosomal counterparts with similar copy numbers, often through enhancer hijacking or 3D architectural rewiring.

11.3.3 ecDNA as a Catalyst of Tumor Heterogeneity

The random and unequal distribution of ecDNA during cell division leads to fluctuations in its copy number among tumor cells, resulting in a genetically diverse tumor population. ecDNA forms spatial clusters "ecDNA hubs"—that colocalize with RNA polymerase II and BRD4, leading to focal regions of hypertranscriptional [27]. The chromatin structure of ecDNA differs markedly from the host genome, favoring open configurations that enhance gene expression.

ecDNA-driven variability alters responses to therapy and immune surveillance, supporting transient phenotypes like drug tolerance. These features align with clonal evolution models, including branching evolution and punctuated equilibrium, where ecDNA plays a critical role in enabling rapid adaptation to therapeutic pressure [28].

11.3.4 Clinical Relevance of ecDNA

The clinical consequences of ecDNA presence are profound. Patients with ecDNA-positive tumors generally experience accelerated disease progression and lower overall survival. Analyses of TCGA datasets indicate a significant correlation between ecDNA and poor prognosis across multiple cancer types [29]. In glioblastoma, ecDNA-amplified EGFR has been associated with earlier relapse and resistance to temozolomide [30], while MYCN ecDNA defines aggressive neuroblastoma requiring intensive treatment [31]. Moreover, the dynamic nature of ecDNA promotes resistance through the selection of subclones harboring distinct resistance-conferring ecDNAs. When one subclone is suppressed by therapy, others may expand, undermining long-term treatment efficacy.

11.3.5 Techniques for Detecting and Visualizing ecDNA

Recent advancements have greatly improved ecDNA detection: Circle-seq and AmpliconArchitect identify circular DNA from short-read data. Hi-C and ATAC-seq reveal regulatory topologies and chromatin accessibility. Live-cell imaging and DNA-FISH allow direct visualization of ecDNA dynamics. Hung et al. (2021) utilized live imaging to observe ecDNA hubs interacting with transcriptional machinery, demonstrating their role in maintaining oncogene expression heterogeneity even in monoclonal cell populations [32].

11.3.6 ecDNA as a Driver of Transcriptional Diversity

11.3.6.1 ecDNA Variability and Intratumoral Divergence

The atypical, non-Mendelian inheritance of extrachromosomal DNA (ecDNA) serves as a powerful driver of intratumoral genetic variation [33]. Research by Turner et al. (2017) demonstrated that glioblastoma cells can eliminate and later reacquire ecDNA carrying the EGFR oncogene following cessation of targeted therapy, highlighting the reversible and adaptable nature of ecDNA-mediated gene amplification [34]. This capacity for rapid modulation fosters the emergence of diverse tumor cell subgroups, each characterized by distinct oncogene copy numbers and transcriptional outputs, thereby enhancing the tumor's evolutionary potential.

11.3.6.2 Mechanistic Basis of ecDNA-Driven Evolution

11.3.6.2.1 Asymmetric Distribution and Clonal Diversification

Because ecDNAs lack centromeres, their segregation during cell division is inherently random, resulting in unequal inheritance among daughter cells. This irregular distribution fuels the formation of heterogeneous clones. Notably, glioblastoma tumors analyzed post-therapy exhibit an increase in ecDNA-driven clonal complexity, especially in populations enriched for EGFR amplification [34, 35].

11.3.6.2.2 Simultaneous Amplification of Oncogenes and Regulatory Sequences

ecDNAs frequently co-amplify oncogenes along with their regulatory elements, such as enhancers and super-enhancers. In malignancies like medulloblastoma and neuroblastoma, ecDNAs harboring MYC or MYCN are often loaded with these powerful regulatory regions, leading to abnormally high gene expression levels and aggressive tumor behavior [36, 37].

11.3.6.2.3 Therapy-Induced Clonal Expansion

Targeted therapy can selectively enrich for ecDNA-bearing clones. For instance, MET-amplified ecDNA in non-small cell lung cancer (NSCLC) mediates resistance to MET inhibitors, and EGFRvIII ecDNA reappears in glioblastoma upon cessation of EGFR-targeted treatment [38].

11.3.6.3 Clinical Implications of ecDNA-Driven Evolution

11.3.6.3.1 Diagnostic Prognostic

Due to its non-uniform distribution, ecDNA may escape detection in small biopsies, posing challenges for molecular diagnostics. In neuroblastoma, MYCN ecDNA may be abundant in tumor cores but undetectable in peripherally sampled regions [39].

11.3.6.3.2 Prognostic Significance

Multiple clinical datasets indicate that ecDNA presence correlates with poorer outcomes, including in glioblastoma and colorectal cancer, where EGFR and MYC amplification via ecDNA predicts shorter survival [36, 40].

11.3.6.3.3 Significance of Therapies

The adaptability of ecDNA under treatment pressure fosters drug resistance. Clones carrying ecDNA-encoded resistance genes often dominate during therapy and recede afterward, re-establishing heterogeneity [40, 41].

11.3.6.3.4 Integration in Oncology

Ongoing trials, such as NCT05155661, are evaluating ecDNA's utility as a biomarker in NSCLC [42]. Therapeutic approaches targeting ecDNA through CRISPR, replication inhibitors, or epigenetic drugs are also being developed [43].

11.3.6.4 Spatial and Temporal Contributions to Heterogeneity

Single-cell sequencing of glioblastoma tumors has demonstrated that ecDNA-bearing clones tend to cluster in specific spatial niches, often at invasive tumor edges, suggesting a role in metastasis. Over time, therapy reshapes ecDNA landscapes, with certain clones emerging or disappearing based on selective pressure [44, 45].

11.3.7 ecDNA as a Driver of Transcriptional Diversity and Spatial-Temporal Heterogeneity

11.3.7.1 ecDNA as a Driver of Transcriptional Diversity

Extrachromosomal DNA (ecDNA) contributes significantly to the complex transcriptional heterogeneity observed in cancer. Unlike chromosomal amplification, ecDNA lacks centromeres and telomeres and adopts a circular configuration, which not only allows for high oncogene copy number but also facilitates unique regulatory interactions that underpin tumor cell diversity (Table 11.1) [46–51].

11.3.7.2 Topological Features and Transcriptional Hyperactivity

ecDNA molecules tend to organize into nuclear hubs enriched in transcriptional machinery. These ecDNA hubs recruit coactivators such as BRD4, RNA Polymerase II, and the Mediator complex, enabling robust transcriptional output. Hoon et al. (2023) demonstrated using Hi-C and ATAC-seq that EGFR-ecDNA in glioblastoma formed transcriptionally active domains with enriched BRD4 binding. Pharmacological inhibition of BRD4 via BET inhibitors (e.g., JQ1) selectively suppressed ecDNA-driven transcription while sparing chromosomal EGFR expression [46].

11.3.7.3 Enhancer Co-Amplification and Trans-Regulatory Activity

The circular nature of ecDNA enables co-amplification of oncogenes with distal enhancers, including super-enhancers located on the same or different chromosomes. In neuroblastoma, MYCN amplification on ecDNA was found to include super-enhancers located hundreds of kilobases away, resulting in transcriptional activation far beyond what linear amplifications achieve. This enhancer hijacking not only amplifies gene expression but enables transcriptional programming unique to ecDNA-bearing clones [47].

11.3.7.4 Epigenetic Adaptability and Transcriptional Switching

ecDNA exhibits flexible chromatin architecture with increased accessibility, low methylation, and active histone modifications, such as H3K27ac. This plasticity allows rapid transcriptional adjustments in response to environmental stressors or therapy. In lung adenocarcinoma models, MET-amplified ecDNA displayed reversible chromatin remodeling under TKI treatment, becoming transcriptionally silent during drug exposure and reactivating upon withdrawal [48]. Such epigenetic plasticity provides a non-genetic basis for therapy resistance and tumor cell survival.

11.3.8 Spatial and Temporal Heterogeneity Induced by ecDNA

ecDNA not only drives transcriptional variance but also contributes to intratumoral heterogeneity across space and time, influencing disease progression and therapy outcomes.

11.3.8.1 Spatial Distribution of ecDNA-Positive Clones

Intratumoral ecDNA localization is often non-random, with enriched distribution in invasive margins or perivascular zones. This spatial bias reflects both microenvironmental selection and proliferative advantages conferred by ecDNA-encoded oncogenes. Using single-cell FISH and transcriptomics in GBM, Flavahan et al. (2023) observed ecDNA-encoded EGFR concentrated in periphery-invading cells, correlating with angiogenesis and tumor dissemination [49]. Such spatial compartmentalization can lead to sampling bias during biopsy, potentially underestimating tumor aggressiveness.

11.3.8.2 Temporal Dynamics and Clonal Evolution

ecDNA populations within tumors exhibit dynamic behavior over time. Under therapeutic pressure, ecDNA-positive subclones may either expand or diminish depending on selective forces. For example, MET or EGFR ecDNA levels fluctuate in NSCLC patients undergoing TKI therapy. Serial biopsies reveal loss of ecDNA under drug pressure, followed by reappearance at relapse, underscoring the reversible nature of ecDNA-mediated resistance [50].

11.3.8.3 Clinical Impact of ecDNA-Mediated Heterogeneity

The spatial and temporal variability introduced by ecDNA complicates diagnosis and treatment. Static biopsies may fail to detect aggressive ecDNA-bearing clones. Furthermore, transient ecDNA loss can enable tumor cells to escape immune surveillance or targeted therapy, only to regain oncogene expression later. Techniques such as ecDNA-specific FISH, plasma-based ecDNA monitoring, and ctDNA sequencing now provide real-time tracking of ecDNA evolution in patients (Table 11.1) [51].

11.3.9 Therapeutic and Diagnostic Implications

Appreciating the role of ecDNA in driving heterogeneity is essential for developing adaptive precision oncology strategies: High ecDNA burden correlates with treatment resistance and poor survival. Detection in tissue or plasma offers a non-invasive prognostic tool. Compounds targeting transcriptional hubs (e.g., BET and CDK7 inhibitors) selectively impair ecDNA-dependent oncogene expression. Genome editing tools like CRISPR/Cas or DNA-damaging agents may disrupt ecDNA maintenance, preventing its reactivation post-therapy.

Clinical studies are currently integrating ecDNA monitoring for patient management:

TABLE 11.1
Mechanisms by Which ecDNA Drives Transcriptional and Spatial-Temporal Heterogeneity

Mechanism	Outcome	Tumor Type	Example Genes	References
ecDNA Transcriptional Hubs	Oncogene hyperactivation	GBM, NSCLC	EGFR, MYC	[46, 47]
Enhancer Hijacking	Aberrant gene regulation	Neuroblastoma	MYCN	[47]
Epigenetic Plasticity	Reversible gene expression	Lung Adenocarcinoma	MET, ALK	[48]
Spatial ecDNA Clustering	Invasive tumor fronts	GBM	EGFR	[49]
Temporal ecDNA Dynamics	Resistance & relapse	NSCLC, GBM	MET, EGFRvIII	[50, 51]

11.4 ECDNA AND TUMOR EVOLUTION

11.4.1 Role of Extrachromosomal DNA (ecDNA) in Adaptive Evolution under Selective Pressures in Adaptive Evolution and Therapy Resistance

Cancer emergence, progression, and adaptation have long been understood through the lens of classical Darwinian principles—inheritance, variation, and selection [52–55].

A crucial player in this evolutionary dynamic is ecDNA, which carries oncogenes, is randomly inherited between daughter cells during mitosis due to the lack of a centromere, unlike chromosomal DNA. This random inheritance of ecDNA is a key contributing factor to intratumoral heterogeneity and accelerated tumor evolution [56].

11.4.1.1 ecDNA Properties Enabling Adaptive Evolution

11.4.1.1.1 Random Segregation of ecDNA

Multiple findings confirm the random inheritance pattern of ecDNA [56].

Lange et al. (2020) reported a Gaussian distribution of ecDNA across daughter cells using a FISH-based method combined with image analysis, supporting the hypothesis of random segregation [57].

This observation was further validated using cancer cell line models and a live-cell imaging system, where CRISPR-Cas9 was employed to insert a TetO array into the intergenic region between MYC and PVT1 on ecDNA in PC3 prostate cancer cells [58].

The ecDNA was tagged with TetR-GFP and tracked during mitosis. Time-lapse imaging of these PC3-TetO cells revealed a distinct pattern of random inheritance, reinforcing the role of ecDNA in promoting variability [59].

These experiments conclusively demonstrated that ecDNA is randomly segregated during cell division, contributing to tumor heterogeneity through Darwinian selection [56].

11.4.1.1.2 ecDNA Boosts Oncogene Expression

Studies using RNA techniques (like allele-specific mRNA analysis) showed that extra gene expression was directly coming from ecDNA, not chromosomes [60].

In various cancers, like glioblastoma and neuroblastoma, ecDNA is commonly found and often carries oncogenes, including MYC, EGFR, PDGFRA, MET, CDK4, and MDM2, which enhance oncogene expression primarily by increasing the gene copy number through focal amplification [61]. Furthermore, higher ecDNA copy numbers were associated with greater oncogene expression [62]. ATAC-seq studies confirmed ecDNA has looser chromatin than chromosomes, making it easier for transcription machinery to access genes [63, 64].

Both increased copy number and looser chromatin on ecDNA lead to higher oncogene expression, fueling cancer growth [63].

11.4.1.1.3 ecDNA Serves as a Reservoir for DNA Recombination

In cancer, ecDNA not only transports oncogenes, but it also serves as a key source of DNA recombination and genomic rearrangement. Studies show that ecDNA can be formed in several ways, such as from extrachromosomal circular elements [64] and by cutting and rejoining DNA segments (excision or ligation) from other ecDNA [65]. Importantly, ecDNA can reintegrate back into chromosomes, leading to Intra-chromosomal (within the same chromosome) rearrangements.

Inter-chromosomal (between different chromosomes) rearrangements [60]. For example, in neuroblastoma and glioblastoma, ecDNA reintegration has been associated with structural genomic alterations and poor clinical outcomes. These rearrangements frequently retain the original ecDNA structure, implying dynamic migration between ecDNA and chromosomal DNA [66]. WGS and Circos plots reveal that numerous chromosomal breakpoints correspond to ecDNA circularization locations [67].

Overall, ecDNA serves as a recombination reservoir, facilitating tumor development and therapeutic resistance.

11.4.1.2 Rapid Tumor Adaptation to Therapy and Stress

Extrachromosomal DNA (ecDNA) enables rapid tumor adaptation to therapeutic and environmental stress. Various findings show that ecDNA supports pre-existing variation, enabling fast selection of resistant clones under drug pressure.

11.4.1.2.1 Rapid Adaptation to Environmental Stress

A discovery by Joshua et al. (2022) highlights the selective growth advantage that ecDNA confers under stressful conditions. The study compared two isogenic glioblastoma (GBM) models: GBM39-EC, which carries approximately 100 copies of EGFRvIII on ecDNA, and GBM39-HSR, which contains the same EGFRvIII copies integrated into chromosomal homogeneously staining regions (HSRs).

Under glucose restriction, GBM39-EC cells sustained growth by rapidly decreasing their ecDNA copy number. In contrast, GBM39-HSR cells, with the same gene amplified on chromosomes, exhibited extreme sensitivity and lacked the capacity to adapt.

This genomic shift in ecDNA occurred within just a few cell cycles, demonstrating ecDNA's unique ability to undergo rapid, reversible changes unlike stable chromosomal alterations [68].

11.4.1.2.2 Rapid Drug Resistance Development

GBM39-EC cells developed resistance to the EGFR inhibitor erlotinib within just two weeks by reversibly decreasing their ecDNA copy number. In contrast, GBM39-HSR cells failed to adapt and remained sensitive to the drug.

This rapid, flexible resistance was further confirmed in tumor samples from glioblastoma patients treated with EGFR inhibitors (erlotinib), which also showed a reduction in EGFR ecDNA copy number, closely reflecting the in vitro findings [68].

Supporting this, a study by Hung et al. (2024) demonstrated that pharmacological stress—such as FGFR2 inhibition using pemigatinib—can induce rapid and coordinated ecDNA copy number changes. This enables tumor cells to adapt swiftly through ecDNA co-segregation and dynamic amplification [69].

11.4.2 Influence on Genomic Instability and Mutation Rate

The combination of random segregation [70], chromothripsis, kataegis, and BFB cycles makes ecDNA hotspots for mutation accumulation and structural variation that support rapid tumor evolution. Uneven ecDNA segregation leads to unequal genetic material distribution, resulting in a fast increase in oncogene copy number and generating extensive intratumoral heterogeneity, which fuels clonal evolution under selective pressure [71].

Structural genomic instability is further amplified by chromothripsis, a condition in which chromosomes break and rejoin chaotically, which leads to circularized ecDNA carrying oncogene-rich fragments [72]. ecDNA is not just amplified but also mutationally diverse, and kataegis contributes to this variability [73]. Additionally, ecDNA formation often involves breakage–fusion–bridge

(BFB) cycles, resulting in secondary breaking and fragmentation, yielding eccDNA by chromosomal breakage, replication, and ligation, resulting in fused sister chromatids. This process increases oncogenes and causes genome instability [74]. In longitudinal studies, ecDNAs carrying clustered mutations are more likely to persist after therapy, underscoring their role as dynamic genomic elements that accelerate mutation rates and contribute to therapeutic resistance [75].

11.5 FUTURE PERSPECTIVES AND RESEARCH DIRECTIONS

11.5.1 Technological Advancements Needed for ecDNA Studies

ecDNA is one of the most fabulous discoveries in the recent history of tumor research. The ecDNA has many extra features to develop tumors and the formation of malignant phenotypes. The high copy number of oncogenes and the high expression levels can improve the ecDNA as a great discovery. So, from the recent study, the ecDNA's formation, mechanism, and role in tumor progression are very effective in acting against the tumor pathogens in the future. Also, the ecDNA is able to diagnose and treat tumors in the future. But there are some drawbacks and limitations with the ecDNAs. The lack of centrosomes, raising the heterogeneity during mitosis because of the bad distribution of ecDNA, can create problems in the diagnosis of tumors in the future. In recent years, though some problems have been resolved, many remain. The primary mechanism of ecDNA cannot develop the treatment. There are no suppressor genes and normal genes in the ecDNA, and if they are present, no function will be found. There is no replication mechanism and no ecDNA elimination process. But the progressive site of ecDNA is very enlightening. Current detection and mapping techniques—such as high-resolution sequencing, long-read sequencing, and single-cell sequencing—have advanced the study of ecDNA. However, further research is required to overcome existing limitations and drawbacks in accurately characterizing ecDNA.

11.6 CONCLUSION

ecDNA plays a pivotal role in driving tumor heterogeneity, rapid genetic evolution, and resistance to therapy. Its unique features, including oncogene amplification and uneven segregation, make it a key contributor to cancer aggressiveness. Understanding ecDNA dynamics offers new opportunities for developing targeted therapies and diagnostic tools, marking a significant advancement in precision oncology and cancer treatment strategies.

REFERENCES

1. Liao Z, Jiang W, Ye L, et al. Classification of extrachromosomal circular DNA with a focus on the role of extrachromosomal DNA (ecDNA) in tumor heterogeneity and progression. Biochimica et Biophysica Acta (BBA)-Reviews on Cancer. 2020 Aug 1;1874(1):188392.
2. Pecorino LT, Verhaak RG, Henssen A, et al. Extrachromosomal DNA (ecDNA): An origin of tumor heterogeneity, genomic remodeling, and drug resistance. Biochemical Society Transactions. 2022 Dec 16;50(6):1911–1920.
3. Zeng X, Wan M, Wu J. ecDNA within tumors: A new mechanism that drives tumor heterogeneity and drug resistance. Signal Transduction and Targeted Therapy. 2020 Nov 24;5(1):277.
4. Dong Y, He Q, Chen X, et al. Extrachromosomal DNA (ecDNA) in cancer: Mechanisms, functions, and clinical implications. Frontiers in Oncology. 2023 Jun 28;13:1194405.
5. Wu S, Bafna V, Mischel PS. Extrachromosomal DNA (ecDNA) in cancer pathogenesis. Current Opinion in Genetics & Development. 2021 Feb 1;66:78–82.
6. Wu S, Tao T, Zhang L, et al. Extrachromosomal DNA (ecDNA): Unveiling its role in cancer progression and implications for early detection. Heliyon. 2023 Nov 1;9(11).
7. Ashique S, Upadhyay A, Garg A, et al. Impact of ecDNA: A mechanism that directs tumorigenesis in cancer drug resistance-a review. Chemico-Biological Interactions. 2022 Aug 25;363:110000.

8. Nguyen DD, Hooper WF, Liu W, et al. The interplay of mutagenesis and ecDNA shapes urothelial cancer evolution. Nature. 2024 Nov 7;635(8037):219–228.
9. Dong Y, He Q, Chen X, et al. Extrachromosomal DNA (ecDNA) in cancer: Mechanisms, functions, and clinical implications. Frontiers in Oncology. 2023 Jun 28;13:1194405.
10. Liao Z, Jiang W, Ye L, et al. Classification of extrachromosomal circular DNA with a focus on the role of extrachromosomal DNA (ecDNA) in tumor heterogeneity and progression. Biochimica et Biophysica Acta (BBA)-Reviews on Cancer. 2020 Aug 1;1874(1):188392.
11. Pecorino LT, Verhaak RG, Henssen A, et al. Extrachromosomal DNA (ecDNA): An origin of tumor heterogeneity, genomic remodeling, and drug resistance. Biochemical Society Transactions. 2022 Dec 16;50(6):1911–1920.
12. Lange JT, Chen CY, Pichugin Y, et al. Principles of ecDNA random inheritance drive rapid genome change and therapy resistance in human cancers. BioRxiv. 2021 Jun 11:2021-06.
13. Malinova A, Schreyer D, Fiorini E, et al. ecDNA amplification of MYC drives intratumor copy-number heterogeneity and adaptation to stress in PDAC. bioRxiv. 2023 Sep 29:2023-09.
14. Huang Q, Zhang S, Wang G, et al. Insight on ecDNA-mediated tumorigenesis and drug resistance. Heliyon. 2024 Mar 30;10(6).
15. Purshouse K, Pollard SM, Bickmore WA. Imaging extrachromosomal DNA (ecDNA) in cancer. Histochemistry and Cell Biology. 2024 Jul;162(1):53–64.
16. Pradella D, Zhang M, Gao R, et al. Engineered extrachromosomal oncogene amplifications promote tumorigenesis. Nature. 2025 Jan 23;637(8047):955–964.
17. Yang QL, Xie Y, Qiao K, et al. Modern biology of extrachromosomal DNA: A decade-long voyage of discovery. Cell Research. 2025 Jan 3:1–2.
18. Yan X, Mischel P, Chang H. Extrachromosomal DNA in cancer. Nature Reviews Cancer. 2024 Apr;24(4):261–273.
19. Ng AW, McClurg DP, Wesley B, et al. Disentangling oncogenic amplicons in esophageal adenocarcinoma. Nature Communications. 2024 May 14;15(1):4074.
20. Obermeyer S, Kapoor H, Markusch H, et al. Transcript elongation by RNA polymerase II in plants: Factors, regulation and impact on gene expression. The Plant Journal. 2024 May;118(3):645–656.
21. Ortolá B, Daròs JA. RNA interference in insects: From a natural mechanism of gene expression regulation to a biotechnological crop protection promise. Biology. 2024 Feb 21;13(3):137.
22. Wang X, Yue F. Hijacked enhancer–promoter and silencer–promoter loops in cancer. Current Opinion in Genetics & Development. 2024 Jun 1;86:102199.
23. Liu S, Dai W, Jin B, et al. Effects of super-enhancers in cancer metastasis: Mechanisms and therapeutic targets. Molecular Cancer. 2024 Jun 7;23(1):122.
24. Tang F, Yang Z, Tan Y, et al. Super-enhancer function and its application in cancer targeted therapy. NPJ Precision Oncology. 2020 Feb 12;4(1):2.
25. Kim H, Nguyen NP, Turner KM, et al. Extrachromosomal DNA is associated with oncogene amplification and poor outcome across multiple cancers. Nature Genetics. 2020;52(9):891–897.
26. Wu S, Turner KM, Nguyen N, et al. Circular ecDNA promotes accessible chromatin and high oncogene expression. Nature. 2019;575(7784):699–703.
27. Hung KL, Yost KE, Xie L, et al. ecDNA hubs drive transcriptional dysregulation in cancer. Nature. 2021;600(7887):283–289.
28. Greaves M, Maley CC. Clonal evolution in cancer. Nature. 2012;481(7381):306–313.
29. Verhaak RGW, Bafna V, Mischel PS. Extrachromosomal oncogene amplification in tumour pathogenesis and evolution. Nature Reviews Cancer. 2019;19(5):283–288.
30. Nathanson DA, Gini B, Mottahedeh J, et al. Targeted therapy resistance mediated by dynamic regulation of extrachromosomal mutant EGFR DNA. Science. 2014;343(6166):72–76.
31. Koche RP, Rodriguez-Fos E, Helmsauer K, et al. Extrachromosomal circular DNA drives oncogenic genome remodeling in neuroblastoma. Nature Genetics. 2020;52(1):29–34.
32. Turner KM, Deshpande V, Beyter D, et al. Extrachromosomal oncogene amplification drives tumour evolution and genetic heterogeneity. Nature. 2017;543(7643):122–125.
33. Verhaak RG, Bafna V, Mischel PS. Extrachromosomal oncogene amplification in tumour pathogenesis and evolution. Nature Reviews Cancer. 2019;19(5):283–288.
34. Nathanson DA, Gini B, Mottahedeh J, et al. Targeted therapy resistance mediated by dynamic regulation of extrachromosomal mutant EGFR DNA. Science. 2014;343(6166):72–76.
35. Kim H, Nguyen NP, Turner K, et al. Extrachromosomal DNA is associated with oncogene amplification and poor outcome across multiple cancers. Nature Genetics. 2020;52(9):891–897.

36. deCarvalho AC, Kowalewski RA, Piekna-Przybylska D, et al. Discordant inheritance of chromosomal and extrachromosomal DNA elements contributes to dynamic disease evolution in glioblastoma. Nature Genetics. 2018;50(5):708–717.
37. Hung KL, Yost KE, Xie L, et al. ecDNA hubs drive cooperative intermolecular oncogene expression. Nature. 2023;614(7949):603–611.
38. Koga T, Blakeley J, Wu JN, et al. Implications of dynamic extrachromosomal EGFR DNA in glioblastoma. Clinical Cancer Research. 2022;28(12):2455–2465.
39. Wu S, Bafna V, Mischel PS. Extrachromosomal DNA in cancer: New insights into oncogene amplification and therapy resistance. Nature Reviews Genetics. 2022;23(9):520–532.
40. Koire A, Katainen R, Siegfried A, et al. Somatic allele-specific expression and extrachromosomal oncogene amplification drive cancer progression. Nature Communications. 2022;13(1):1545.
41. Luebeck J, Corces MR, Kundaje A, et al. Structural evolution of chromatin and transcriptional landscapes via ecDNA. Nature Genetics. 2022;54(5):782–791.
42. Makar K, Wróbel A, Antczak A, Tworek D. Clinical utility of ctDNA analysis in lung cancer—A review. Advances in Respiratory Medicine. 2025;93(3):17.
43. Kim H, Nguyen NP, Luebeck J, et al. CRISPR targeting of ecDNA loci as a therapeutic strategy in oncogene-amplified tumors. Nature Biotechnology. 2023;41(2):240–249.
44. Lv W, Zeng Y, Li C, et al. Spatial–temporal diversity of extrachromosomal DNA shapes urothelial carcinoma evolution and the tumor immune microenvironment. Cancer Discovery. 2025 Jun 3;15(6):1225–46.
45. Noorani I, Haughey M, Luebeck J, et al. Extrachromosomal DNA driven oncogene spatial heterogeneity and evolution in glioblastoma. bioRxiv. 2024 Oct:2024-10.
46. Hung KL, Yost KE, Xie L, et al. ecDNA hubs drive cooperative intermolecular oncogene expression. Nature. 2021 Dec 23;600(7890):731–736.
47. Helmsauer K, Valieva ME, Ali S, et al. Enhancer hijacking determines extrachromosomal circular MYCN amplicon architecture in neuroblastoma. Nature Communications. 2020 Nov 16;11(1):5823.
48. Kobayashi K, Tan AC. Unraveling the impact of intratumoral heterogeneity on EGFR tyrosine kinase inhibitor resistance in EGFR-mutated NSCLC. International Journal of Molecular Sciences. 2023 Feb 18;24(4):4126.
49. Webb MG, Chow F, McCullough CG, et al. Resolving spatial subclonal genomic heterogeneity of loss of heterozygosity and extrachromosomal DNA in gliomas. Nature Communications. 2025 Jun 13;16(1):1–7.
50. Lange JT, Chen CY, Pichugin Y, et al. Principles of ecDNA random inheritance drive rapid genome change and therapy resistance in human cancers. BioRxiv. 2021 Jun 11:2021-06.
51. Arshadi A, Tolomeo D, Venuto S, et al. Advancements in focal amplification detection in tumor/liquid biopsies and emerging clinical applications. Genes. 2023 Jun 20;14(6):1304.
52. Merlo LM, Pepper JW, Reid BJ, et al. Cancer as an evolutionary and ecological process. Nature Reviews Cancer. 2006 Dec 1;6(12):924–935.
53. Greaves M, Maley CC. Clonal evolution in cancer. Nature. 2012 Jan 19;481(7381):306–313.
54. McGranahan N, Swanton C. Clonal heterogeneity and tumor evolution: Past, present, and the future. Cell. 2017 Feb 9;168(4):613–628.
55. McGranahan N, Swanton C. Biological and therapeutic impact of intratumor heterogeneity in cancer evolution. Cancer Cell. 2015 Jan 12;27(1):15–26.
56. Watkins TB, Lim EL, Petkovic M, et al. Pervasive chromosomal instability and karyotype order in tumour evolution. Nature. 2020 Nov 5;587(7832):126–132.
57. Lange JT, Rose JC, Chen CY, et al. The evolutionary dynamics of extrachromosomal DNA in human cancers. Nature Genetics. 2022 Oct;54(10):1527–1533.
58. Tasan I, Sustackova G, Zhang L, et al. CRISPR/Cas9-mediated knock-in of an optimized TetO repeat for live cell imaging of endogenous loci. Nucleic Acids Research. 2018 Sep 28;46(17):e100.
59. Grimm JB, Tkachuk AN, Xie L, et al. A general method to optimize and functionalize red-shifted rhodamine dyes. Nature Methods. 2020 Aug;17(8):815–821.
60. Morton AR, Dogan-Artun N, Faber ZJ, et al. Functional enhancers shape extrachromosomal oncogene amplifications. Cell. 2019 Nov 27;179(6):1330–1341.
61. Decarvalho AC, Kim H, Poisson LM, et al. Discordant inheritance of chromosomal and extrachromosomal DNA elements contributes to dynamic disease evolution in glioblastoma. Nature Genetics. 2018 May;50(5):708–717.
62. Wu S, Turner KM, Nguyen N, et al. Circular ecDNA promotes accessible chromatin and high oncogene expression. Nature. 2019 Nov 28;575(7784):699–703.
63. Kim H, Nguyen NP, Turner K, et al. Extrachromosomal DNA is associated with oncogene amplification and poor outcome across multiple cancers. Nature Genetics. 2020 Sep;52(9):891–897.

64. Windle B, Draper BW, Yin YX, et al. A central role for chromosome breakage in gene amplification, deletion formation, and amplicon integration. Genes & Development. 1991 Feb 1;5(2):160–174.
65. Burrell RA, McGranahan N, Bartek J, et al. The causes and consequences of genetic heterogeneity in cancer evolution. Nature. 2013 Sep 19;501(7467):338–345.
66. Alt FW, Kellems RE, Bertino JR, et al. Selective multiplication of dihydrofolate reductase genes in methotrexate-resistant variants of cultured murine cells. Journal of Biological Chemistry. 1978 Mar 10;253(5):1357–1370.
67. McClintock B. Chromosome organization and genic expression. In Cold Spring Harbor Symposia on Quantitative Biology 1951 Jan 1 (Vol. 16, pp. 13–47). Cold Spring Harbor Laboratory Press.
68. Lange JT, Chen CY, Pichugin Y, et al. Principles of ecDNA random inheritance drive rapid genome change and therapy resistance in human cancers. BioRxiv. 2021 Jun 11:2021-06.
69. Shapiro JA. How chaotic is genome chaos? Cancers. 2021 Mar 17;13(6):1358.
70. Gu X, Yu J, Chai P, et al. Novel insights into extrachromosomal DNA: Redefining the onco-drivers of tumor progression. Journal of Experimental & Clinical Cancer Research. 2020 Dec;39:1.
71. Andor N, Graham TA, Jansen M, et al. Pan-cancer analysis of the extent and consequences of intratumor heterogeneity. Nature Medicine. 2016 Jan;22(1):105–113.
72. Simovic-Lorenz M, Ernst A. Chromothripsis in cancer. Nature Reviews Cancer. 2025 Feb;25(2):79–92.
73. Lv W, Pan X, Han P, et al. Extrachromosomal circular DNA orchestrates genome heterogeneity in urothelial bladder carcinoma. Theranostics. 2024 Aug 19;14(13):5102.
74. Dong Y, He Q, Chen X, et al. Extrachromosomal DNA (ecDNA) in cancer: Mechanisms, functions, and clinical implications. Frontiers in Oncology. 2023 Jun 28;13:1194405.
75. Berezovsky A. Platelet Derived Growth Factor Receptor Alpha (PDGFRA) Oncogene Dependency in Glioblastoma (Doctoral dissertation, Wayne State University).

12 Role of ecDNA in Cancer Metastasis and Disease Recurrence

Anas Islam,[1] *Asad Ahmad,*[1] *Badruddeen,*[1] *Mohammad Irfan Khan,*[1] *and Yusuf Asad*[2]

1 Faculty of Pharmacy
Integral University
Lucknow, India

2 Department of Biomedical Engineering
Rowan University
Glassboro, NJ, USA

12.1 INTRODUCTION

Extrachromosomal DNA (ecDNA) comprises large, circular fragments excised from chromosomal loci that persist as autonomous nuclear elements. First observed as "double-minute" chromosomes more than half a century ago, ecDNAs were long regarded as rare curiosities. High-coverage whole-genome sequencing and dedicated analytics such as AmpliconArchitect have now revealed that ecDNAs are widespread in human tissues, harboring coding and regulatory segments that span hundreds of kilobases to several megabases. Their topological independence from centromeres and telomeres allows ecDNAs to replicate during S phase yet evade the orderly segregation rules that govern chromosomes, permitting free three-dimensional roaming within the nucleus and frequent encounters with transcriptional machinery [1, 2]. Because ecDNAs lack centromeres, they are apportioned stochastically to daughter cells at mitosis. This random assortment, coupled with the capacity for rapid rolling-circle replication, drives explosive copy-number expansion and seeds profound intratumor heterogeneity. Chromatin profiling shows that ecDNA is packaged into a uniquely open configuration that favors enhancer-promoter contact over long distances, creating "ecDNA hubs" where multiple oncogene copies—for example, MYC, EGFR, and CDK4—are transcribed at levels far higher than those achieved by chromosomal amplification. Such transcriptional hyperactivation fuels accelerated proliferation, promotes survival under stress, and endows tumors with remarkable evolutionary plasticity [2, 3].

Large-scale genomic surveys underscore the clinical significance of these elements. A pan-cancer analysis of more than 8,000 tumors found ecDNA in roughly 20% of newly diagnosed cancers and in nearly one-third of metastatic or heavily pre-treated lesions; ecDNA positivity correlated with higher histological grade, shorter progression-free survival, and multidrug resistance. Similar enrichment has been documented in glioblastoma, esophageal carcinoma, and sarcoma, where ecDNA-bearing tumors belong to the most aggressive molecular subtypes and display particularly poor outcomes. Mechanistically, ecDNA can co-amplify oncogenes and immunomodulatory loci, simultaneously driving growth signaling and blunting anti-tumor immunity, thereby compounding its adverse prognostic impact [4, 5]. ecDNA has therefore re-shaped modern cancer biology by uncoupling oncogenic dosage from the constraints of the linear genome. Its fluid copy-number dynamics create a "liquid"

 DOI: 10.1201/9781003655299-12

genetic layer that tumors can gain, discard, or reshuffle in response to environmental pressure or therapeutic challenge. Single-cell lineage-tracing and CRISPR-based capture experiments demonstrate preferential retention of ecDNA in disseminating clones and its persistence in minimal residual disease after therapy, linking ecDNA directly to early metastatic seeding and late relapse. Collectively, these insights position ecDNA as a central nexus connecting genome instability, clonal evolution, therapy resistance, and metastatic competence—and they highlight the urgent need to devise strategies that selectively eliminate or disable these formidable circles [3, 6].

12.2 BASICS OF CANCER METASTASIS AND DISEASE RECURRENCE

Cancer metastasis is classically defined as the multistep dissemination of malignant cells from a primary tumor to anatomically distant organs where they initiate secondary growths. The process begins with local invasion driven by epithelial-to-mesenchymal transition (EMT) and matrix remodeling, proceeds through intravasation into blood or lymphatic vessels, survival as circulating tumor cells, arrest and extravasation at permissive sites, and culminates in colonization that may require years of dormancy before overt outgrowth. Conceptually, this cascade still reflects Stephen Paget's "seed-and-soil" hypothesis: only tumor cells (the seeds) equipped with the right molecular cargo can exploit receptive microenvironments (the soil) to establish metastases. Contemporary reviews integrating genomics, biomechanics, and immunology have refined this framework, revealing how tumor cell plasticity, metabolic rewiring, mechanical force sensing, and immune evasion cooperate to complete each step of the cascade. Key signaling axes—such as TGF-β/SMAD, PI3K-AKT, Rho-ROCK, and chemokine networks (CXCL12-CXCR4, CCL21-CCR7)—orchestrate EMT, motility, and niche preparation, while tumor-secreted exosomes and neutrophil extracellular traps pre-condition future metastatic sites. These insights emphasize that metastasis is not an inevitable late event but an early and genetically programmed trait that evolves under selective pressure [7–9].

Disease recurrence represents the re-emergence of clinically detectable cancer after a period of apparent remission. It may stem from residual micrometastatic deposits that survive surgery or radiotherapy, from therapy-tolerant persister cells within the primary bed, or from minimal residual disease (MRD) persisting in the circulation or bone marrow. Advances in liquid biopsy have shown that even a few hundred tumor cells—or fragmented tumor DNA—can forecast impending relapse months before radiological evidence. Mechanistically, recurrent tumors often derive from subclones selected by adjuvant therapy: they carry drug-resistance mutations, epigenetic plasticity, or reside in protective niches such as perivascular spaces. Stress-induced quiescence, enhanced DNA-damage repair, overexpression of drug efflux pumps, and rewired apoptotic thresholds all contribute to survival under therapeutic assault. Importantly, the biology of recurrence interweaves with that of metastasis: the same cellular programs that enable distant spread—stem-like phenotypes, immune suppression, metabolic flexibility—also confer the capacity to lie dormant and later re-ignite tumor growth [10].

Clinically, the twin challenges of metastasis and recurrence account for roughly 90% of cancer-related mortality worldwide. Large epidemiological audits confirm that patients whose tumors are metastatic at diagnosis have five-year survival rates below 30% for most solid malignancies, compared with rates exceeding 70% for their non-metastatic counterparts. Analysis of >18 million US cases projects that although the incidence of de novo metastatic cancer may decline modestly, it will still impose a rising absolute burden as survivorship improves for indolent subtypes; by 2040, the number of individuals living with metastatic disease is expected to rise by nearly 47%. Moreover, registry data indicate that relapse remains common even after apparently curative therapy: up to one-third of breast-cancer deaths and half of colorectal-cancer deaths occur more than five years after initial treatment, underscoring the need for durable surveillance strategies. Financially and psychosocially, metastatic recurrence multiplies treatment costs, hospitalizations, and quality-of-life decrements, and it complicates trial design because heterogeneous, pre-treated tumors show unpredictable drug responses. Collectively, these statistics highlight metastasis and recurrence as the principal obstacles to cancer cure and the critical benchmarks for evaluating emerging therapeutic strategies [8, 11].

12.3 ECDNA CHARACTERISTICS RELEVANT TO METASTASIS AND RECURRENCE

Extrachromosomal DNA (ecDNA) is a circular, non-chromosomal DNA structure found in many cancers, distinguished by its ability to carry amplified oncogenes and regulatory elements outside the standard chromosomal framework. Unlike linear chromosomal DNA, ecDNA replicates autonomously, is highly accessible for transcription, and segregates unevenly during cell division, which increases genetic diversity and intratumoral heterogeneity—key factors in metastasis and recurrence [4, 12–14]. ecDNA is more frequently observed in advanced, metastatic, and treatment-resistant tumors, and its presence correlates with poor prognosis and reduced survival [15, 16]. Mechanistically, ecDNA can rapidly amplify oncogenes and facilitate structural rearrangements, driving clonal evolution and enabling tumors to adapt to selective pressures such as chemotherapy [12, 16]. Its dynamic nature allows for the swift emergence of drug resistance and supports immune evasion, as ecDNA-positive cells often downregulate immune recognition molecules [15]. Compared to chromosomal DNA, ecDNA is more prone to localized hypermutation and structural complexity, further fueling tumor evolution. Its dynamic nature allows for the swift emergence. The unique physical and genetic properties of ecDNA—such as its circularity, high copy number, and mobility—make it a potent driver of cancer progression, multifocality, and recurrence [17]. These features also present promising diagnostic and therapeutic targets, as ecDNA can be detected in bodily fluids and may be selectively targeted to disrupt tumor growth [14].

12.4 ECDNA AND CANCER CELL BEHAVIOR

Extrachromosomal DNA exerts a direct, gene-dosage-driven influence on cell states that underwrite metastatic competency. By clustering dozens of oncogene copies—most famously MYC, EGFR, or CCND1—within "ecDNA hubs," cancer cells achieve transcriptional outputs that dwarf those produced by linear amplifications. In pancreatic ductal adenocarcinoma, single-cell lineage tracing has shown that variable MYC-bearing ecDNA copy number toggles cells between epithelial and mesenchymal programs in real time, thereby enhancing motility, invasion, and anoikis resistance without fixed genetic mutation of EMT regulators [18]. Such plasticity extends to stem-like features: ecDNA-high subclones preferentially up-regulate SOX2, KLF4, and ALDH1 signatures, seed circulating tumor-cell pools early and dominate experimental metastases in xenograft models. Concomitantly, enhancer hijacking events created by circular re-arrangements couple metastasis-related promoters to super-enhancers, further boosting expression of Rho-GTPase and chemokine axes that guide tissue homing [19].

Stochastic segregation of acentromeric circles at mitosis accelerates clonal diversification, giving tumors a rolling catalogue of phenotypes on which selection can act. Multi-regional sampling in urothelial carcinoma revealed sharply discordant ecDNA landscapes between spatially separated foci within a single lesion; copy-number "jackpot" clones carrying high ecDNA burden preferentially expanded at the invasive front and were over-represented in lymph-node metastases [20]. Chromatin-accessibility profiling indicates that these circles maintain an enhancer-rich, H3K27ac-laden architecture even under therapeutic stress, enabling rapid transcriptional rebound after drug withdrawal and facilitating outgrowth of persister populations that later seed relapse [21]. Beyond intrinsic cell programs, ecDNA remodels the tumor microenvironment to favor dissemination and survival. Transcriptomic deconvolution of 870 human tumors demonstrated broad suppression of antigen-processing and MHC-I/II display in ecDNA-positive cancers, accompanied by enrichment of DNA-damage-response and cell-cycle signatures that foster immune-excluded phenotypes [22, 23].

Single-cell atlases reveal that ecDNA-bearing clones secrete higher levels of CXCL8, CSF1, and VEGFA, recruiting neutrophils and tumor-associated macrophages that construct pre-metastatic niches. In Barrett's esophagus-to-adenocarcinoma evolution, circles co-amplifying immunomodulatory genes such as PD-L1 and CD274 emerge early, correlate with reduced cytotoxic-T-cell infiltration and persist through malignant transformation, illustrating how ecDNA-encoded immune-evasion programs are embedded before overt metastasis [16]. Metabolically, circular amplification of SLC2A1 and LDHA increases glycolytic flux, acidifying the extracellular milieu and impairing dendritic-cell activation, while ecDNA-encoded NRF2 variants enhance oxidative-stress tolerance, supporting colonization of hypoxic metastatic beds. Collectively, these adaptations weave ecDNA into every layer of the metastatic cascade—cell-intrinsic plasticity, clonal heterogeneity, and niche conditioning—and explain why circles are consistently retained in minimal residual disease after therapy.

12.5 MECHANISMS LINKING ECDNA TO METASTASIS

Extrachromosomal DNA confers metastatic capacity primarily by massively amplifying oncogenes whose products govern motility, invasion, survival in circulation, and colonization of distant tissues. Because circles can carry dozens to hundreds of intact gene copies, their transcriptional output dwarfs that of chromosomal gains and can be modulated instantaneously through copy-number fluctuation. Glioblastoma provides a canonical example: drug-resistant sub-clones harbor ecDNA loaded with the constitutively active receptor variant EGFRvIII. Exposure to the small-molecule inhibitor erlotinib rapidly selects cells with higher EGFR-ecDNA dosage; when drug pressure is removed, those circles are lost, restoring drug sensitivity and illustrating the reversible dosage plasticity that fuels invasive escape and therapeutic relapse [24]. Similar dynamics are seen in neuroblastoma, where MYCN-containing ecDNAs not only drive proliferation but also up-regulate CXCR4, ITGA5, and metalloproteinases, collectively enhancing chemotaxis toward bone-marrow stroma and degradation of extracellular matrix barriers—two essential steps for bone metastasis [25].

ecDNA enables rapid and high-level amplification of oncogenes, which drives aggressive tumor behavior, including enhanced metastatic potential. This amplification increases intratumoral genetic heterogeneity, allowing cancer cells to adapt quickly to new environments and resist therapies, both of which are critical for successful metastasis [26, 27]. The presence of ecDNA is significantly higher in metastatic and pretreated tumors compared to primary, untreated cancers, highlighting its role in cancer progression and metastatic spread [4, 26].

ecDNA not only amplifies oncogenes but also alters chromatin accessibility and gene regulation, leading to hyperactivation of signaling pathways that promote tumor invasiveness and metastasis [13, 28, 29]. ecDNA can function as mobile enhancers, forming hubs that interact with chromosomal DNA to globally amplify transcription of genes involved in cell migration, invasion, and survival [28, 30, 31]. These interactions facilitate the activation of pathways that are directly relevant to metastatic behavior.

Studies have shown that ecDNA is retained and even enriched in advanced and metastatic tumors, with features such as localized hypermutation and structural complexity that further drive tumor evolution and adaptation [4, 32]. For example, in glioblastoma and prostate cancer models, ecDNA hubs have been observed to promote cooperative oncogene expression, supporting aggressive and metastatic phenotypes [28, 31]. Additionally, ecDNA's uneven segregation during cell division increases genetic diversity, providing a pool of variants that can be selected for metastatic traits [26, 32].

12.6 ECDNA IN DISEASE RECURRENCE

Extrachromosomal DNA (ecDNA) plays a significant role in cancer by promoting tumor recurrence, therapy resistance, and the survival of minimal residual disease. ecDNA is frequently found in cancer cells and is associated with poor clinical outcomes and increased tumor aggressiveness. Its non-chromosomal inheritance leads to high oncogene copy numbers and substantial intra-tumoral genetic heterogeneity, which accelerates tumor evolution and adaptation. This heterogeneity enables cancer cells to survive therapeutic pressures, contributing to drug resistance and tumor recurrence even after initial treatment responses. The circular structure of ecDNA allows for altered gene regulation and elevated oncogene expression, further driving malignant behavior and enabling rapid genomic remodeling. Additionally, ecDNA can integrate into or detach from chromosomal DNA, causing structural changes and genomic rearrangements that support tumor progression and resistance to chemotherapy. The presence of ecDNA also results in increased transcription-replication conflicts and replication stress, which can be exploited for targeted therapies but also underlie the resilience of ecDNA-positive tumors to conventional treatments. Overall, ecDNA's ability to amplify oncogenes, promote genetic diversity, and facilitate rapid adaptation underlies its central role in therapy resistance, tumor recurrence, and the persistence of minimal residual disease in cancer [13, 27, 33, 34].

Targeting extrachromosomal DNA (ecDNA) in recurrent cancer is an emerging therapeutic focus, as ecDNA drives oncogene amplification, therapy resistance, and rapid tumor evolution. The most advanced strategy involves exploiting the unique replication stress and transcription-replication conflicts present in ecDNA-positive tumors, with promising results from selective CHK1 inhibitors. Several approaches are being explored:

***CHK1 Inhibition*:** ecDNA-positive tumors experience high replication stress and depend heavily on the CHK1 checkpoint kinase for survival. Selective CHK1 inhibitors, such as BBI-2779 and BBI-cmpd1, have shown potent and preferential killing of ecDNA-containing cancer cells in preclinical models, suppressing tumor growth and preventing resistance to targeted therapies. This approach exploits a synthetic lethality unique to ecDNA-driven cancers [34, 35].

***Enhancing Transcription-Replication Conflict*:** By further increasing the already high transcription-replication conflict in ecDNA-positive cells, researchers can induce lethal DNA damage and cell death specifically in these tumor cells, sparing normal cells [34, 35].

***Precision Medicine and Patient Selection*:** Detecting ecDNA in tumors can help identify patients most likely to benefit from novel ecDNA-targeted therapies, allowing for more effective clinical trial design and personalized treatment strategies [14, 33, 36].

Ongoing research aims to develop additional agents that target ecDNA biogenesis, maintenance, or its unique chromatin structure, as well as to integrate ecDNA profiling into routine clinical practice for better therapeutic stratification [12–14, 21, 33].

12.7 CHALLENGES AND FUTURE DIRECTIONS

Despite rapid advancements in the understanding of extrachromosomal DNA, several technical and scientific challenges hinder its full characterization and therapeutic targeting. One of the primary obstacles is the accurate detection and structural reconstruction of ecDNA, which often requires high-depth sequencing and sophisticated computational tools that are not yet standard in clinical settings. Traditional short-read sequencing struggles to distinguish ecDNA from chromosomal rearrangements, while single-cell approaches, though powerful, remain expensive and technically demanding. Additionally, the stochastic segregation and structural heterogeneity of ecDNA pose difficulties in consistent tracking across tumor samples and over time, making it challenging to assess its dynamic role in tumor evolution, especially in metastatic and recurrent disease contexts.

From a therapeutic perspective, ecDNA represents a unique yet complex target. Its circular nature and high oncogene content offer potential points of intervention, such as disrupting ecDNA replication, transcriptional activity, or chromatin accessibility. However, selectively targeting ecDNA without affecting normal chromosomal DNA remains a major hurdle. Many of the regulatory elements on ecDNA, such as enhancers and promoters, are shared with linear DNA, increasing the risk of off-target effects. Furthermore, the rapid copy-number changes and reintegration capabilities of ecDNA mean that even successful interventions may be temporary unless followed by sustained pressure or combination therapies.

Future research needs to address both mechanistic and translational gaps. Mechanistically, deeper insights are needed into how ecDNA forms, persists, and interacts with nuclear architecture, as well as how it influences gene regulation beyond simple dosage effects. Understanding the timing and triggers of ecDNA emergence in the metastatic cascade and during treatment-induced stress could uncover critical windows for intervention. On the translational front, developing robust ecDNA-specific biomarkers for early detection, minimal residual disease monitoring, and prediction of therapy resistance is essential. There is also a need for innovative drug-delivery systems that can exploit ecDNA-specific vulnerabilities with minimal toxicity. Ultimately, integrating ecDNA profiling into clinical decision-making and tailoring therapies based on ecDNA dynamics holds promise for transforming the management of metastatic and recurrent cancers.

12.8 CONCLUSION

Extrachromosomal DNA (ecDNA) has emerged as a central player in cancer progression, metastasis, and recurrence due to its unique structural, genetic, and regulatory features. By amplifying oncogenes and enhancing transcriptional activity beyond chromosomal constraints, ecDNA drives aggressive tumor phenotypes, promotes intratumoral heterogeneity, and facilitates escape from therapeutic pressure. Its capacity to support the survival of minimal residual disease and fuel relapse makes it a formidable barrier to durable cancer remission. Despite these challenges, ecDNA also offers novel opportunities for therapeutic targeting, as its distinct biology reveals specific vulnerabilities absent in chromosomal DNA. Advances in sequencing technologies, chromatin mapping, and functional genomics have begun to unravel the complexity of ecDNA, yet many aspects of its formation, regulation, and clinical impact remain incompletely understood. Future research focused on ecDNA-specific biomarkers, targeted interventions, and real-time monitoring strategies may enable precision oncology approaches that preempt metastasis and recurrence. Ultimately, integrating ecDNA biology into cancer diagnostics and treatment paradigms holds significant promise for improving patient outcomes in some of the most treatment-resistant and lethal cancer types.

REFERENCES

[1] Paulsen, T., Kumar, P., Koseoglu, M.M., & Dutta, A. Discoveries of extrachromosomal circles of DNA in normal and tumor cells. Trends in Genetics. 2018 Apr;34(4):270–278. https://doi.org/10.1016/j.tig.2017.12.010. Epub 2018 Jan 9. PMID: 29329720; PMCID: PMC5881399.

[2] Turner, K.M., Deshpande, V., Beyter, D., Koga, T., Rusert, J., Lee, C., Li, B., Arden, K., Ren, B., Nathanson, D.A., & Kornblum, H.I. Extrachromosomal oncogene amplification drives tumour evolution and genetic heterogeneity. Nature. 2017 Mar 2;543(7643):122–125.

[3] Wu, S., Turner, K.M., Nguyen, N., Raviram, R., Erb, M., Santini, J., Luebeck, J., Rajkumar, U., Diao, Y., Li, B., & Zhang, W. Circular ecDNA promotes accessible chromatin and high oncogene expression. Nature. 2019 Nov 28;575(7784):699–703.

[4] Kim, H., Kim, S., Wade, T., Yeo, E., Lipsa, A., Golebiewska, A., Johnson, K.C., An, S., Ko, J., Nam, Y., & Lee, H.Y. Mapping extrachromosomal DNA amplifications during cancer progression. Nature Genetics. 2024 Nov;56(11):2447–2454.

[5] Bailey, C., Pich, O., Thol, K., Watkins, T.B., Luebeck, J., Rowan, A., Stavrou, G., Weiser, N.E., Dameracharla, B., Bentham, R., & Lu, W.T. Origins and impact of extrachromosomal DNA. Nature. 2024 Nov 7;635(8037):193–200.

[6] Lange, J.T., Rose, J.C., Chen, C.Y., Pichugin, Y., Xie, L., Tang, J., Hung, K.L., Yost, K.E., Shi, Q., Erb, M.L., & Rajkumar, U. The evolutionary dynamics of extrachromosomal DNA in human cancers. Nature Genetics. 2022 Oct;54(10):1527–1533.

[7] Mathot, L., & Stenninger, J. Behavior of seeds and soil in the mechanism of metastasis: A deeper understanding. Cancer Science. 2012 Apr;103(4):626–631. https://doi.org/10.1111/j.1349-7006.2011.02195.x. Epub 2012 Jan 29. PMID: 22212856; PMCID: PMC7659190.

[8] Shi, X., Wang, X., Yao, W., Shi, D., Shao, X., Lu, Z., Chai, Y., Song, J., Tang, W., & Wang, X. Mechanism insights and therapeutic intervention of tumor metastasis: Latest developments and perspectives. Signal Transduction and Targeted Therapy. 2024 Aug 2;9(1):192.

[9] Liu, Z., Chen, J., Ren, Y., Liu, S., Ba, Y., Zuo, A., Luo, P., Cheng, Q., Xu, H., & Han, X. Multi-stage mechanisms of tumor metastasis and therapeutic strategies. Signal Transduction and Targeted Therapy. 2024 Oct 11;9(1):270.

[10] Song, M., Pan, W., Yu, X., Ren, J., Tang, C., Chen, Z., Wang, Z., Deng, Y., He, N., Liu, H., & Li, S. Minimal residual disease detection: Implications for clinical diagnosis and cancer patient treatment. MedComm (2020). 2025 May 15;6(6):e70193. https://doi.org/10.1002/mco2.70193. PMID: 40384986; PMCID: PMC12079024.

[11] Hudock, N.L., Mani, K., Khunsriraksakul, C., Walter, V., Nekhlyudov, L., Wang, M., Lehrer, E.J., Hudock, M.R., Liu, D.J., Spratt, D.E., & Zaorsky, N.G. Future trends in incidence and long-term survival of metastatic cancer in the United States. Communications Medicine. 2023 May 27;3(1):76.

[12] Huang, Q., Zhang, S., Wang, G., & Han, J. Insight on ecDNA-mediated tumorigenesis and drug resistance. Heliyon. 2024;10(10): e27733. https://doi.org/10.1016/j.heliyon.2024.e27733.

[13] Li, Z., Wang, B., Liang, H., & Han, L. Pioneering insights of extrachromosomal DNA (ecDNA) generation, action and its implications for cancer therapy. International Journal of Biological Sciences. 2022;18:4006–4025. https://doi.org/10.7150/ijbs.73479.

[14] Mao, X., Rao, G., Li, G., & Chen, S. Insights into extrachromosomal DNA in cancer: Biogenesis, methodologies, functions, and therapeutic potential. Advanced Biology. 2025;9. https://doi.org/10.1002/adbi.202400433.

[15] Xiang, X., Gao, Q., Verhaak, R., Xu, Z., Li, Y., Li, Y., Sun, C., Wang, Y., Hansen, J., Prasad, G., Teng, H., Qin, T., Zhou, R., Li, C., Lv, W., Jiang, S., Han, P., Henssen, A., Luo, Y., . . . Lin, L. Spatial-temporal diversity of extrachromosomal DNA shapes urothelial carcinoma evolution and tumor-immune microenvironment. Cancer Discovery. 2025. https://doi.org/10.1158/2159-8290.CD-24-1532.

[16] Luebeck, J., Ng, A., Galipeau, P., Li, X., Sanchez, C., Katz-Summercorn, A., Kim, H., Jammula, S., He, Y., Lippman, S., Verhaak, R., Maley, C., Alexandrov, L., Reid, B., Fitzgerald, R., Paulson, T., Chang, H., Wu, S., Bafna, V., & Mischel, P. Extrachromosomal DNA in the cancerous transformation of Barrett's oesophagus. Nature. 2023;616:798–805. https://doi.org/10.1038/s41586-023-05937-5.

[17] Gu, X., Yu, J., Chai, P., Ge, S., & Fan, X. Novel insights into extrachromosomal DNA: Redefining the oncodrivers of tumor progression. Journal of Experimental & Clinical Cancer Research. 2020;39(1):Article 215. https://doi.org/10.1186/s13046-020-01726-4.

[18] Fiorini, E., Malinova, A., Schreyer, D., Pasini, D., Bevere, M., Alessio, G., Rosa, D., D'Agosto, S., Azzolin, L., Milite, S., Andreani, S., Lupo, F., Veghini, L., Grimaldi, S., Pedron, S., Castellucci, M., Nourse, C., Salvia, R., Malleo, G., . . . Corbo, V. MYC ecDNA promotes intratumour heterogeneity and plasticity in PDAC. Nature. 2025;640(8059):811–820. https://doi.org/10.1038/s41586-025-08721-9.

[19] Mortenson, K.L., Dawes, C., Wilson, E.R., Patchen, N.E., Johnson, H.E., Gertz, J., Bailey, S.D., Liu, Y., Varley, K.E., & Zhang, X. 3D genomic analysis reveals novel enhancer-hijacking caused by complex structural alterations that drive oncogene overexpression. Nature Communications. 2024 Jul 20;15(1):6130. https://doi.org/10.1038/s41467-024-50387-w. PMID: 39033128; PMCID: PMC11271278.

[20] Lv, W., Zeng, Y., Li, C., Liang, Y., Tao, H., Zhu, Y., Sui, X., Li, Y., Jiang, S., Gao, Q., Rodriguez-Fos, E., Prasad, G., Wang, Y., Zhou, R., Xu, Z., Pan, X., Chen, L., Xiang, X., Teng, H., . . . Han, P. Spatial-temporal diversity of extrachromosomal DNA shapes urothelial carcinoma evolution and tumor-immune microenvironment. Cancer Discovery. 2025 Mar 11. https://doi.org/10.1158/2159-8290.CD-24-1532. Epub ahead of print. PMID: 40062518.

[21] Yang, Q.L., Xie, Y., Qiao, K., Lim, J.Y., & Wu, S. Modern biology of extrachromosomal DNA: A decade-long voyage of discovery. Cell Research. 2025 Jan 3:1–2.

[22] Lin, M.S., Jo, S.Y., Luebeck, J., Chang, H.Y., Wu, S., Mischel, P.S., & Bafna, V. Transcriptional immune suppression and up-regulation of double-stranded DNA damage and repair repertoires in ecDNA-containing tumors. Elife. 2024 Jun 19;12:RP88895.

[23] Wu, T., Wu, C., Zhao, X., Wang, G., Ning, W., Tao, Z., Chen, F., & Liu, X.S. Extrachromosomal DNA formation enables tumor immune escape potentially through regulating antigen presentation gene expression. Scientific Reports. 2022 Mar 4;12(1):3590.

[24] Nathanson, D.A., Gini, B., Mottahedeh, J., Visnyei, K., Koga, T., Gomez, G., Eskin, A., Hwang, K., Wang, J., Masui, K., & Paucar, A. Targeted therapy resistance mediated by dynamic regulation of extrachromosomal mutant EGFR DNA. Science. 2014 Jan 3;343(6166):72–76.

[25] Koche, R.P., Rodriguez-Fos, E., Helmsauer, K., Burkert, M., MacArthur, I.C., Maag, J., Chamorro, R., Munoz-Perez, N., Puiggròs, M., Dorado Garcia, H., & Bei, Y. Extrachromosomal circular DNA drives oncogenic genome remodeling in neuroblastoma. Nature Genetics. 2020 Jan 1;52(1):29–34.

[26] Zeng, X., Wan, M., & Wu, J. ecDNA within tumors: A new mechanism that drives tumor heterogeneity and drug resistance. Signal Transduction and Targeted Therapy. 2020;5. https://doi.org/10.1038/s41392-020-00403-4.

[27] Ashique, S., Upadhyay, A., Garg, A., Mishra, N., Hussain, A., Negi, P., Hing, G., Bhatt, S., Ali, M., Gowthamarajan, K., Singh, S., Gupta, G., Chellappan, D., & Dua, K. Impact of ecDNA: A mechanism that directs tumorigenesis in cancer drug resistance-a review. Chemico-Biological Interactions. 2022;110000. https://doi.org/10.1016/j.cbi.2022.110000.

[28] Zhu, Y., Gujar, A., Wong, C., Tjong, H., Ngan, C., Gong, L., Chen, Y., Kim, H., Liu, J., Li, M., Mil-Homens, A., Maurya, R., Kuhlberg, C., Sun, F., Yi, E., deCarvalho, A., Ruan, Y., Verhaak, R., & Wei, C. Oncogenic extrachromosomal DNA functions as mobile enhancers to globally amplify chromosomal transcription. Cancer Cell. 2021. https://doi.org/10.1016/j.ccell.2021.03.006.

[29] Wu, S., Bafna, V., & Mischel, P. Extrachromosomal DNA (ecDNA) in cancer pathogenesis. Current Opinion in Genetics & Development. 2021;66:78–82. https://doi.org/10.1016/j.gde.2021.01.001.

[30] Zhu, Y., Gong, L., & Wei, C. Guilt by association: EcDNA as a mobile transactivator in cancer. Trends in Cancer. 2022. https://doi.org/10.1016/j.trecan.2022.04.011.

[31] Hung, K., Yost, K., Xie, L., Shi, Q., Helmsauer, K., Luebeck, J., Schöpflin, R., Lange, J., González, C., Weiser, N., Chen, C., Valieva, M., Wong, I., Wu, S., Dehkordi, S., Duffy, C., Kraft, K., Tang, J., Belk, J., . . . Chang, H. ecDNA hubs drive cooperative intermolecular oncogene expression. Nature. 2021;600:731–736. https://doi.org/10.1038/s41586-021-04116-8.

[32] Xing, J., Ning, Q., Tang, D., Mo, Z., Lei, X., & Tang, S. Progress on the role of extrachromosomal DNA in tumor pathogenesis and evolution. Clinical Genetics. 2020;99:503–512. https://doi.org/10.1111/cge.13896.

[33] Yan, X., Mischel, P., & Chang, H. Extrachromosomal DNA in cancer. Nature Reviews Cancer. 2024. https://doi.org/10.1038/s41568-024-00669-8.

[34] Tang, J., Weiser, N., Wang, G., Chowdhry, S., Curtis, E., Zhao, Y., Wong, I., Marinov, G., Li, R., Hanoian, P., Tse, E., Mojica, S., Hansen, R., Plum, J., Steffy, A., Milutinovic, S., Meyer, S., Luebeck, J., Wang, Y., . . . Chang, H. Enhancing transcription–replication conflict targets ecDNA-positive cancers. Nature. 2024;635:210–218. https://doi.org/10.1038/s41586-024-07802-5.

[35] Chowdhry, S., Milutinovic, S., Tse, E., Garcia, S., Perusse, D., Ritland, M., Ko, J., Wilkinson, D., Turner, K., Steffy, A., Plum, J., Norman, B., Pferdekamper, A., Meyer, T., Liao, D., Elsdon, R., Lange, J., Pinkerton, A., Hansen, R., . . . Kasibhatla, S. Abstract 1626: Tumors driven by oncogene amplified extrachromosomal DNA (ecDNA) demonstrate enhanced sensitivity to cell cycle checkpoint kinase 1 (CHK1) inhibition. Cancer Research. 2023. https://doi.org/10.1158/1538-7445.am2023-1626.

[36] Noorani, I., Mischel, P., & Swanton, C. Leveraging extrachromosomal DNA to fine-tune trials of targeted therapy for glioblastoma: Opportunities and challenges. Nature Reviews Clinical Oncology. 2022;19:733–743. https://doi.org/10.1038/s41571-022-00679-1.

13 Computational Approaches to Study ecDNA in Cancer Genomics

Monika Kaushik,[1] Priya Chaudhary,[2] Mohini Mondal,[3] Rohini Armo,[4] Nikunj Solanki,[5] and Sonal Solanki[5]

1 Amity Institute of Pharmacy
Amity University Gwalior
Maharajpura, Gwalior, India
2 Centre for Herbal Pharmacology and Environmental Sustainability, Chettinad Hospital and Research Institute
Chettinad Academy of Research and Education
Kelambakkam, India
3 Department of Pharmaceutical Technology
JIS University
Kolkata, India
4 Department of Pharmaceutical Chemistry
Shri Rawatpura Sarkar Institute of Pharmacy
Durg Chhattisgarh, India
5 Department of Pharmaceutics
HSBPVT's, Group of Institutions, Faculty of Pharmacy
Kashti, Ta-Shrigonda, India

13.1 INTRODUCTION

13.1.1 Overview of Cancer Genomics

Extrachromosomal components like extrachromosomal circular DNA are included in the whole cancer genome in addition to changes in gene sequences, epigenetic modifications, and chromosome spatial arrangement [1, 2]. In both normal and malignant tissues, there is extrachromosomal circularization of DNA, which shows up as tiny, circular, nonamplified DNAs, including microDNAs (called eccDNA) [1]. The term "ecDNA" refers to clonal, massive, copy number amplified circular extrachromosomal DNAs, which are frequently observed in cancer cells [1]. Rapid tumor progression is facilitated by ecDNA, which is a key component of the cancer genome since it may carry complete oncogenes and induce their heightened expression [1].

ecDNA was first discovered in the 1960s. The use of low-throughput imaging-based approaches, such as FISH, slowed down the advancement of ecDNA research in cancer since then. High-throughput, universal methods for examining ecDNA have been developed as a result of the introduction of new sequencing and computational technologies. The study of ecDNA in malignancies has been greatly enhanced by these sequencing-based methodologies. Currently, around 50% of all human cancer types include ecDNA [2]. As shown before, ecDNA does not have any natural chromosomal characteristics, including centromeres or telomeres, and its only function is to boost

 DOI: 10.1201/9781003655299-13

oncogene expression. Rapid carcinogenesis and treatment resistance are common outcomes of cancer cells being able to evolve rapidly in such a situation [1].

Limited research on ecDNA in hematological malignancies has been conducted via case reports, series of cases, and small cohort studies of certain illnesses, including acute leukemia and lymphoma [1]. In most cases, the amplified oncogenes are MYC or MLL, and the emphasis is on acute myeloid leukemia. The variety of illnesses studied and the sample sizes in hematological cancer research are still rather small, even though they have been included in several larger pan-cancer studies in the past few years [3]. The frequency of occurrence, structural features, roles, and underlying processes of ecDNA production in hematological malignancies are thus important factors that need more thorough research.

13.1.2 Emergence of Extrachromosomal DNA (ecDNA) as a Genomic Feature

As a primary carrier of amplified oncogenes, extrachromosomal DNA (ecDNA)—circular DNA made from nucleosomal chromatin—is less well understood in cancer. It is more open than chromosomal DNA, which leads to enhanced transcriptional activity [4]. Some tumor cells possess large copy numbers of oncogenes, which may significantly impact tumor heterogeneity and treatment resistance, since ecDNA is inherited in a non-Mendelian fashion during mitosis, because it lacks a centromere. ecDNA was discovered in a variety of cancer types but seldom in healthy tissues in recent investigations using whole-genome sequencing (WGS) and cytogenetics techniques. Additionally, research has shown a correlation between ecDNA and more aggressive tumors as well as a worse clinical outcome. Therefore, it is expected that medications that target the creation and maintenance of ecDNA might be effective cancer treatments, since the processes of ecDNA dynamics and their function in cancer are slowly becoming understood [4].

With its distinct inheritance mechanisms that allow it to adapt rapidly to changing contexts, ecDNA-based oncogene amplification may undoubtedly contribute to faster cancer evolution. Nevertheless, is there a difference in how ecDNA and chromosomal DNA work specifically? Determining the genome's function relies on its spatial organization [5]. The nucleus is hierarchically structured and divides DNA into 23 pairs of chromosomes, which are further separated into separate territories. This division occurs in human cells. Chromatin accessibility to the cell's transcriptional and regulatory machinery is determined by the location of nucleosomes, which are arranged into histone octamers that wind around the DNA. The interplay between regulatory and transcription factors determines the expression of genes inside insulated areas generated by this dynamic process, which is further structured into chromatin loops. Recently, groundbreaking research has shown that the chromatin landscape in cancer is significantly different from normal, and changes in the spatial architecture, such as the disruption of topologically related domain boundaries, are involved in tumor pathogenesis in cancer. One of the most basic questions that needs answering is how the structure of ecDNA affects epigenetic regulation, chromatin accessibility, and gene transcription [5].

Recently, researchers used CRISPR/Cas9 editing to create circular DNA fragments of kilobase size that were found to have a gene that encodes an improved green fluorescent protein. These fragments are comparable to the tiny circular eccDNAs mentioned earlier. Higher expression levels of oncogene transcription have previously been demonstrated7,8 for the big (105–107 bp) ecDNAs comprising oncogenes and regulatory elements that are the subject of this study, which vary significantly from this model. To better understand the pathogenic role of ecDNAs in cancer and how their chromosomal or extrachromosomal context impacts the impact of an oncogene on tumor development, progression, and drug resistance, new methods are being developed to probe the chromatin structure, transcriptional landscape, and functional contribution of ecDNAs in cancer pathogenesis [5].

13.1.3 Importance of Computational Analysis in ecDNA Research

Oncogene amplification and variability in tumors are two areas where research into cancer-related issues using extrachromosomal circular DNA (ecDNA) has recently picked up pace. Standard laboratory methods have a hard time detecting ecDNA due to its complicated structure, which is often big, heterogeneous, and circular. Consequently, the detection and evaluation of ecDNA from sequencing data requires computational analysis [5]. Sequence read density between a tumor genomic and a corresponding normal human genome reference may be utilized to derive tumor DNA copy number patterns using high-throughput short-read DNA sequencing.

An untested assumption behind this analytical method is that it can identify the chromosomal loci in normal tissue that include amplified genes seen in cancer cells. Thus, the ecDNA might potentially go unnoticed by the conventional methods of analyzing high-throughput short-read DNA sequencing, according to Mischel, Bafna, and colleagues. They created an objective method for ecDNA identification by combining whole genome sequencing, computational, and cytogenetic imaging studies [5]. They found that amplification on ecDNA is common in cancer, but oncogene amplification on native chromosomal sites is rather rare. Their findings also show that ecDNA-based gene amplification leads to an increase in copy number and intratumoral genetic heterogeneity, indicating that ecDNA plays a crucial role in the development of cancer. Building on earlier research showing that oncogenes on ecDNAs are often also located on chromosomes as HSRs, the AmpliconArchitect tool for precise structural mapping of the amplicons shows that some ecDNAs may "jump" onto chromosomal HSRs. It is unclear what mechanism underlies this procedure [6].

The graphical representation of circular DNA is made possible by computer visualization methods such as Ribbon and Circos, which aid in understanding and formulation of hypotheses [5]. These methods cover a wide range of applications, including non-invasive cancer diagnostics, real-time monitoring of treatment resistance, and ecDNA detection based on liquid biopsies [6]. Functional implications can only be uncovered by computational analysis when dealing with raw sequencing data. Due to its sensitivity, scalability, and depth, ecDNA is useful for research and clinical translation for complete investigations.

13.2 BIOLOGICAL BASIS OF ECDNA IN CANCER

13.2.1 Formation and Structure of ecDNA

Circular ecDNA comprised of chromosomal fibers with nucleosomes was discovered in the 1980s through TEM research [7]. Wu et al. discovered in 2019 that ecDNA was more accessible and less compact than chromosomal DNA when they combined WGS analysis, optical imaging, and ultrasound imaging. This led to an increase in locus expression. As this study demonstrated, oncogenes and other genes may be overexpressed in ecDNA with high copy numbers. Compared to linear DNA, ecDNA amplification boosted oncogene expression, according to a WGS study by Kim et al. [8]. And, hence, gene expression is enhanced by circular ecDNA molecules that include considerably decompacted chromatin.

Several molecular processes related to chromosomal damage may initiate ecDNA production, which is greatly related to genomic instability [7]. Chromothripsis, DNA religation, fork stalling, template flipping, and breakage–fusion–bridge (BFB) cycles are all examples of such processes.

13.2.1.1 The Chromothripsis

When tens to hundreds of locally localized DNA breaks happen at the same time, causing DNA fragment rearrangements and a broad loss of DNA sequences, this is referred to as chromothripsis [7]. This procedure could include cancer-causing genes that might promote cell proliferation.

13.2.1.2 DNA Religation

Additional data supports the chromothripsis concept, which suggests that DNA ligation is a mechanism for ecDNA production. Using the CRISPR-Cas9 technique, scientists have created ecDNA in human cell lines by removing DNA segments from chromosomes [7]. Additional study is required to identify the molecular processes that may explain the ligation of deleted chromosomal DNA to produce ecDNA in the absence of chromothripsis, according to an alternative scenario.

13.2.1.3 Cycles of Breakage–Fusion–Bridge

DNA breaks, telomeres shorten, and chromosomes end-to-end unite to create an anaphase bridge with two centromeres; this is one of the traditional processes of ecDNA production, and it is called the breakage–fusion–bridge (BFB) cycle [7]. As a consequence of the cycle's effects on gene amplification and chromosomal rearrangement, cancer can develop.

13.2.1.4 Template Switching and Fork Stalling

Breakpoint analysis of ecDNA has shown that fork stalling and template flipping are additional processes of ecDNA synthesis. Based on the roles of ecDNA, which promote tumor development and treatment resistance, new ecDNA might be produced from either chromosomal DNA or secondary change of existing ecDNA [7].

13.2.1.5 Combination of Pre-Existed ecDNA

ecDNA, once created, continues to develop quickly, even when combined with pre-existing ecDNA. By joining two smaller ecDNA, a bigger ecDNA might be created, as shown in the case of ecDNA harboring KRAS in esophageal cancer [7].

13.2.2 Functional Role in Oncogene Amplification and Heterogeneity

Encouraging oncogene expression, enhancing copy number as well, and boosting chromatin accessibility are three important roles that ecDNA plays in the genesis and evolution of cancer (Figure 13.1) [7].

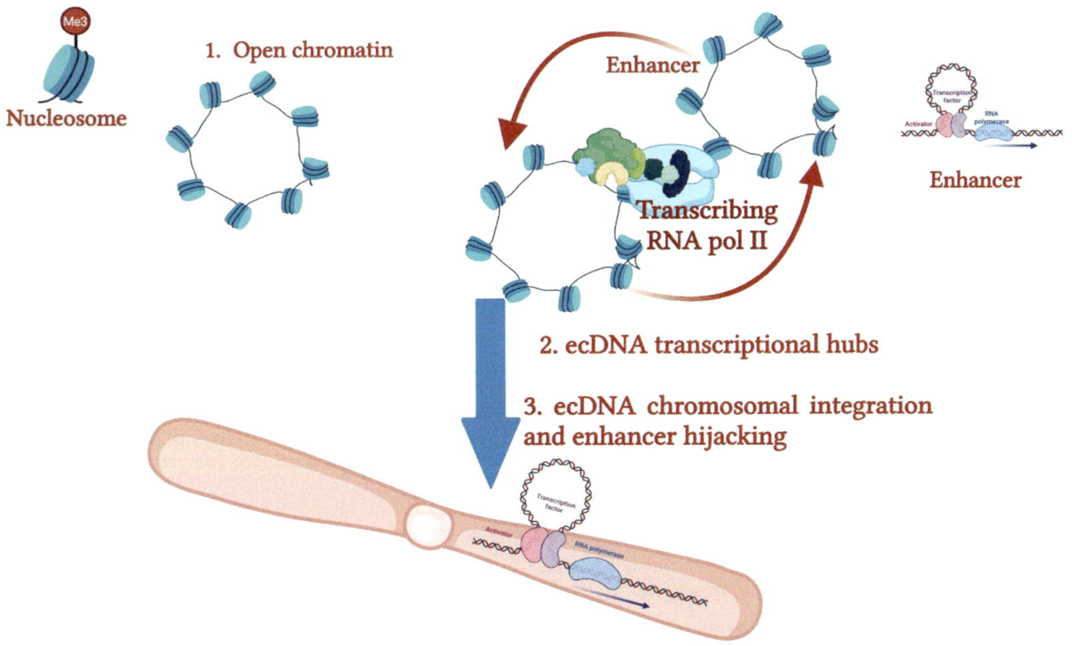

FIGURE 13.1 Functional dynamics of ecDNA: open chromatin, transcriptional hubs, and enhancer hijacking.

13.2.2.1 ecDNA Increases Copy Number and Chromatin Accessibility to Boost Oncogene Expression

Many cancers, including glioblastoma, amplify oncogenes on ecDNA [9]. The expression of oncogenes may be enhanced by increasing the copy number of ecDNA by focused amplification. Unlike chromosomes, ecDNA lacks higher-order compaction, making chromatin more accessible [7].

13.2.2.2 ecDNA Promotes Oncogene Expression by Hijacking Enhancers

ecDNA may hijack proximal or peripheral enhancers to boost oncogene activation via ultra-long chromatin interactions [7]. The enhancer and oncogene cooperate to promote expression. Chromatin capture analysis showed that distal enhancer e4 interacted with MYCN promoter to increase MYCN expression in neuroblastoma cell lines [10].

13.2.2.3 ecDNA Regulates Oncogene Expression by Interacting with Chromosomal DNA or Forming Hubs

ecDNA may influence gene expression by interacting with chromosomal DNA or forming hubs [7]. It interacts with chromosomal DNA to promote transcription as a mobile transacting component. Connecting to generate ecDNA hubs may modulate oncogene expression [11].

13.2.2.4 DNA Recombination Reservoir ecDNA

Serving as a reservoir for DNA recombination is one of ecDNA's other crucial roles. Extrachromosomal circular elements or DNA fragments excision or ligation create ecDNA, which stores DNA recombination [7]. Reintegration into linear chromosomes causes intra- and inter-chromosomal rearrangements. WGS data showed that most intra- and inter-chromosomal rearrangements in neuroblastoma genomes correlated with extrachromosomal circularization, indicating that ecDNA reintegration remodels the genome [12]. ecDNA promotes oncogene expression, chromatin accessibility, and DNA recombination, contributing to cancer initiation and evolution [9].

13.2.3 Challenges in Distinguishing ecDNA from Chromosomal DNA

Although it originates from chromosomes, extrachromosomal circular DNA (eccDNA) is different from chromosomal DNA [13]. They are present in humans, nematodes, fruit flies, and yeast. Humans have three sizes and copies of extrachromosomal circular DNA. Traditional sequencing and mapping techniques failed to identify ecDNA from chromosomal DNA. Sequence similarity is very important in ecDNA because it comprises perfect, or near-perfect copies of genomic areas amplified from chromosomes. Traditional short-read sequencing cannot differentiate circular and linear DNA in linear reference genomes. High-copy-number ecDNA patches in sequencing data may be misinterpreted as chromosomal amplifications without computational tools. ecDNA breakpoint junctions cannot be resolved by traditional mapping methods because they cannot recognize circular topologies [14].

The lack of centromeres and telomeres in ecDNA is another concern. ecDNA lacks these features, making it harder to find in the nucleus during imaging or karyotyping tests, despite its relevance for chromosomal identification and segregation. Even within a single tumor, ecDNA varies in shape, dimension, and amount across cell populations [3]. Computational reconstruction and comparisons of datasets are made more difficult by biological variability. Interfering with the distinction between extrachromosomal as well as chromosomal components, ecDNA can either integrate back into the genome or generate concatenated multimers. ecDNA identification and verification are being improved by long-read sequencing, Circle-Seq, and high-resolution imaging; nonetheless, analytical difficulties still exist. To distinguish ecDNA from chromosomal DNA, sequencing, enhancement, and computational workflows must be enhanced [15].

13.3 HIGH-THROUGHPUT SEQUENCING TECHNIQUES FOR ECDNA PROFILING

To better understand how extrachromosomal circular DNA (ecDNA) is formed, structured, and what functions it plays in cancer, high-throughput sequencing tools have become crucial. Oncogenes and regulatory elements are often found in ecDNAs, which are acentric DNA elements that are circular in shape. Sequencing techniques that can identify variations in structure and unusual genomic characteristics are necessary for their recognition and evaluation [3].

13.3.1 Whole Genome Sequencing (WGS)

Initial discovery of ecDNA is most often accomplished using whole genome sequencing (WGS). It gives a complete picture of the genome without bias, which helps scientists find structural changes, copy number variations, and localized amplifications. Focused areas of high copy number with acute breakpoints, frequently surrounded by complicated junctions, are common features of ecDNAs. Circular DNA is reconstructed and classified from WGS datasets using specialized computational tools like ECdetect, Circle-Map, and AmpliconArchitect. Despite its extensive coverage and high resolution, short-read WGS may not be able to decipher intricate ecDNA structures without computational assistance [16].

13.3.2 ATAC-Seq and RNA-Seq

Another robust approach that measures chromatin accessibility is ATAC-seq, which stands for Assay for Transposase-Accessible Chromatin utilizing sequencing. ecDNAs are very transcriptionally active because they often have open chromatin regions. Active enhancers and promoters inside ecDNA molecules may be identified using ATAC-seq, which reveals areas of enhanced accessibility [17]. This method gives epigenomic context, which is useful for WGS. The functional effects of ecDNA can only be understood with the use of RNA-seq, also known as transcriptome sequencing. The increased copy number and open chromatin state of oncogenes housed on ecDNA, including CCND1, MYC, or EGFR, causes them to be produced at much greater levels than their chromosomal counterparts5. Science has made great strides in connecting structural presence with gene activity via the integration of RNA-seq and WGS [18].

13.3.3 Single-Cell Genomics and Long-Read Sequencing

The ability to conduct analysis at the cellular level has allowed single-cell sequencing technologies to further develop ecDNA profiling. This clonal diversity and heterogeneity may be uncovered by using single-cell DNA-seq to show how ecDNA composition differs among single tumor cells. Similarly, ecDNA's cell-by-cell transcriptional effects may be evaluated using single-cell RNA-seq. Due to ecDNA-driven oncogene amplification, this method is essential for identifying cell groups that may be more resistant to treatment [19].

A lot of the problems with short-read approaches are addressed by long-read sequencing systems like PacBio and Oxford Nanopore Technologies' (ONT) offerings. These systems provide readings that cover many kilobases, allowing for the direct viewing of whole ecDNA molecules in a single read. ecDNA multimers, concatemers, and complicated rearrangements are easier to identify using long-read sequencing than with conventional short-read technologies [19]. Another cutting-edge approach developed for circular DNA enrichment is Circle-Seq. With this method, circular DNA is enriched for sequencing by selectively degrading linear DNA with exonucleases. Circle-Seq works well with high-throughput sequencing to improve the efficiency of detecting ecDNA fragments with low abundance. There are merits and limits to each sequencing strategy. As compared to WGS, ATAC-seq and RNA-seq provide functional information rather than structural detail; single-cell techniques have limitations due to coverage depth; and long-read platforms, while informative, are often more costly and have greater error rates. For that reason, an improved and more comprehensive ecDNA profile is produced by integrating these methods into processes. The capacity to identify, characterize, and comprehend ecDNA in cancer has been greatly enhanced by high-throughput sequencing. Unraveling the biochemical and clinical consequences of ecDNA will continue to rely on integrative sequencing methodologies, even as technology marches forward [20].

13.4 COMPUTATIONAL TOOLS AND PIPELINES FOR ECDNA DETECTION

13.4.1 Circle-Map

In order to identify genomic rearrangements that support the circular DNA shape, Circle-Map [1] uses the alignment of the read sequence to the reference genome as input. This alignment is then utilized to detect the splitting of the read into two halves. The loss of multiple split alignments may occur when using Circle-MAP [16]. This usually happens when the split segment is either too short or the sequence identity is too strong to align to more than two spots. A soft-clipping alignment was used to describe each of these profiles. To create a circular DNA breakpoint map, circle-map first employs posterior probability to realign the soft-spliced segment with the reference genome. The precision of identifying circular DNA junction locations is improved by this improved method [21].

13.4.2 AmpliconArchitect

For the purpose of detecting and evaluating amplicon fine structure, AmpliconArchitect [1] was created. A wide variety of human-viral heterozygous extrachromosomal amplicon structures inside various tumor types have been effectively identified by this method. The technology achieves extensive study of these structures by combining the use of PacBio sequencing with metaphase fluorescence in situ hybridization (FISH) [22]. In the UPCI: SCC090 cell line, the fine structure of an extrachromosomal heterozygous amplicon containing FOXE1 (Forkhead box E1) was mapped. AmpliconArchitect operates based on two aspects of ecDNA: (1) there was a significant increase in the copy number, and (2) it is an SV, or special ring structure. One such tool is AmpliconArchitect, which can recreate ecDNA Amplicon structures using data from whole-genome sequencing. In order to restore the actual eccDNA structure to the best of our abilities, AmpliconArchitect offers a universal Amplicon model that encodes all supporting structures and an algorithmic framework for rebuilding the potential amplicon structures [1, 22].

13.4.3 EcSeg, Circle-Seq, and Other Emerging Tools

The use of ecDNA in cancer therapy requires thorough, reliable, and economical research. These days, advancements rely on widefield optical imaging and gene sequencing, and their immediate uses are being demonstrated [21]. With widefield images, ecDNA may be visualized, revealing the focal magnification's geographic location [22]. According to earlier studies, ecDNA may be discolored using the fluorescein DAPI (4,6-diamidino-2-phenylindole), and it can be verified using centromere-specific FISH (fluorescence in situ hybridization) markers and genetic coding. FISH probing off-target proto-oncogenes typically reveal advantageous sites for both genomic and extrachromosomal genes in resting cells [23]. The compressed grid of genetic code permits the noticeable restriction of specific genes within the chromosomes or ecDNA at the same time as metaphase. However, the combination of DAPI with FISH prevents an unbiased evaluation of the number of repetitions in ecDNA, which is necessary to analyze the variation. In order to address this problem, scientists developed an image editing tool called ECdetect [24], which provides a thorough, repetitive, and accurate evaluation for

DAPI-labeled cells' ecDNA in a partly controlled way. Therefore, the AmpliconReconstructor (AR) tool combines NGS data with distant optical projection (OM) of a wider sequenced tract to provide insight into the reformation of the breakdown graph. Prior to the breakdown graph similarities, AR transforms the sequencing contig with the OM information. It then reliably transforms ecDNA, which can provide precise ecDNA quantification and aid in understanding the ecDNA synthesis procedure [25].

13.4.4 Strengths, Limitations, and Comparative Analysis

The growing reliance on laboratory models and genome-wide association procedures with laboratory animals of disease is another aspect of data fusion and the "omics" combo. Up to now, there have existed several successful instances of integrative genomes that duplicate or confirm signals using integrated omics [26]. For example, in vitro studies of lymphoblastoid cells have demonstrated success in identifying transcriptional and genome development outputs that support clinical sequencing consequences. Results from other organisms, such as domestic dogs, have additionally been shown to be a great way to combine omics [27]. A comparison method using the canine's clasp showed potential in the study of cancer. Comparing the genomic copy-number dispersion between humans and dogs has given researchers invaluable insight into the development of bone marrow cancer lymphoma, intracranial malignancies, and other kinds of tumors [28].

13.5 INTEGRATIVE BIOINFORMATICS APPROACHES

13.5.1 Multi-Omics Integration for ecDNA Function Prediction

To identify individuals with ecDNA and to restore ecDNA organization, the AmpliconArchitect (AA) program was used to analyze genome-wide association data from gastrointestinal cancer cell lines [29]. Additionally, full-exome sequencing (WES) and RNA analysis were performed on patients with cancer of the gastric region. The abundance of ecDNA was assessed using the GCAP statistics, and its relationship to mortality rates was examined [30]. When it comes to DNA damages and immune-associated chain reactions, ecDNA appears normal in gastric cancer. The development of SNU216-TR and SNU216 demonstrated that rather than merely producing treatment-resistant paralogs, the unique ecDNA in SNU216-TR may be responsible for resistance to medications via introns. Both large-scale ecDNA decoding and separation using the recently chosen ecDNA purifying technique CRISPR-CATCH have made use of the TGS application nanopore. Furthermore, TGS-dependent scWGS processes that enable SV and ecDNA identification have implicitly evolved [31].

13.5.2 Machine Learning Applications in ecDNA Classification

The resource information CytoCellDB, which guides the classification of ecDNA in lines of cancer cells, is demonstrated in this work. Five hundred and seventy-seven cell lines queryable distinguish between types of datasets, including information on chromosomal abnormalities, HSRs, and additional ecDNA, are included in CytoCellDB [32]. According to the information accessible, CytoCellDB is probably the biggest library of its kind, with 438 genuine negative (TN; ecDNA-) and 139 actually positive (TP; ecDNA+) specimens. This indicates a 400% increase in cell types with ecDNA labels and a 200% increase in total ecDNA+ data. Most of our mathematical divinations are verified by investigators using FISH tests and G-banding karyotyping. As a benefit, CytoCellDB opens up a new avenue for investigating the molecular impact of ecDNA globally and will come before the development of personalized treatments, indicators, and treatment plans for malignancies driven by ecDNA-dependent proliferation [33].

13.5.3 Network Biology and Pathway Enrichment Analysis

Scientists employed RNA sequencing to ascertain the expression of genes along with the patterns of circular DNA over identical specimens in order to thoroughly characterize the typical epithelium-adenoma-adenocarcinoma pattern on many different levels [34]. Interestingly, transcriptome analysis reveals two distinct rising clusters and reveals particular gene sequencing. As of right now, over 3,000 genes have been identified from these clusters and have been found to be distinct in pathways linked to NF-kB communication, immune response-associated channels, and cell cycle regulation, which includes the receptor pathway for T-cells signaling process, antigen presentation, and digestion. Building a robust matrix of expression patterns through the use of microarray analysis has been useful for assessing gene expression datasets from rare, unavailable forms of thyroid cancer, such as ATC. In comparison to over 60 normal thyroid tissues, global gene expression was assessed in roughly 100 PTC and 20 ATC specimens [35].

13.6 VISUALIZATION AND STRUCTURAL RECONSTRUCTION OF ECDNA

13.6.1 Cytogenetic Tools—DNA Fluorescence in Situ Hybridization (FISH)

Conventional microscope technology was used to first detect ecDNA. DNA luminescence in situ hybridization (DNA-FISH) allows examination of genetic locus, including cancer-promoting genes and regulatory sequences visible on ecDNA, and they are primarily visible on the phase of metaphase propagating via 4′,6-diamidino-2-phenylindole (DAPI) stains. RNA-FISH can detect ecDNAs' nascent translations of RNA by utilizing probes that target intronic parts. These morphological techniques continue to be the most effective means of identifying individual ecDNA in cells, as well as their regional makeup and ecDNA/HSR interactions [36]. FISH relies on the highest grade epifluorescence, or gradually confocal-based, imaging, particularly when combined with quantification commitment, and requires some prior knowledge of the nucleotides that are present on ecDNA. It is possible to compare the transcriptional efficiency between DNA- and RNA-FISH information, but it is difficult to combine them without causing any signal attenuation. To manage this restriction, research has previously used repeated screening of nuclei to halt DNA- and RNA-FISH findings [37].

13.6.2 Live-Cell Imaging

Live-cell imaging would enable the observation of ecDNA progress in the interim, in opposition to DNA-FISH, which requires cell preservation. Although there are evolutionary programs for this, it is still difficult for scientists to observe single-copy loci in living cells [38]. The Casilio setup directs different fluorescence reporter genes to a single region using a dead-Cas9, guiding RNAs (gRNAs) that pick the ecDNA limit, and an RNA-aptamer. This technique raises some concerns regarding the buildup of fluorescent materials by producing large foci of fluorescent RNA-binding polypeptide of varying sizes that appear noticeably larger and more irregular than the messages produced by FISH. Two-hued ecDNA labeling of the ecEGFR limit is employed to customize call cases of co-localization, and it has been suggested that the huge foci of signal from probes directed to ecDNA result from centers of aggregate ecDNA. However, it was necessary to turn on the difference between the information produced by DNA-FISH and actual cell viewing. Other techniques, which rely on the orientation of fluorescent fusion proteins with respect to non-segregated bacterium series, could possibly be modified to visualize ecDNA in cells that operate; however, this has not yet been attempted and involves an organization of bacterial lineages into ecDNA [39].

13.6.3 Combining DNA and Protein Imaging

The analysis of loci placement proportionate to nuclear protein molecules, such as RNA polymerase II, is guided by the combination of fluorescence and genetic site labeling. The "hubs" or condensation concept, which postulates that the primary RNA Pol II, co-activators, and transcription variables will converge alongside and split off from the shared nucleoplasm, makes this intriguing [40]. However, because condensed matter is lighter and more active, further research into this theory requires upscaling or, alternatively, lone individual scanning. There is an ongoing debate on the role that these regulatory regions play in regulating genomic activation. It was established that ecDNA may assemble in a rotating manner, enhancing its translated item, because it contains cancer-promoting genes and desired enhancement factors, as well as larger patches of accessible chromatin. Scientists failed to identify a close spatial association among ecDNA and big RNA Pol II foci (>500 nm) in glioblastoma cell lines, despite the fact that some imaging methodologies have revealed a more-than-anticipated overlap between RNA Pol II and ecDNA markers [41].

13.6.4 3D Genomics and Chromatin Interaction Data

The dynamic and intricate process of three-dimensional (3D) chromatin activity in the nucleus has a significant impact on a variety of processes in cells, including DNA replication, genetic control, and modification. Given that malignancies are the primary focus of present research, it is critical to comprehend the geographic spread and exchange of genomic information for the sake to disclose the complications of diseases [42]. The capacity of Hi-C technology to measure and record the geographical proximity to genetic areas within a nuclear domain forms its basis. In order to build these structures, this influential approach guides the comprehensive study of crucial chromatin restriction, which includes chromatin cycles, topologically associating domains (TADs), and the role of insulating proteins like cohesin and CCCTC-binding factor (CTCF). Nevertheless, Hi-C can be used to detect more advanced levels of chromatin alteration in addition to primary chromatin tightness, offering a thorough understanding of how the genetic material works as a single, cohesive system [43, 45].

13.7 APPLICATIONS IN CANCER DIAGNOSIS AND THERAPY

13.7.1 ecDNA-Based Biomarkers for Precision Oncology

Real-time surveillance of cancer features and changes is made possible by the ability of liquid biopsies to collect and identify ecDNA present in the blood of cancer survivors using Circa-seq. EccDNA is a powerful biochemical marker for a number of illnesses associated with chromosome mutations and changes [44]. Acute mixed immunosuppression condition has been shown to benefit from the incorporation of eccDNA derived from T-cell channel genes. For prenatal evaluation, early detection, estimation, and treatment of gynecologic malignancies, extrachromosomal circular DNA (ecDNA) can serve as a useful target for treatment and a noninvasive diagnostic marker. ecDNA with MYCN in a biallelic form may be a powerful biochemical marker for establishing risk assessment in neoplasm patients [45].

13.7.2 Computationally Guided Therapeutic Targeting of ecDNA

In their investigation of the role of genomic circular DNA in neurological tumors, researchers combined transcriptomics and genomics and discovered a variety of hitherto unidentified extrachromosomal circular DNA [46]. Additionally, it has been revealed that the main component of somatic chromosomal combination is extrachromosomal circular DNA. The capacity of circular DNA to undergo chimerism, reintegrate into the linear genome, and improve the development of genes that cause cancer is demonstrated. This study showed that extrachromosomal circular DNA can control aberrant tumor-linked variations in gene expression in addition to driving genomic recombination. This study has only focused on neuroma; future investigations can expand to include different types of tumors. Additionally, it offers new insights that can guide the exploration of alternative functions carried out by extrachromosomal circular genetic DNA [47].

13.7.3 Role in Treatment Monitoring and Resistance Prediction

Short segments of circulatory DNA that encode cancer-causing genes were found; these segments might be typical of carcinoma cells. These pieces are essential for creating heterogeneity between cells, which aids in the difficult treatment of aggressive malignancies. This study demonstrates that as tumor-promoting genes proliferate in ecDNA, malignancies become more diverse or shifted, allowing them to keep on expressing more cancer-causing genes [48]. When tumor cells split, ecDNA will randomly enter the progeny cells; hence, a single carcinoma cell may not have ecDNA in the center of the cell. This is how cancer individuals differ from genomes. Surprisingly, variability among cells in a tumor increases with the degree of genetic changes in ecDNA. Cellular variance can increase tumor adaptability as it progresses toward external constraints, primarily medication [49].

13.8 CURRENT CHALLENGES AND FUTURE DIRECTIONS

Medical predictability is the main obstacle to diagnosing, predicting, and treating disorders. It affects the use of machine learning algorithms as well as human (physicians) medical professionals. According to this concept, there is no known cure for an indicator that a person presents for evaluation. In this instance, the patterns and processes associated with biological investigative exploration, as well as the degree of uncertainty identified, have been shown through the use of statistical and mathematical representations. There have been several algorithms using deep learning introduced that quantify the variation in the classification results [47]. For a more reliable and context-dependent evaluation of the results, the wet-lab research and the data it produces can then be integrated with the computerized modeling approach. The most promising technological system for the study of multi-omics data combinations has been the use of deep learning. The complete instruction architecture and the system's improved feature presentation nature have contributed to its increased success. The most crucial step is creating associations between the information from omics [49].

13.9 CONCLUSION

Human disorders feature composite metabolic networks. The developing aspects of biological information are essential for fully understanding the molecular structure and appreciating the biological processes involved in illnesses. In order to clarify the significance of multi-omics in disease categorization, diagnosis, forecasting, and therapy, this chapter presents an overview of algorithmic organization. Among the conditions that have been studied are Alzheimer's, cancer, and Down syndrome. We give a comprehensive overview of the data held in omics libraries and talk about the execution constraints. The chapter discussed that deep learning systems may accurately diagnose illnesses at a level comparable to that of medical professionals. To improve the accuracy and comprehension of the next research, the drawbacks of machine learning, such as the complexity of the designs, must be tackled.

REFERENCES

1. Wu, S., Tao, T., Zhang, L., Zhu, X., & Zhou, X. (2023). Extrachromosomal DNA (ecDNA): Unveiling its role in cancer progression and implications for early detection. Heliyon, 9(11).
2. Luebeck, J., Ng, A. W., Galipeau, P. C., Li, X., Sanchez, C. A., Katz-Summercorn, A. C., . . . & Verhaak, R. G. (2023 Apr 27). Extrachromosomal DNA in the cancerous transformation of Barrett's oesophagus. Nature, 616(7958), 798–805.
3. Verhaak, R. G., Bafna, V., & Mischel, P. S. (2019). Extrachromosomal oncogene amplification in tumour pathogenesis and evolution. Nature Reviews Cancer, 19(5), 283–288.
4. Dong, Y., He, Q., Chen, X., Yang, F., He, L., & Zheng, Y. (2023). Extrachromosomal DNA (ecDNA) in cancer: Mechanisms, functions, and clinical implications. Frontiers in Oncology, 13, 1194405.
5. Shachar, S., Voss, T. C., Pegoraro, G., Sciascia, N., & Misteli, T. (2015). Identification of gene positioning factors using high-throughput imaging mapping. Cell, 162(4), 911–923.

6. Corces, M. R., Granja, J. M., Shams, S., Louie, B. H., Seoane, J. A., Zhou, W., . . . & Chang, H. Y. (2018). The chromatin accessibility landscape of primary human cancers. Science, 362(6413), eaav1898.
7. Yan, X., Mischel, P., & Chang, H. (2024). Extrachromosomal DNA in cancer. Nature Reviews Cancer, 24(4), 261–273.
8. Kim, H., Nguyen, N. P., Turner, K., Wu, S., Gujar, A. D., Luebeck, J., . . . & Verhaak, R. G. (2020). Extrachromosomal DNA is associated with oncogene amplification and poor outcome across multiple cancers. Nature Genetics, 52(9), 891–897.
9. Morton, A. R., Dogan-Artun, N., Faber, Z. J., MacLeod, G., Bartels, C. F., Piazza, M. S., . . . & Scacheri, P. C. (2019). Functional enhancers shape extrachromosomal oncogene amplifications. Cell, 179(6), 1330–1341.
10. Helmsauer, K., Valieva, M. E., Ali, S., Chamorro González, R., Schöpflin, R., Röefzaad, C., . . . & Koche, R. P. (2020). Enhancer hijacking determines extrachromosomal circular MYCN amplicon architecture in neuroblastoma. Nature Communications, 11(1), 5823.
11. Windle, B., Draper, B. W., Yin, Y. X., O'Gorman, S., & Wahl, G. M. (1991). A central role for chromosome breakage in gene amplification, deletion formation, and amplicon integration. Genes & Development, 5(2), 160–174.
12. Smith, J. P., & Sheffield, N. C. (2020). Analytical approaches for ATAC-seq data analysis. Current Protocols in Human Genetics, 106(1), e101.
13. Yuan, X. Q., Zhou, N., Song, S. J., Xie, Y. X., Chen, S. Q., Yang, T. F., . . . & Peng, L. (2024). Decoding the genomic enigma: Approaches to studying extrachromosomal circular DNA. Heliyon, 10(17).
14. Thompson, P. M., Hibar, D. P., Stein, J. L., & Jahanshad, N. (2016). Imaging genomics and ENIGMA. In Genomics, Circuits, and Pathways in Clinical Neuropsychiatry (pp. 101–115). Academic Press.
15. Shoshani, O., Brunner, S. F., Yaeger, R., Ly, P., Nechemia-Arbely, Y., Kim, D. H., . . . & Cleveland, D. W. (2021). Chromothripsis drives the evolution of gene amplification in cancer. Nature, 591(7848), 137–141.
16. Deshpande, V., Luebeck, J., Nguyen, N. P. D., Bakhtiari, M., Turner, K. M., Schwab, R., . . . & Bafna, V. (2019). Exploring the landscape of focal amplifications in cancer using amplicon architect. Nature Communications, 10(1), 392.
17. Grandi, F. C., Modi, H., Kampman, L., & Corces, M. R. (2022). Chromatin accessibility profiling by ATAC-seq. Nature Protocols, 17(6), 1518–1552.
18. Fang, R., Preissl, S., Li, Y., Hou, X., Lucero, J., Wang, X., . . . & Ren, B. (2021). Comprehensive analysis of single cell ATAC-seq data with SnapATAC. Nature Communications, 12(1), 1337.
19. Gupta, P., O'Neill, H., Wolvetang, E. J., Chatterjee, A., & Gupta, I. (2024). Advances in single-cell long-read sequencing technologies. NAR Genomics and Bioinformatics, 6(2), lqae047.
20. Tian, L., Jabbari, J. S., Thijssen, R., Gouil, Q., Amarasinghe, S. L., Voogd, O., . . . & Ritchie, M. E. (2021). Comprehensive characterization of single-cell full-length isoforms in human and mouse with long-read sequencing. Genome Biology, 22, 1–24.
21. Turner, K. M., Deshpande, V., Beyter, D., Koga, T., Rusert, J., Lee, C., . . . & Mischel, P. S. (2017). Extrachromosomal oncogene amplification drives tumour evolution and genetic heterogeneity. Nature, 543(7643), 122–125.
22. Prada-Luengo, I., Krogh, A., Maretty, L., & Regenberg, B. (2019). Sensitive detection of circular DNAs at single-nucleotide resolution using guided realignment of partially aligned reads. BMC Bioinformatics, 20, 1–9.
23. Kumar, P., Dillon, L. W., Shibata, Y., Jazaeri, A. A., Jones, D. R., & Dutta, A. (2017). Normal and cancerous tissues release extrachromosomal circular DNA (eccDNA) into the circulation. Molecular Cancer Research, 15(9), 1197–1205.
24. Mann, L., Seibt, K. M., Weber, B., & Heitkam, T. (2022). ECCsplorer: A pipeline to detect extrachromosomal circular DNA (eccDNA) from next-generation sequencing data. BMC Bioinformatics, 23(1), 40.
25. Rajkumar, U., Turner, K., Luebeck, J., Deshpande, V., Chandraker, M., Mischel, P., & Bafna, V. (2019). EcSeg: Semantic segmentation of metaphase images containing extrachromosomal DNA. Iscience, 21, 428–435.
26. Thomas, R., Duke, S. E., Wang, H. J., Breen, T. E., Higgins, R. J., Linder, K. E., . . . & Breen, M. (2009). 'Putting our heads together': Insights into genomic conservation between human and canine intracranial tumors. Journal of Neuro-Oncology, 94, 333–349.
27. Rotroff, D. M., & Motsinger-Reif, A. A. (2016). Embracing integrative multiomics approaches. International Journal of Genomics, 2016(1), 1715985.
28. Brown, C. C., Havener, T. M., Medina, M. W., Auman, J. T., Mangravite, L. M., Krauss, R. M., . . . & Motsinger-Reif, A. A. (2012). A genome-wide association analysis of temozolomide response using lymphoblastoid cell lines shows a clinically relevant association with MGMT. Pharmacogenetics and Genomics, 22(11), 796–802.

29. Yong, J. X., & Zeng, Z. L. (2024). IDDF2024-ABS-0202 Multi-omics profiling reveals characteristics and associations of gastric cancer with extrachromosomal DNA (ecDNA) and drug resistance. BMJ Publishing Group Ltd & British Society of Gastroenterology.
30. Fan X., Yang, C., Li, W., Bai, X., Zhou, X., Xie, H., Wen, L., Tang, F. (2021). SMOOTH-seq: Single-cell genome sequencing of human cells on a third-generation sequencing platform. Genome Biology, 22(1), 195. https://doi.org/10.1186/s13059-021-02406-y.
31. Van Dijk, E. L., Jaszczyszyn, Y., Naquin, D., & Thermes, C. (2018). The third revolution in sequencing technology. Trends in Genetics, 34(9), 666–681.
32. Fessler, J., Ting, S., Yi, H., Haase, S., Chen, J., Gulec, S., . . . & Brunk, E. (2024). CytoCellDB: A comprehensive resource for exploring extrachromosomal DNA in cancer cell lines. NAR Cancer, 6(3), zcae035.
33. Sun, H., Lu, X., & Zou, L. (2023 Jan 1). EccBase: A high-quality database for exploration and characterization of extrachromosomal circular DNAs in cancer. Computational and Structural Biotechnology Journal, 21, 2591–2601.
34. Lu, L., Chen, M., Zhang, G., Liu, Y., Xu, X., Jiang, Z., . . . & Xu, H. (2024). Comprehensive profiling of extrachromosomal circular DNAs in colorectal cancer progression. Scientific Reports, 14(1), 28519.
35. Geraldo, M. V., & Kimura, E. T. (2015). Integrated analysis of thyroid cancer public datasets reveals role of post-transcriptional regulation on tumor progression by targeting of immune system mediators. PLoS One, 10(11), e0141726.
36. Purshouse, K., Friman, E. T., Boyle, S., Dewari, P. S., Grant, V., Hamdan, A., . . . & Bickmore, W. A. (2022). Oncogene expression from extrachromosomal DNA is driven by copy number amplification and does not require spatial clustering in glioblastoma stem cells. Elife, 11, e80207.
37. Purshouse, K., Pollard, S. M., & Bickmore, W. A. (2024). Imaging extrachromosomal DNA (ecDNA) in cancer. Histochemistry and Cell Biology, 162(1), 53–64.
38. Alexander, J. M., Guan, J., Li, B., Maliskova, L., Song, M., Shen, Y., . . . & Weiner, O. D. (2019). Live-cell imaging reveals enhancer-dependent Sox2 transcription in the absence of enhancer proximity. Elife, 8, e41769.
39. Clow, P. A., Du, M., Jillette, N., Taghbalout, A., Zhu, J. J., & Cheng, A. W. (2022). CRISPR-mediated multiplexed live cell imaging of nonrepetitive genomic loci with one guide RNA per locus. Nature Communications, 13(1), 1871.
40. Zhu, Y., Gujar, A. D., Wong, C. H., Tjong, H., Ngan, C. Y., Gong, L., . . . & Wei, C. L. (2021). Oncogenic extrachromosomal DNA functions as mobile enhancers to globally amplify chromosomal transcription. Cancer Cell, 39(5), 694–707.
41. Mir, M., Bickmore, W., Furlong, E. E., & Narlikar, G. (2019). Chromatin topology, condensates and gene regulation: Shifting paradigms or just a phase? Development, 146(19), dev182766.
42. Yoon, I., Kim, U., Jung, K. O., Song, Y., Park, T., & Lee, D. S. (2024). 3C methods in cancer research: Recent advances and future prospects. Experimental & Molecular Medicine, 56(4), 788–798.
43. Wu, S., Turner, K. M., Nguyen, N., Raviram, R., Erb, M., Santini, J., . . . & Mischel, P. S. (2019). Circular ecDNA promotes accessible chromatin and high oncogene expression. Nature, 575(7784), 699–703.
44. Luebeck, J., Coruh, C., Dehkordi, S. R., Lange, J. T., Turner, K. M., Deshpande, V., . . . & Bafna, V. (2020). Amplicon reconstructor integrates NGS and optical mapping to resolve the complex structures of focal amplifications. Nature Communications, 11(1), 4374.
45. Luo, X., Zhang, L., Cui, J., An, Q., Li, H., Zhang, Z., . . . & Xiao, F. (2023). Small extrachromosomal circular DNAs as biomarkers for multi-cancer diagnosis and monitoring. Clinical and Translational Medicine, 13(9), e1393.
46. Chiu, R. W., Dutta, A., Henssen, A. G., Lo, Y. D., Mischel, P., & Regenberg, B. (2020). What is extrachromosomal circular DNA and what does it do? Clinical Chemistry, 66(6), 754–759.
47. Arco Martín, J. E., Ramírez Pérez De Inestrosa, J., Martínez Murcia, F. J., & Gorriz Sáez, J. M. (2022). Uncertainty-driven ensembles of multi-scale deep architectures for image classification. Elsevier.
48. Wu, N., Wei, L., Zhu, Z., Liu, Q., Li, K., Mao, F., . . . & Zhao, X. (2024). Innovative insights into extrachromosomal circular DNAs in gynecologic tumors and reproduction. Protein & Cell, 15(1), 6–20.
49. Gong, P., Cheng, L., Zhang, Z., Meng, A., Li, E., Chen, J., & Zhang, L. (2023). Multi-omics integration method based on attention deep learning network for biomedical data classification. Computer Methods and Programs in Biomedicine, 231, 107377.

14 ecDNA and Immune Modulation

Progress and Challenges of Cancer Immunotherapy

Debasmita Paul,[1] Arijit Das,[2] Abhijit Ghosh,[2] and Biplab Debnath[1]

1 Department of Pharmaceutical Technology, Bharat Technology Uluberia, India
2 Department of Pharmaceutical Technology Bharat Technology West Bengal, India

14.1 INTRODUCTION

14.1.1 Overview of Cancer Immunotherapy and Extrachromosomal DNA (ecDNA)

Cancer immunotherapy improved the field of oncology by using the body's immune system to find and kill cancer cells. Chemotherapy and radiotherapy, common cancer treatments, damage both healthy and cancerous tissues without being selective. Immunotherapy changes immune responses to tumor-specific antigens as a more targeted approach. This type of treatment includes a lot of different methods, such as immune checkpoint inhibitors (ICIs), chimeric antigen receptor (CAR) T-cell therapy, cancer vaccines, and oncolytic viruses [1]. These treatments have long-lasting clinical benefits for many types of cancer, especially melanoma, non-small cell lung cancer (NSCLC), and renal cell carcinoma [2]. Cancer cell evolution and tumor heterogeneity are two of the most difficult aspects of cancer therapy, as they drive natural and adaptive responses to anti-cancer medications. Gain-of-function mutations of oncogenes and drug resistance genes, as well as downregulation of tumor suppressor genes, are examples of genetic changes that contribute to adaptability. Tumors frequently exhibit large-scale genomic changes, such as amplifications, deletions, and loss of heterozygosity. Dysregulation at the epigenetic, transcriptional, translational, and posttranslational levels has also been shown to have a significant impact on cancer cell survival and tumor development. Though its significance is rapidly becoming apparent, the role of extrachromosomal DNA (ecDNA)—circular DNA derived from nucleosomal chromatin—in cancer is still in the research phase. More open than chromosomal DNA, ecDNA has been discovered to be a key carrier of amplified oncogenes [3], resulting in increased transcriptional activity [4]. ecDNA is inherited in a non-Mendelian fashion during mitosis [5] because it lacks a centromere, causing some tumor cells to carry high copy numbers of oncogenes, which can have a significant impact on tumor heterogeneity and drug resistance. ecDNA was discovered in several cancer types in recent studies using cytogenetics and whole-genome sequencing (WGS) but rarely in normal tissues [6]. Studies have found that ecDNA may be associated with more aggressive cancers and poorer clinical outcomes [7]. As the mechanisms of ecDNA dynamics and their role in cancer become better understood, it is expected that medications targeting ecDNA formation and maintenance will be beneficial as cancer treatments [8]. The use of immune

DOI: 10.1201/9781003655299-14

checkpoint inhibitors, especially monoclonal antibodies that target cytotoxic T-lymphocyte antigen-4 (CTLA-4), programmed death receptor-1 (PD-1), and its ligand PD-L1, has opened new ways to treat cancer [9]. But only some patients have long-lasting responses, which show complicated tumor-immune interactions and the importance of understanding resistance mechanisms. Recent research has focused on the tumor microenvironment (TME), which consists of immunosuppressive cells (like regulatory T-cells and myeloid-derived suppressor cells), cytokines (like IL-10 and TGF-β), and metabolic byproducts (like adenosine and lactate) that weaken immune effector functions [10, 11]. So, for immunotherapy to work, it needs to boost the body's ability to fight tumors and block signals that suppress the immune system in the TME [12, 13]

14.1.2 Objective and Scope of the Review

The primary objective of this review is to explore the interplay between extrachromosomal DNA and immune modulation in the context of cancer immunotherapy. As the field of ecDNA research is expanding, it is important to synthesize emerging evidence on how these extrachromosomal elements influence immune recognition, response, and resistance. Understanding the bidirectional relationship between ecDNA and the immune system could reveal novel biomarkers for immune responsiveness and aid in the creation of effective, personalized cancer therapies.

14.2 EXTRACHROMOSOMAL DNA (ECDNA) IN CANCER

14.2.1 Role of ecDNA in Oncogene Amplification and Tumor Evolution

The amplification of oncogenes via ecDNA is a mechanism by which cancer cells attain unregulated proliferation and persistence. In contrast to linear chromosomal amplification, ecDNA-mediated oncogene amplification results in significant increases in copy number without disturbing adjacent genes or regulatory regions. ecDNA can contain numerous oncogenes or enhancer elements in proximity, creating "super enhancers" that promote aggressive transcriptional programs.

The MYC oncogene, a master regulator of cell proliferation, is frequently amplified on ecDNA in various cancers, including medulloblastoma and glioblastoma multiforme (GBM). This leads to MYC overexpression and supports a stem-like phenotype in tumor cells, causing resistance against conventional therapies. Similarly, ecDNA-mediated amplification of EGFR, CDK4, and MDM2 is frequently observed in glioblastomas, supporting rapid cell proliferation, bypassing cell cycle checkpoints, and resisting apoptosis [14, 15].

Tumor evolution is influenced by ecDNA. Their random segregation during mitosis creates a non-Mendelian inheritance pattern, resulting in genetically diverse daughter cells within the same tumor mass. This heterogeneity allows tumor cells to rapidly adapt to selective pressures, such as chemotherapy or immune surveillance. ecDNA can integrate back into chromosomes or form hybrid structures with chromosomal DNA, which have complex rearrangements and further enhance tumor evolution [16].

14.2.2 Mechanisms of ecDNA Formation and Propagation in Cancer Cells

Multiple mechanisms of genome instability are involved in the complex process of ecDNA formation. The breakage–fusion–bridge (BFB) cycle is the most prevalent model for ecDNA genesis. This process commences with chromosomal double-strand breaks (DSBs), which are the result of oncogenic signaling or replication stress. These breaks can result in end-to-end chromosome fusions, which are followed by bridge formation during anaphase and subsequent breakage. Ultimately, DNA fragments are released, which circularize to form ecDNA [17]. Chromothripsis, a catastrophic one-step genomic rearrangement event, is a substantial contributor to the formation of ecDNA. In the process of chromothripsis, chromosomes are fragmented into numerous pieces and subsequently reassembled in a random manner. The production of circular DNA molecules that are enriched with oncogenic sequences and regulatory elements. The ligation and circularization of these fragments are significantly influenced by DNA repair pathways, particularly non-homologous end joining (NHEJ), which facilitates the synthesis of ecDNA. The dynamics of ecDNA propagation during cell division are distinctive. ecDNA is distributed unequally among daughter cells and lacks centromeres, contrary to chromosomes that segregate evenly. This leads to stochastic variations in ecDNA content across tumor cells. This uneven inheritance is essential for the selection of clones with a favorable oncogene dosage under therapeutic stress and the generation of cellular heterogeneity. Amplification can occur in a brief period of time as ecDNA undergoes rolling circle replication. ecDNA demonstrates open chromatin conformation, high levels of histone acetylation, and increased accessibility to transcriptional machinery, which are potent hubs of gene expression, from an epigenetic perspective [18]. ecDNA functions as a mobile regulatory element, containing enhancers that regulate the expression of chromosomal genes at distant loci, thereby complicating cancer epigenomics (Figure 14.1).

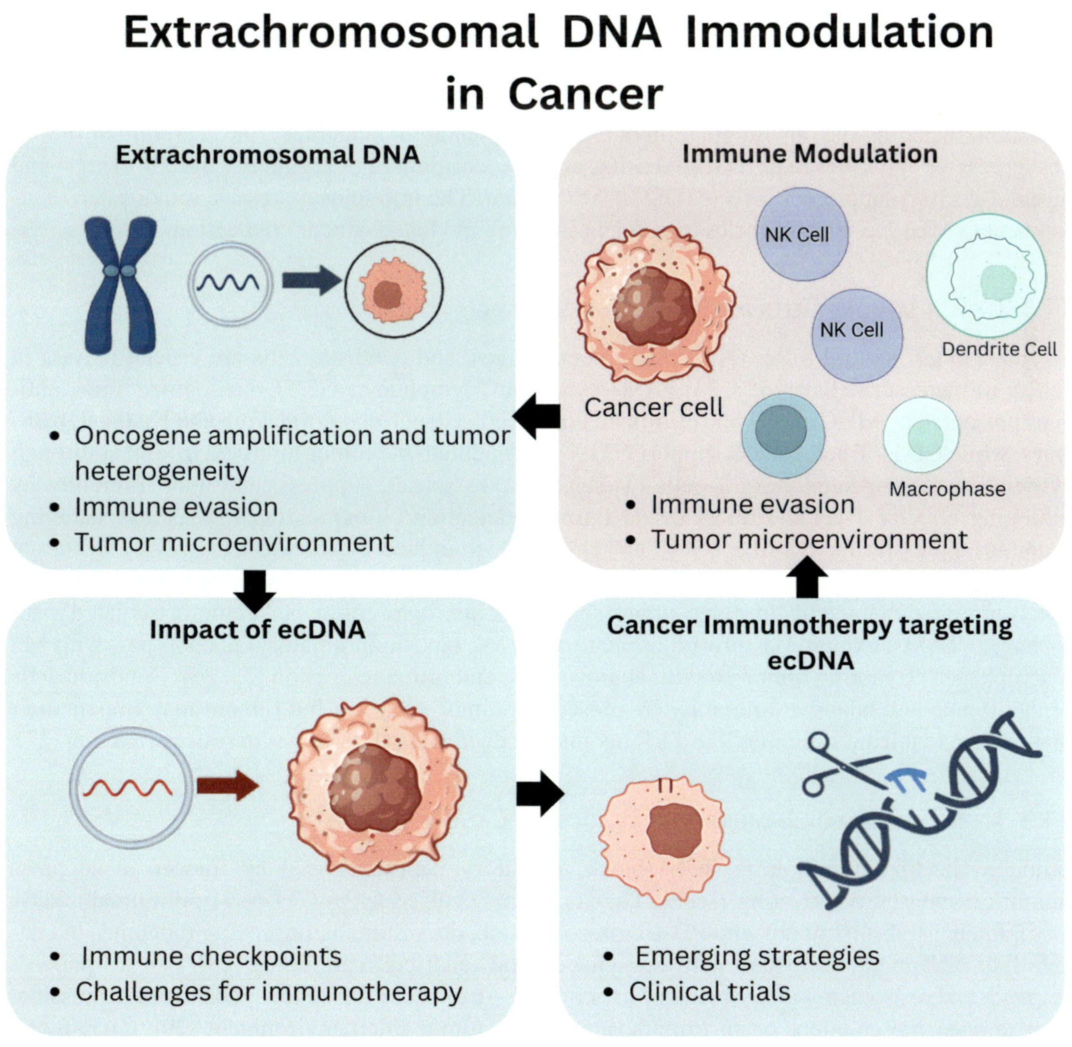

FIGURE 14.1 Extrachromosomal DNA (ecDNA) immodulation in cancer.

14.3 IMMUNE SYSTEM MODULATION IN CANCER THERAPY

14.3.1 Basics of the Immune Response against Cancer

The immune system is a network of cells and molecules responsible for detecting and eradicating abnormal cells, including malignant ones. Cancer immunosurveillance refers to the immune system's capacity to identify and inhibit tumor progression. The identification of tumor-associated antigens (TAAs) and neoantigens—proteins primarily expressed by cancer cells as a result of genetic alterations, including mutations, translocations, or epigenetic modifications, such as extrachromosomal DNA (ecDNA) [19]. Intrinsic tumor characteristics, including ecDNA-mediated oncogene amplification, enhance the expression of immune checkpoint molecules (e.g., PD-L1), disrupt antigen presentation pathways, or stimulate the release of immunosuppressive cytokines [20, 21]. Efficient modulation of the immune system is crucial for reinstating antitumor immunity. The immune surveillance hypothesis proposes that the immune system observes and eradicates emerging tumor cells. Cancer cells frequently exhibit neoantigens—peptides originating from mutated proteins—that are absent in normal cells, enabling the immune system to identify and target them. The primary immune response commences with antigen presentation by dendritic cells (DCs), which sequester tumor antigens and migrate to lymph nodes to activate naive T-cells. Activated T-cells, especially cytotoxic CD8+ T lymphocytes, infiltrate tumors and induce cell lysis by releasing perforin and granzymes. Helper CD4+ T-cells deliver crucial cytokine support, augmenting the efficacy of cytotoxic T-cells and additional immune cells [22]. Mechanisms to circumvent immune detection, including the downregulation of antigen presentation machinery, the secretion of immunosuppressive cytokines (e.g., TGF-β, IL-10), and the recruitment of regulatory T-cells (Tregs) and myeloid-derived suppressor cells (MDSCs), are present. The immunosuppressive tumor microenvironment (TME) presents obstacles to efficient immune-mediated cancer eradication.

14.3.2 Key Immune Cells in Cancer Surveillance

T lymphocytes, natural killer (NK) cells, macrophages, and dendritic cells are crucial players in cancer immune surveillance [23, 24]. CD8+ cytotoxic T lymphocytes (CTLs) recognize tumor antigens presented by MHC class I and eliminate tumor cells via perforin/granzyme and Fas/FasL pathways, while CD4+ T helper cells support CTLs and modulate the tumor microenvironment through cytokine secretion. Regulatory T-cells (Tregs), a CD4+ subset, suppress antitumor immunity by inhibiting effector T-cells. Tumors evade immune detection by expressing PD-L1 and secreting immunosuppressive factors like TGF-β and IL-10, which induce T-cell exhaustion. NK cells target tumor cells lacking MHC I, often downregulated by ecDNA-related gene silencing, but their function is inhibited in hypoxic, immunosuppressive tumor microenvironments. Tumor-associated macrophages (TAMs) exhibit M1 (pro-inflammatory) or M2 (anti-inflammatory) phenotypes, with M2 macrophages promoting tumor growth, angiogenesis, and immune evasion [25, 26]. Dendritic cells bridge innate and adaptive immunity by presenting tumor antigens, but tumors may impair their maturation, requiring strategies like TLR agonists or dendritic cell vaccines to restore function [27].

14.3.3 Strategies for Immune Modulation in Cancer Therapy

Immune checkpoint inhibitors (ICIs) block regulatory pathways used by tumors to suppress immune responses, reactivating T-cells. They target molecules such as CTLA-4 (ipilimumab), PD-1 (nivolumab, pembrolizumab), and PD-L1 (atezolizumab, durvalumab), improving outcomes in cancers like melanoma, non-small cell lung cancer, and renal cell carcinoma [28]. Not all patients respond, and resistance—either primary or acquired—can occur due to low neoantigen expression, poor antigen presentation, or an immunosuppressive tumor microenvironment [29]. CAR-T cell therapy involves engineering a patient's T-cells to express synthetic receptors that recognize tumor antigens independently of MHC, with success against CD19 in B cell malignancies. Its effectiveness in solid tumors is limited due to toxicity and antigen escape, prompting research into enhancing Chimeric Antigen Receptor T-Cell (CAR-T) cell function and tumor targeting [30]. Cytokine therapies, such as IL-2 and IFN-α, boost immune cell activity but are hampered by systemic toxicity, leading to the exploration of targeted delivery methods to improve therapeutic impact [31].

14.4 IMPACT OF ECDNA ON IMMUNE MODULATION

14.4.1 How ecDNA Affects Immune Evasion

Immune evasion causes malignant cells to evade immune surveillance. ecDNA contributes through several interrelated mechanisms, including aberrant gene expression, antigen presentation interference, and modulation of interferon signaling pathways. ecDNA contains amplified oncogenes, including MYC, EGFR, and MDM2, which promote tumor growth and inhibit immune recognition. MYC amplification on ecDNA diminishes major histocompatibility complex (MHC) class I molecules, hindering antigen presentation to cytotoxic T lymphocytes (CTLs) and promoting immune evasion [32]. EGFR amplification increases the release of immunosuppressive cytokines, including IL-6 and TGF-β, thereby inhibiting immune cell activation [33]. ecDNA interferes with the cyclic GMP-AMP synthase-stimulator of interferon genes (cGAS-STING) pathway. ecDNA activates the cGAS-STING axis, triggering Type I interferon responses and pro-inflammatory cytokine production [34]. Tumor cells develop mechanisms to suppress this through mutations in STING or downregulation of cGAS, causing ecDNA to accumulate without activating immune responses [35]. Immunogenic cell death (ICD) is critical for initiating anti-tumor immunity, and ecDNA can alter this. ecDNA may encode anti-apoptotic genes or modulate stress response pathways to prevent ICD, enabling immune evasion [36]. ecDNA facilitates immune escape by shielding tumor cells from immune surveillance and impairing innate and adaptive immune responses.

14.4.2 ecDNA and Tumor Heterogeneity: Challenges for Immunotherapy

Tumor heterogeneity is a challenge in immunotherapy, as the diverse genetic and phenotypic characteristics among tumor cells hinder immune recognition and response. ecDNA contributes to heterogeneity owing to its structural characteristics and mode of inheritance. In contrast to chromosomal DNA, ecDNA is devoid of centromeres, resulting in random and uneven segregation during mitosis. Daughter cells acquire distinct ecDNA copies, fostering genetic mosaicism within the tumor. Induces subclonal diversification and fluctuating oncogene expression, modifying immunogenicity among tumor cells, enabling some subpopulations to evade immune targeting. Gene amplification and deletion occur by ecDNA in response to environmental stimuli, such as immunotherapy. Tumor cells temporarily enhance PD-L1 or other immune checkpoint regulators on ecDNA to evade immune responses. This plasticity complicates treatment as conventional immunotherapeutic targets may exhibit variable expression across tumor regions or be absent in resistant clones. ecDNA-driven heterogeneity is spatially and temporally dynamic. Research employing single-cell sequencing and imaging has demonstrated that ecDNA distribution differs across various tumor regions and temporal stages. This variability undermines T-cells' capacity to generate a reliable and effective anti-tumor response and diminishes the effectiveness of CAR-T or TCR-T therapies aimed at specific antigens. The increase in intratumoral heterogeneity mediated by ecDNA restricts the efficacy of immune-based therapies and underscores the necessity for dynamic monitoring of treatment strategies [37].

14.4.3 Influence of ecDNA on Immune Checkpoints and Tumor Microenvironment

The tumor microenvironment (TME) consists of immune cells, fibroblasts, endothelial cells, extracellular matrix (ECM), and signaling molecules. ecDNA plays a crucial role in altering this microenvironment, influencing the effectiveness of immunotherapy. ecDNA influences the expression of immune checkpoints, including PD-L1 and CTLA-4, via gene amplification or epigenetic modification. In glioblastoma and non-small cell lung cancer, ecDNA-mediated amplification of PD-L1 resulted in immune suppression and suboptimal responses to checkpoint inhibitors. Extrachromosomal DNA (ecDNA) may modulate enhancer activity near checkpoint genes, resulting in their overexpression without direct amplification [38]. ecDNA-mediated oncogene expression impacts cytokine and chemokine synthesis, affecting the tumor microenvironment. Oncogenes such as EGFR, CDK4, and MYC enhance the secretion of IL-10, VEGF, and CCL2, which suppress dendritic cell maturation, recruit regulatory T-cells (Tregs), and facilitate M2 macrophage polarization [39]. These create an immunosuppressive microenvironment that obstructs cytotoxic T-cell infiltration and functionality. ecDNA-induced gene expression results in stromal reprogramming. Extrachromosomal DNA containing HIF1A or VEGF genes augments angiogenesis and hypoxia, inhibiting T-cell function and fostering immune exhaustion [40]. Elevated ECM deposition obstructs immune cell migration within tumors, forming physical barriers to immune infiltration. ecDNA can activate cGAS-STING signaling, causing a pro-inflammatory tumor microenvironment. The liberation of tumor-derived ecDNA into the cytoplasm or extracellular vesicles functions as a danger-associated molecular pattern (DAMP), activating dendritic cells and facilitating T-cell recruitment. This is transient or counteracted by immunosuppressive mechanisms derived from tumors [41].

14.5 CURRENT PROGRESS IN CANCER IMMUNOTHERAPY TARGETING ECDNA

14.5.1 Emerging Strategies to Target ecDNA-Driven Cancers

Recent studies have concentrated on advanced therapeutics—cell and gene therapy, RNA interference, and oligonucleotide treatments—with conventional approaches such as targeted medicines and immunotherapies. ecDNA is a big circle of DNA that has important oncogenes. It is increased in invasive malignancies and not found in normal cells, which makes it a possible target for treatment. Its unique traits may facilitate non-invasive biopsy methods for cancer diagnosis, resistance assessment, and treatment forecasting, improving cancer management and monitoring [42, 43].

14.5.2 Role of CRISPR and Gene-Editing Techniques

Recent advancements have transformed nucleic acid editing by directing CRISPR/Cas9 to specific genomic locations, enabling accurate gene alterations. This technique has made it possible to conduct things like targeted ablation, domestication, amplification, transcriptional regulation, and chromatin modification. It does have problems, like alternative splicing, low percentages of targeted cells, mixing of targeting and reporter RNA, and impacts on the transcriptome that are not targeted. Lipid nanoparticles can transfer mRNA directly to bigger ECMs, but this doesn't work for many brain applications since there aren't enough surface indicators to target cells. Solid tumors can lose genes that are important for targeting monotherapy, including checkpoints. This makes it harder to find or fix unexpected cellular states or signaling pathways when editing is done with transgenic viruses or electroporation. Multiplexed CRISPRa technique employs a diverse array of gRNAs for a specific library size, optimizing coverage while minimizing noise. In this architecture, gRNA controls against genes can be biochemically extracted with only three poolings. In this multiplexed screening setup, CRISPRa gRNAs that target TAA sets that are often overexpressed have been used. Mass spectrometry-based proteomics helps confirm whether multiplexed gRNA designs really do cause protein translation. Contamination from other multiplexed gRNAs or qPCR primers may elevate false positive rates in CRISPRa; this can be mitigated through meticulous oligonucleotide design. The structures of different gRNA configurations are still unknown because their structural parts haven't been investigated without Cas proteins. On-target disruptions by gRNAs targeting non-repeated sequences confuse biological conclusions regarding CRISPR specificity. Simpler gRNA designs that test different CRISPRs or zFN perturbations against well-defined motifs in the gRNA sliding window may help prevent off-target effects. This is different from Cas9-directed gRNAs that target transcribed gene TSSs. Improving on-target potency through exact gRNA changes may uncover concealed on-target hits usually produced by alternative CRISPR designs, perhaps elucidating regulatory DNA-protein interactions in mammalian cells [44–46].

14.5.3 Potential Biomarkers for ecDNA in Immunotherapy Response

Recent research has commenced investigations into the function of ecDNA as a tumor-specific biomarker for immune profiling and the assessment of immunotherapy responses. One study demonstrated that particular oncogenic ecDNAs could elucidate the contributions of cancer cells to the tumor immune microenvironment (TIME) by distinguishing encoded proteins. Targeting ecDNA3922+F79 cells with particular antibodies indicates a novel strategy for identifying carcinogenesis and treatment responses. A barrier persists in differentiating ecDNAs when numerous similar protein-coding variants are present in TIME and organs. Looking into other signals, including various splicing sites from ecDNA-encoded proteins, could make diagnoses more accurate [42, 47].

14.5.4 Clinical Trials Involving ecDNA

Extrachromosomal DNA (ecDNA) has emerged as a pivotal element in modulating immune responses in cancer. Recent clinical trials have begun exploring ecDNA's role in immune evasion and resistance to immunotherapy, especially immune checkpoint inhibitors (ICIs). A notable example is the NCT04107168 trial, which assessed tumor evolution and immune modulation in glioblastoma, emphasizing ecDNA-driven oncogene amplification (such as EGFR) and its impact on T-cell exclusion and immune checkpoint resistance. The study found that tumors with ecDNA exhibited heightened immune evasion by modulating the tumor microenvironment (TME), decreasing antigen presentation, and increasing immunosuppressive signals. These findings underscore the complexity ecDNA introduces to cancer immunotherapy, revealing potential resistance mechanisms to ICIs. As a result, targeting ecDNA formation or its downstream signaling could represent a novel therapeutic strategy to overcome immune escape in solid tumors, thereby enhancing the efficacy of current immunotherapies [48].

14.6 SAFETY AND EFFICACY CONCERNS IN TARGETING ECDNA

The discovery of drugs has resulted in the approval of 61 immune checkpoint inhibitors (ICIs), such as monoclonal antibodies targeting PD-1, PD-L1, and CTLA-4, revolutionizing cancer therapy. Nonetheless, merely 11% of patients attained a complete response for a duration exceeding two years, with some thereafter exhibiting resistance and recurrence. Low amounts of tumor-infiltrating lymphocytes, "cold" tumors, and large levels of immune suppressors are some of the things that make treatments fail. The quick approval and usage of ICIs have made many worry about their safety and effectiveness. This has led to more immune-related adverse events (irAEs) like pneumonitis, colitis, and dermatitis. There is proof that the beginning and severity of irAE are related to a good response to treatment. Disrupting immunological checkpoints to boost the immune system can lead to serious autoimmune diseases, especially in people who already have them. The association between severe irAEs and enhanced prognosis indicates a comprehensive, non-selective immune response [47].

14.7 FUTURE PERSPECTIVES AND RESEARCH DIRECTIONS

14.7.1 Potential Novel Therapeutic Targets

Cancers use a number of different ways to avoid being recognized by the immune system, which makes tumor immunotherapy less effective. Comprehending these evasion methods, especially in ecDNA-driven malignancies, necessitates comprehensive data and insights that are still insufficiently investigated. Immunotherapy, particularly immune checkpoint blockade (ICB) and CAR-T cell therapy, exhibits potential for tumor treatment; yet, the emphasis on ecDNA has uncovered a scarcity of research on innovative therapeutic targets. ICB has been utilized for many tumors, including MS-MLH1-deficient colorectal cancer, whereas CAR-T cell treatment has demonstrated considerable efficacy in B cell malignancies, although it encounters difficulties with solid tumors. Clinical successes suggest the possibility of reprogramming the immune system to target cancerous cells. Investigations should focus on novel immune recognition proteins that may propel the field forward. Effective cancer immunotherapy boosts the immune system's ability to fight tumors by making them more immunogenic. Tumor cells, which come from normal somatic cells, gather mutations and epimutations, creating proteins that are not normal. These aberrant proteins, including neo-antigens (NAs), exhibit tumor selectivity, underscoring the necessity for additional investigation in this domain [49, 50].

14.7.2 Combination Therapies Involving ecDNA Targeting and Immune Modulation

Even though ecDNA is known to be a cancer biomarker, there aren't many investigations on ecDNA-targeting reagents in immune-modulated treatment. Clamped and pp65 mRNA metAA complexes with TLR7/8a can boost antitumor immune responses in T and NK cells. In melanoma (B16), the combination of ecDE-NEF-HCV and TLR3a injection led to elevated levels of antigen-specific IgG4 and IgA. However, the combination with PD-L1 antibodies for tumor growth inhibition requires additional investigation. It is essential to combine immune-stimulated therapies with ecDNA-targeting medicines, including TLR agonists, infiltration analysis, and transcriptome profiling for the purpose of monitoring evasion and progression. Nanoproligomers that target ecDNA can be used as biopharmaceutical medications made just for patients. Two main techniques have been created to target tumor ecDNAs: advanced therapeutic biologics that focus on fusion genes in ecDNAs and chemotherapies that target oncogenic proteins from eDNA-encoded transgenes. These fusions can change some microenvironments, which can change the nitrogen pools in tumors and cause telomeres to shorten. Tumor-infiltrating lymphocytes (TIL) are crucial for monitoring agents. Engineered single-chain variable fragment antibodies targeting ecDNA-encoded polypeptides, along with mu and plx RNA aptamers, can identify TLRs. Novel nanobodies that target complement fragments on tumors improve the infiltration of immune cells and encourage a T-cell-mediated abscopal effect. Ongoing attempts aim to create indirect approaches for ecDNA detection without isolation, studying eyeless DNA and RNA/DNA hybrids through in situ staining patterns or amplification and sequencing-based identification [42, 50].

14.7.3 Personalized Medicine Approaches in ecDNA-Associated Cancers

This article summarizes what we currently know about ecDNA biology, including its topology, expression, evolution, intercellular transfer, and effect on cancer hallmarks. It also suggests new research areas for personalized treatment in ecDNA-linked tumors. ecDNA sequencing, together with phylogenetics, can be used to study the evolutionary route of a tumor and the diversity of its clones. Responses of ecDNA-related genes to targeted medicines and immunological checkpoints can inform personalized cancer treatment. ecDNA is a new area of cancer research, but it's already changing the field. Historically, the evolution of tumors and responses to treatment were examined using genetic modifications in primary or metastatic tissues. Understanding ecDNA biology indicates that somatic mutations could arise in normal tissues prior to tumor initiation or in the early stages of tumor progression. This review seeks to stimulate additional research about ecDNA's function in cancer and its utilization in personalized therapy. Recent progress in ecDNA biology and its ramifications for innovative personalized medicine approaches targeting tumor evolution and medication resistance are delineated [43, 47].

14.8 CONCLUSION

The finding of ecDNA has motivated scholars to investigate it as a possible novel field of inquiry. Recent literature has substantially improved our comprehension of the molecular mechanisms underlying ecDNA production, motility, and their roles in cancer progression and therapeutic resistance. New ways are being found to use DNA as a cancer biomarker to find it early, keep an eye on treatment, and figure out how bad it is. Most ecDNA research focuses on cellular issues, and there are not many studies that look at its involvement in making a TME that suppresses the immune system. It is still hard to get ecDNA out of BM samples without changing its shape. Present NGS data mostly concentrates on sequencing the coding region, which constitutes merely 1.2% of the human genome, thus constraining our comprehension of crucial regulatory, non-coding, and epigenetic components. Numerous genomic areas are duplicated and co-evolved, influencing GC composition and targeted LM alteration, which impedes our capacity to elucidate the functions of ecDNA in cancer evolution and diversification.

REFERENCES

1. Mellman I, Coukos G, Dranoff G. Cancer immunotherapy comes of age. Nature (2011) 480 (7378):480–489. Published 2011 Dec 21. https://doi.org/10.1038/nature10673
2. Sharma P, Allison JP. The future of immune checkpoint therapy. Science (2015) 348 (6230):56–61. https://doi.org/10.1126/science.aaa8172
3. Wu S, Bafna V, Chang HY, Mischel PS. Extrachromosomal DNA: An emerging hallmark in human cancer. Annu Rev Pathol (2022) 17:367–86. https://doi.org/10.1146/annurevpathmechdis-051821-114223
4. Xing J, Ning Q, Tang D, Mo Z, Lei X, Tang S. Progress on the role of extrachromosomal DNA in tumor pathogenesis and evolution. Clin Genet (2021) 99 (4):503–12. https://doi.org/10.1111/cge.13896
5. Levan A, Levan G. Have double minutes functioning centromeres? Hereditas (1978) 88(1):81–92. https://doi.org/10.1111/j.1601-5223.1978.tb01606.x
6. Pecorino LT, Verhaak RGW, Henssen A, Mischel PS. Extrachromosomal DNA (ecDNA): An origin of tumor heterogeneity, genomic remodeling, and drug resistance. Biochem Soc Trans (2022) 50(6):1911–20. https://doi.org/10.1042/BST20221045
7. Turner KM, Deshpande V, Beyter D, Koga T, Rusert J, Lee C, Li B, Arden K, Ren B, Nathanson DA, Kornblum HI, Taylor MD, Kaushal S, Cavenee WK, Wechsler-Reya R, Furnari FB, Vandenberg SR, Rao PN, Wahl GM, Bafna V, . . . Mischel PS. Extrachromosomal oncogene amplification drives tumour evolution and genetic heterogeneity. Nature (2017) 543 (7643):122–5. https://doi.org/10.1038/nature21356
8. Kim H, Nguyen NP, Turner K, Wu S, Gujar AD, Luebeck J, Liu J, Deshpande V, Rajkumar U, Namburi S, Amin SB, Yi E, Menghi F, Schulte JH, Henssen AG, Chang HY, Beck CR, Mischel PS, Bafna V, Verhaak RGW. Extrachromosomal DNA is associated with oncogene amplification and poor outcome across multiple cancers. Nat Gen (2020) 52 (9):891–897. https://doi.org/10.1038/s41588-020-0678-2
9. Topalian SL, Drake CG, Pardoll DM. Targeting the PD-1/B7-H1(PD-L1) pathway to activate anti-tumor immunity. Curr Opin Immunol (2012) 24 (2):207–12. https://doi.org/10.1016/j.coi.2011.12.009
10. Joyce JA, Fearon DT. T cell exclusion, immune privilege, and the tumor microenvironment. Science (2015) 348 (6230):74–80. https://doi.org/10.1126/science.aaa6204
11. Yi E, Chamorro González R, Henssen AG, Verhaak RG. Extrachromosomal DNA amplifications in cancer. Nat Rev Genet (2022 Dec) 23 (12):760–71. https://doi.org/10.1038/s41576-022-00521-5
12. Bafna V, Mischel PS. Extrachromosomal DNA in cancer. Annu Rev Genomics Hum Genet (2022 Aug 31) 23 (1):29–52. https://doi.org/10.1146/annurev-genom-120821-100535
13. Hung KL, Yost KE, Xie L, et al. ecDNA hubs drive cooperative intermolecular oncogene expression. Nature (2021) 600 (7890):731–6. https://doi.org/10.1038/s41586-021-04116-8
14. Schumacher TN, Schreiber RD. Neoantigens in cancer immunotherapy. Science (New York, N.Y.) (2015) 348 (6230):69–74. https://doi.org/10.1126/science.aaa4971
15. Vivier E, Tomasello E, Baratin M, Walzer T, Ugolini S. Functions of natural killer cells. Nat Immunol (2008) 9 (5):503–10. https://doi.org/10.1038/ni1582
16. Qian BZ, Pollard JW. Macrophage diversity enhances tumor progression and metastasis. Cell (2010) 141 (1):39–51. https://doi.org/10.1016/j.cell.2010.03.014
17. Farhood B, Najafi M, Mortezaee K. CD8+ cytotoxic T lymphocytes in cancer immunotherapy: A review. J Cell Physiol (2019) 234 (6):8509–21. https://doi.org/10.1002/jcp.27782

18. Ribas A, Wolchok JD. Cancer immunotherapy using checkpoint blockade. Science (New York, N.Y.) (2018) 359 (6382):1350–55. https://doi.org/10.1126/science.aar4060
19. Maude SL, Laetsch TW, Buechner J, Rives S, Boyer M, Bittencourt H, Bader P, Verneris MR, Stefanski HE, Myers GD, Qayed M, De Moerloose B, Hiramatsu H, Schlis K, Davis KL, Martin PL, Nemecek ER, Yanik GA, Peters C, Baruchel A, . . . Grupp SA. Tisagenlecleucel in children and young adults with B-cell lymphoblastic leukemia. The New England Journal of Medicine (2018) 378 (5):439–48. https://doi.org/10.1056/NEJMoa1709866
20. Ljunggren HG, Kärre K. In search of the 'missing self': MHC molecules and NK cell recognition. Immunology Today (1990) 11 (7):237–244. https://doi.org/10.1016/0167-5699(90)90097-s
21. Mantovani A, Marchesi F, Malesci A, Laghi L, Allavena P. Tumour-associated macrophages as treatment targets in oncology. Nat Rev Clin Oncol (2017) 14 (7):399–416. https://doi.org/10.1038/nrclinonc.2016.217
22. Carroll SM, DeRose ML, Gaudray P, et al. Double minute chromosomes can be produced from precursors derived from a chromosomal deletion. Mol Cell Biol (1988) 8 (4):1525–33. https://doi.org/10.1128/mcb.8.4.1525-1533.1988
23. Verhaak RGW, Bafna V, Mischel PS. Extrachromosomal oncogene amplification in tumour pathogenesis and evolution. Nat Rev Cancer (2019) 19 (5):283–288. https://doi.org/10.1038/s41568-019-0128-6
24. Nikolaev S, Santoni F, Garieri M, et al. Extrachromosomal driver mutations in glioblastoma and low-grade glioma. Nat Commun (2014) 5:5690. Published 2014 Dec 4. https://doi.org/10.1038/ncomms6690
25. Nathanson DA, Gini B, Mottahedeh J, Visnyei K, Koga T, Gomez G, Eskin A, Hwang K, Wang J, Masui K, Paucar A, Yang H, Ohashi M, Zhu S, Wykosky J, Reed R, Nelson SF, Cloughesy TF, James CD, Rao PN, . . . Mischel PS. Targeted therapy resistance mediated by dynamic regulation of extrachromosomal mutant EGFR DNA. Science (New York, N.Y.) (2014) 343 (6166):72–6. https://doi.org/10.1126/science.1241328
26. Shoshani O, Brunner SF, Yaeger R, et al. Chromothripsis drives the evolution of gene amplification in cancer [published correction appears in Nature. 2021 Mar;591(7850):E19. https://doi.org/10.1038/s41586-021-03379-5.]. Nature (2021) 591 (7848):137–41. https://doi.org/10.1038/s41586-020-03064-z
27. Bignell GR, Greenman CD, Davies H, et al. Signatures of mutation and selection in the cancer genome. Nature (2010) 463 (7283):893–8. https://doi.org/10.1038/nature08768
28. Yi E, Gujar AD, Guthrie M, et al. Live-cell imaging shows uneven segregation of extrachromosomal DNA elements and transcriptionally active extrachromosomal DNA hubs in cancer. Cancer Discov (2022) 12 (2):468–83. https://doi.org/10.1158/2159-8290.CD-21-1376
29. Robert M, Crasta K. Breaking the vicious circle: Extrachromosomal circular DNA as an emerging player in tumour evolution. Sem Cell Dev Bio (2022) 123:140–150. https://doi.org/10.1016/j.semcdb.2021.11.015
30. Sanborn JZ, Salama SR, Grifford M, et al. Double minute chromosomes in glioblastoma multiforme are revealed by precise reconstruction of oncogenic amplicons. Cancer Res (2013) 73 (19):6036–45. https://doi.org/10.1158/0008-5472.CAN-13-0186
31. Kohl NE, Kanda N, Schreck RR, et al. Transposition and amplification of oncogene-related sequences in human neuroblastomas. Cell (1983) 35 (2 Pt 1):359–67. https://doi.org/10.1016/0092-8674(83)90169-1
32. Qiu H, Shao ZY, Wen X, Zhang LZ. New insights of extrachromosomal DNA in tumorigenesis and therapeutic resistance of cancer. Am J Cancer Res (2020) 10 (12):4056–65. Published 2020 Dec 1.
33. Miao D, Margolis CA, Gao W, Voss MH, Li W, Martini DJ, Norton C, Bossé D, Wankowicz SM, Cullen D, Horak C, Wind-Rotolo M, Tracy A, Giannakis M, Hodi FS, Drake CG, Ball MW, Allaf ME, Snyder A, Hellmann MD, . . . Van Allen EM. Genomic correlates of response to immune checkpoint therapies in clear cell renal cell carcinoma. Science (New York, N.Y.) (2018) 359 (6377):801–6. https://doi.org/10.1126/science.aan5951
34. Hanahan D, Weinberg RA. Hallmarks of cancer: The next generation. Cell (2011) 144 (5):646–74. https://doi.org/10.1016/j.cell.2011.02.013
35. Ablasser A, Chen ZJ. cGAS in action: Expanding roles in immunity and inflammation. Science (New York, N.Y.) (2019) 363 (6431):eaat8657. https://doi.org/10.1126/science.aat8657
36. Harding SM, Benci JL, Irianto J, Discher DE, Minn AJ, Greenberg RA. Mitotic progression following DNA damage enables pattern recognition within micronuclei. Nature (2017) 548 (7668):466–70. https://doi.org/10.1038/nature23470
37. Kroemer G, Galluzzi L, Kepp O, Zitvogel L. Immunogenic cell death in cancer therapy. Annu Rev Immunol (2013) 31:51–72. https://doi.org/10.1146/annurev-immunol-032712-100008
38. Cooper WA, Tran T, Vilain RE, Madore J, Selinger CI, Kohonen-Corish M, Yip P, Yu B, O'Toole SA, McCaughan BC, Yearley JH, Horvath LG, Kao S, Boyer M, Scolyer RA. PD-L1 expression is a favorable prognostic factor in early stage non-small cell carcinoma. Lung Cancer (Amsterdam, Netherlands) (2015) 89 (2):181–8. https://doi.org/10.1016/j.lungcan.2015.05.007

39. Zhu Y, Gujar AD, Wong CH, Tjong H, Ngan CY, Gong L, Chen YA, Kim H, Liu J, Li M, Mil-Homens A, Maurya R, Kuhlberg C, Sun F, Yi E, deCarvalho AC, Ruan Y, Verhaak RGW, Wei CL. Oncogenic extrachromosomal DNA functions as mobile enhancers to globally amplify chromosomal transcription. Cancer Cell (2021) 39 (5):694–707.e7. https://doi.org/10.1016/j.ccell.2021.03.006
40. Liu J, Lichtenberg T, Hoadley KA, Poisson LM, Lazar AJ, Cherniack AD, Kovatich AJ, Benz CC, Levine DA, Lee AV, Omberg L, Wolf DM, Shriver CD, Thorsson V, Cancer Genome Atlas Research Network, & Hu H. An integrated TCGA pan-cancer clinical data resource to drive high-quality survival outcome analytics. Cell (2018) 173 (2):400–16.e11. https://doi.org/10.1016/j.cell.2018.02.052
41. Semenza GL. Hypoxia-inducible factors: Mediators of cancer progression and targets for cancer therapy. Trends Pharmacol Sci (2012) 33 (4):207–14. https://doi.org/10.1016/j.tips.2012.01.005
42. Buenrostro JD, Wu B, Litzenburger UM, Ruff D, Gonzales ML, Snyder MP, Chang HY, Greenleaf WJ. Single-cell chromatin accessibility reveals principles of regulatory variation. Nature (2015) 523 (7561):486–90. https://doi.org/10.1038/nature14590
43. Huang Q, Zhang S, Wang G, Han J. Insight on ecDNA-mediated tumorigenesis and drug resistance. Heliyon (2024) 10 (6):e27733. https://doi.org/10.1016/j.heliyon.2024.e27733
44. McCarty NS, Graham AE, Studená L, Ledesma-Amaro R. Multiplexed CRISPR technologies for gene editing and transcriptional regulation. Nature Communications [Internet]. 2020 Mar 9;11(1). Available from: https://doi.org/10.1038/s41467-020-15053-x
45. Biederstädt A, Manzar GS, Daher M. Multiplexed engineering and precision gene editing in cellular immunotherapy. Front Immunol (2022) 13:1063303. https://doi.org/10.3389/fimmu.2022.1063303
46. Aussel C, Cathomen T, Fuster-García C. The hidden risks of CRISPR/Cas: structural variations and genome integrity. Nature Communications [Internet]. 2025 Aug 5;16(1). Available from: https://doi.org/10.1038/s41467-025-62606-z
47. Xu L, Zou C, Zhang S, Chu TSM, Zhang Y, Chen W, Zhao C, Yang L, Xu Z, Dong S, Yu H, Li B, Guan X, Hou Y, Kong FM. Reshaping the Systemic Tumor Immune Environment (STIE) and Tumor Immune Microenvironment (TIME) to enhance immunotherapy efficacy in solid tumors. J Hematol Oncol (2022) 15 (1):87. https://doi.org/10.1186/s13045-022-01307-2
48. Thompson NA, Stewart GD, Welsh SJ, Doherty GJ, Robinson MJ, Neville BA, Vervier K, Harris SR, Adams DJ, Dalchau K, Bruce D, Demiris N, Lawley TD, Corrie PG. The MITRE trial protocol: A study to evaluate the microbiome as a biomarker of efficacy and toxicity in cancer patients receiving immune checkpoint inhibitor therapy. BMC Cancer (2022) 22 (1):99. https://doi.org/10.1186/s12885-021-09156-x
49. Wu T, Wu C, Zhao X, Wang G, Ning W, Tao Z, Chen F, Liu XS. Extrachromosomal DNA formation enables tumor immune escape potentially through regulating antigen presentation gene expression. Sci Rep (2022) 12 (1):3590. https://doi.org/10.1038/s41598-022-07530-8
50. Gu X, Yu J, Chai P, Ge S, Fan X. Novel insights into extrachromosomal DNA: Redefining the oncodrivers of tumor progression. J Exp Clin Cancer Res: CR (2020) 39 (1):215. https://doi.org/10.1186/s13046-020-01726-4

15 Extrachromosomal DNA (ecDNA) and Clinical Implications in Cancer

Yumna Khan,[1] *Mudasir Maqbool,*[2] *Somya Sharma,*[3] *Amita Joshi Rana,*[4] *Ayesha Sultana,*[5] *and Md. Sadique Hussain*[6]

1 Institute of Biotechnology and Genetic Engineering (Health Division)
The University of Agriculture
Khyber Pakhtunkhwa, Pakistan
2 Department of Pharmacology
Government Medical College Baramulla
Jammu and Kashmir, India
3 School of Pharmaceutical Sciences
Jaipur National University
Jaipur, India
4 College of Pharmacy
Graphic Era Hill University
Bhimtal, India
5 Department of Pharmaceutics
Yenepoya Pharmacy College & Research Centre
Yenepoya (Deemed to be University)
Mangalore, India
6 Uttaranchal Institute of Pharmaceutical Sciences
Uttaranchal University
Prem Nagar, Dehradun, India
School of Pharmaceutical Sciences
Lovely Professional University
Punjab, India

15.1 INTRODUCTION

Extrachromosomal DNA (ecDNA) is a type of non-linear DNA, forming a circle, separate from and outside the linear chromosomes of the cells that comprise the nucleus. Unlike bacterial plasmids, ecDNA in human cells is not a part of the normal genome and is usually only found in a pathological context, most often in cancer. The size of these circular DNA molecules varies between hundreds of kilobases to a few megabases and lacks centromeres and telomeres, a structural characteristic that sets the DNA in chromosomes apart [1]. ecDNAs can form autonomous copies and may contain several oncogene copies and are thus subject to overexpression. This increased expansion permits tumor cells to quickly adjust to stressful conditions in the environment, including drug exposure, to improve the heterogeneity of the tumor and therapeutic tolerance [2, 3].

Among the most unique features of ecDNA is the capability to enable high-level gene expression through its open chromatin structure and the absence of repressing histone regulation, thus facilitating transcriptional plasticity [4]. Besides, ecDNAs may integrate into the chromosomal DNA

DOI: 10.1201/9781003655299-15

or form an extended ecDNA in the presence of elaborate genomic rearrangements, which further exacerbates their role in genomic instability and cancer development [5].

The idea of extrachromosomal components in cancer first emerged in the 1960s when the phenomenon of exceptionally small acentric chromatin bodies was visualized in cytogenetic preparations known as double minutes [6]. These structures have long been regarded as cytogenetic oddities whose meaning remains unclear. It was only with the arrival of next-generation sequencing (NGS) and more powerful computational resources that have helped establish the circular character of ecDNA and its contribution to the amplification of oncogenes were established. A critical study by Turner et al. (2017) characterized circular amplicons in several types of cancer and pinpointed *MYC*, *EGFR*, and *CDK4* as examples of oncogenic genes in circular amplicons, making ecDNA a cause of intratumoral heterogeneity and aggressive phenotypes [2].

Later, wide-scale genomic analyses and those involving The Cancer Genome Atlas (TCGA) data showed that ecDNA is common in more than one in every four human cancers, especially those found in glioblastoma (GB), neuroblastoma (NB), and lung squamous cell carcinoma [7, 8]. Additional utilization of methods, including Circle-Seq, optical mapping, and single-cell whole genome sequencing, has better defined the mechanisms of ecDNA biogenesis as well as the dynamics of ecDNA in cancer cells [9].

The clinical and biological consequences of ecDNA in cancer are profound, as ecDNA facilitates oncogene amplification in cancer cells and leads to tumor heterogeneity through reversible and rapid oncogene amplification independent of underlying chromosomal architecture [10]. Such plasticity allows cancer cells to rapidly respond to stressful conditions, such as therapy and immune monitoring. As an example, GB resistance to *EGFR*-targeted drugs is linked to amplification of *EGFR* on ecDNA, and dynamic ecDNA element gain or loss is a source of variable drug resistance [11, 12].

Moreover, by increasing genetic plasticity, ecDNA stimulates clonal evolution. It may be re-incorporated into chromosomes, resulting in complicated rearrangements, or may act itself by producing more ecDNA molecules by breakage–fusion–bridge processes and maintenance of the condition of constant genomic remodeling [13]. In a clinical setting, ecDNA is linked to an aggressive tumor behavior, dismal prognosis, and resistance to treatment in many malignancies [14]. Notably, the lack of chromosomal mapping of the ecDNA provides a major obstacle to standard gene-targeting applications, meaning there is a need to come up with new therapeutic strategies targeting ecDNA-relevant mechanisms [15].

The purpose of the chapter is to ensure a detailed review of the contemporary knowledge concerning the ecDNA in cancer. The role of ecDNA in the oncogene amplification, intratumoral heterogeneity, therapeutic resistance, and cancer evolution will be critically discussed. Recent progress relative to unanswered questions in the area has also been presented in the chapter, focusing on the prospect of ecDNA as a diagnostic, prognostic, and therapeutic biomarker. This chapter will attempt to explain these complex but exciting interconnections between the aspects of the multifaceted role of ecDNA in cancer biology and highlight its importance as a new target in precision oncology by incorporating the results of genomic, epigenetic, and clinical experiments.

15.2 BIOLOGICAL CHARACTERISTICS OF ECDNA

ecDNA is a novel and dynamic constituent of the cancer genome, which is a form of DNA existing outside the chromosome. ecDNA is normally circular, acentric, and different from and not equal to chromosomal DNA in form and activity. It does not bind with the centromere, and this causes its random segregation in the process of cell division [16]. This uneven dispensation results in daughter cells inheriting moreecDNA than their parent cells, increasing genetic variation and cancer heterogeneity. This property makes the tumor cells more fit, and they are able to adjust quickly to new environments and evade treatment. The ecDNA has a circular shape that enables autonomous replication, which makes it more likely to amplify oncogenes [17]. Research studies have revealed that ecDNA tends to host genes that play a vital role in tumorigenesis, such as oncogenes such as

MYCN, *EGFR*, *HER2*, and *MDM2*. This upregulation of oncogenes is important in cancer development and has been one of the reasons why cancer proliferates at a high rate [18]. ecDNA is usually found in tumors, especially in aggressive cancers such as GB and NB, and increases genes that foster cancerous actions, including *MYCN* in NB and *EGFR* in GB [19]. The circularity of ecDNA causes an open chromatin structure relative to chromosomal DNA. In contrast to the chromosomal DNA, ecDNA is less dense and can be more actively transcribed [12]. This accessibility leads to increased expression of oncogenes present on the ecDNA, leading to the aggressive nature of some cancers. As an example, MYCN amplification on ecDNA has been linked to poor outcomes in NB, where overexpression of the gene is a driver of tumorigenesis and chemoresistance [20]. Likewise, *EGFR* amplification on ecDNA in GB causes continued activation of growth signaling pathways, a survival advantage to tumor cells, especially during therapeutic stress. The acentric nature of ecDNA makes it unstable, and this can be used in developing therapeutic approaches [21]. Also in mitosis, ecDNA may be mis-segregated and end up in micronuclei, membrane-enclosed structures that contain fragmented ecDNA. These micronuclei may be removed or reincorporated into the nucleus, and there is a possibility that amplified oncogenes can be lost [22]. This is a process called micronucleus formation, which has been reported in NB cells whereby amplification of MYCN on ecDNA is spontaneously lost, resulting in a reversal of the malignant phenotype and a subsequent loss of tumorigenicity. This elimination is not, however, absolute, and ecDNA may also insert into chromosomal loci, resulting in long-term amplification of oncogenes, especially in treatment-resistant tumors [23]. ecDNA is vital in drug resistance. The capacity of ecDNA to promote numerous copies of oncogenes enables tumor cells to escape the impacts of chemotherapy and targeted agents [15]. As an example in colon cancer, methotrexate is a chemotherapy drug that is resistant due to amplification of the DHFR gene on ecDNA. Amplified DHFR makes tumor cells escape the effect of inhibition of methotrexate inhibition, allowing them to survive. This does not only occur in colon cancer but also in other types of cancer, such as GB and breast cancer, in which ecDNA is known to amplify certain genes, such as *EGFR* and HER2, respectively, to mediate targeted therapy resistance [24]. The capacity of ecDNA to copy genes resistant-related and allow their dynamic control highlights its importance as a major participant in resistance to cancer therapy. In GB, for example, a mutant form of *EGFR*, EGFRvIII is often amplified on ecDNA, causing resistance to *EGFR* tyrosine kinase inhibitors (TKIs) such as Erlotinib [18]. It has been observed that when EGFRvIII amplification is present in ecDNA, the tumor is aggressive and resistant to treatment, and this could be worsened by MDM2 amplification, all leading to persistent tumor growth even after drugs are involved. Likewise, the ecDNA frequently harbors the ABCG2 gene that encodes a drug transporter protein, which facilitates resistance to numerous chemotherapeutic agents in the case of GB [12]. The function of ecDNA in drug resistance is also demonstrated in cervical carcinoma, where methotrexate resistance is bestowed by DHFR amplification on ecDNA. The overexpression of a gene that causes resistance to multiple drugs-MDR1 has also been found in oral squamous cell carcinoma and human epidermoid carcinoma, which causes resistance to vinblastine and colchicine, respectively [21]. The fluidity of the ecDNA enables cancer cells to be quickly exposed to environmental pressures such as drug treatment. The copy number of ecDNA that tumor cells possess can also be altered, which can potentially happen due to chromothripsis, at which massive genomic rearrangement leads to the genesis of ecDNA composed of amplified oncogenes [24]. This adaptability to treatment stresses allows tumors to avoid treatment efforts, so ecDNA is an essential determinant of tumor development and treatment failure [16].

15.3 ROLE OF ECDNA IN ONCOGENE AMPLIFICATION

Recent systematic studies have shown that ecDNA is an effective mechanism allowing high-level amplification of oncogenes, contributing to tumor heterogeneity, immune evasion, and drug resistance in various cancers. In small-cell lung cancer (SCLC), ecDNA was found to be the main driver of MYC amplification, forming transcription-amplifying structures that facilitate spatial proximity between enhancers and oncogene loci. This structural reorganization supported MYC overexpression stability, which induced profound tumor plasticity and increased aggressive biological behavior with worse patient survival rates [25]. Another substantial study quantified ecDNA-mediated *MYC* paralogue amplification (*MYC*, *MYCN*, and *MYCL*) and found significantly higher copy numbers of ecDNA and greater ecMYC^{+} cell line frequency in SCLC (22/47) than in other cancers (15/282). Spatial single-cell profiling of more than 100,000 cells identified that ecMYC^{+} tumors showed increased proliferation markers (Ki67, VEGFA) and immunosuppressive regulators (FAP, FOXP3) with low T-cell infiltration, together developing an immunologically "cold" tumor microenvironment [26]. Furthermore, in ROS1-positive non-small cell lung cancer (NSCLC), ecDNA enabled dynamic amplification of MET under TKI selective pressure, in a way that allowed bypass signaling and strong resistance regardless of MET gene mutations or fusions. The adaptive amplification highlights the function of ecDNA to maintain oncogenic signaling under targeted therapy pressure [27].

Expanding this paradigm to gastrointestinal malignancies, ecDNA was shown to induce maximum oncogene dosage in advanced esophagogastric cancer (EGC) and, specifically, for ERBB2. Tumors with ecDNA-mediated ERBB2 amplification exhibited a remarkable median fold-change in copy number of 13.5, significantly higher than that of ERBB2-amplified/ecDNA^{-} tumors (median 2.7), which was associated with more aggressive clinical behaviors and variably responsive treatments for HER2-targeting therapies [28]. In gastric cancer, single-cell analyses further showed that ecDNA not only amplified oncogenic drivers such as FGFR2 and MYC but also maintained intraclonal heterogeneity by dynamically structurally remodeling. After drug treatment with the FGFR2 inhibitor AZD4547, resistant subclones increased in ecDNA copy number, showed chimeric ecDNA morphology, and had elevated levels of FGFR2. Most importantly, all these changes were eliminated upon drug removal, which restored the sensitivity of the cells to the drug and emphasized the dynamic and reversible nature of ecDNA-dependent resistance [29]. In addition, ecDNA was present in 32% of gastric cancer and was linked to distinct molecular signatures characterized by low PD-L1 expression and low scores of tumor inflammation, constituting an immunologically evasive phenotype likely resistant to anti-PD-1 immunotherapy [30].

15.4 ECDNA-MEDIATED THERAPY RESISTANCE

Emerging evidence identifies ecDNA as a key mediator of therapeutic resistance in many cancers by facilitating oncogene amplification, bypass signaling activation, upregulation of efflux transporters, induction of autophagy, and promotion of DNA repair mechanisms (Figure 15.1). In lung adenocarcinoma, MET, HER2, and RET ecDNA-mediated amplification enhanced resistance to EGFR TKIs by an epigenetic factor at ecDNA promoters that maintained high-level oncogene expression and bypassed signal activation [31]. Likewise, in gastric cancer, constitutive oncogenic signaling activation by ecDNA has been reported. For instance, in FGFR2-amplified KATOIII cells, resistance to infigratinib developed through increased ecDNA-mediated amplification of FGFR2, and upon withdrawal of the drug, rapid loss of ecDNA with resensitization was observed, demonstrating the reversible and dynamic nature of this phenomenon [32]. Moreover, trastuzumab-resistant gastric cancer cells (SNU216-TR) contained ecDNA with HER2, thus maintaining aberrant HER2 signaling and potentially leading to immune evasion through the coexistence of non-oncogenic ecDNA elements [33]. In methotrexate-resistant colorectal cancer, amplification of DHFR driven by ecDNA and elevated FAM151B expression enabled enhanced folate metabolism and induced extensive drug resistance. Additionally, proteins like MSH3 and ZFYVE16 contributed to the stability of ecDNA even at a later stage of resistance, although they expressed downregulated expression profiles [34].

Further supporting this, functional experiments in methotrexate-resistant HT29 colorectal cancer cells showed that MSH3 played a key role in ecDNA-mediated DHFR amplification. Loss of MSH3 destabilized ecDNA maintenance, reduced DHFR copy number, promoted ecDNA expulsion, and restored methotrexate sensitivity, indicating a DNA repair-associated mechanism essential

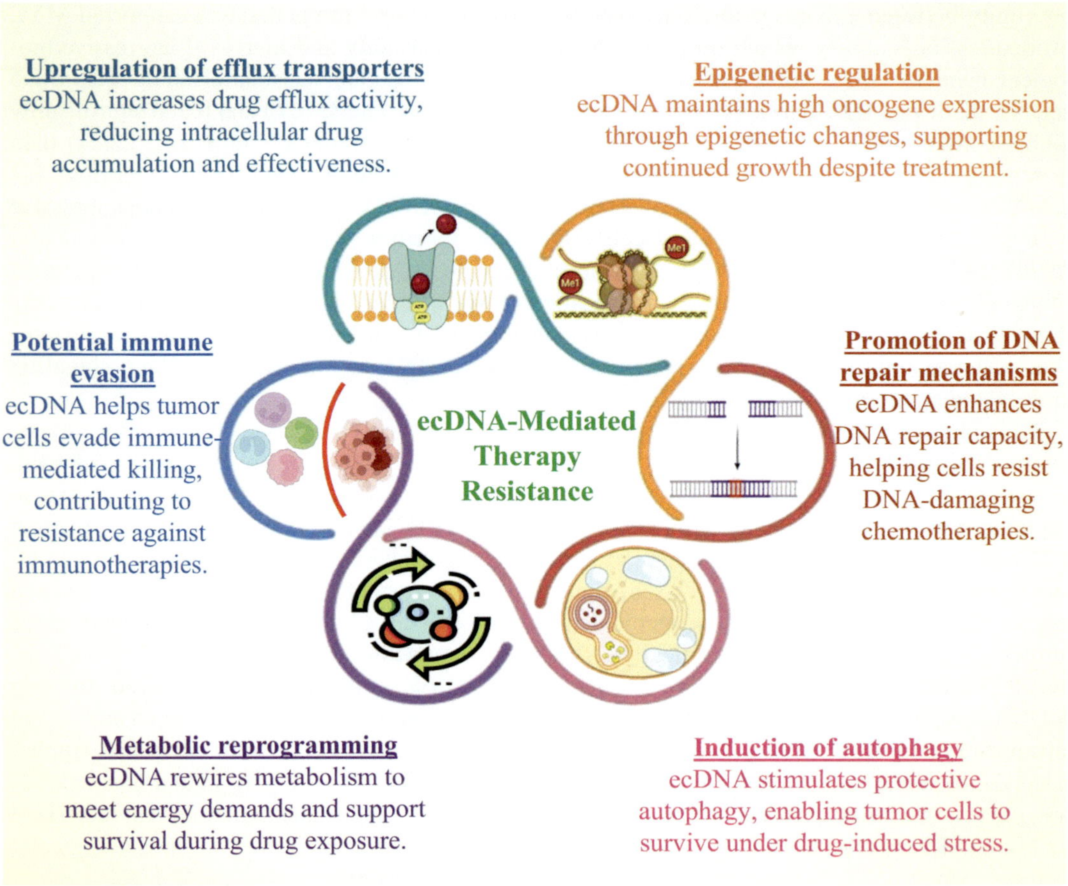

FIGURE 15.1 ecDNA-mediated therapy resistance.

for ecDNA-driven resistance maintenance [35]. In high-grade serous ovarian cancer, ecDNA components with HIF1A facilitated cisplatin resistance through induction of DNA repair, enhanced drug efflux capability, and epithelial-mesenchymal transformation, whose synergistic action facilitated survival of cells under genotoxic stress [36]. Additionally, in hypopharyngeal squamous cell carcinoma, eccDNA-associated RAB3B amplification triggered autophagy, establishing a cytoprotective response to cisplatin treatment, as confirmed through the application of functional assays and organoid models [37]. Furthermore, in several cell lines treated with paclitaxel, ecDNA-induced ABCB1 amplification led to enhanced P-glycoprotein levels, promoting drug efflux and dose-dependent, reversible drug resistance. Removal of paclitaxel in COLO320DM and COLO320HSR cells resulted in a sudden loss of ABCB1 ecDNA and resensitization, emphasizing the plasticity of ecDNA-based resistance mechanisms [38].

15.5 ECDNA AND TUMOR EVOLUTION

Accumulation of ecDNA has become an important and dynamic mechanism driving tumor evolution, facilitating rapid oncogene amplification and profound intratumoral heterogeneity by non-Mendelian inheritance. Significantly, integrative DNA and RNA sequencing analysis comparisons among GB patient tumors and corresponding neurosphere and xenograft models revealed that ecDNA-mediated oncogene amplifications were widely common and stabilized under disease progression, thus ensuring clonal heterogeneity and permitting adaptive evolutionary paths [39]. Moreover, a detailed whole-genome sequencing, structural modeling, and cytogenetic analysis in 17 types of cancers established that ecDNA was present in about 50% of tumors but was still rare in normal cells, strengthening its tumor-specific origin. Such ecDNA structures with very high copy numbers and transcript levels were selectively amplified by oncogenic drivers over the amplification of chromosomes, and thus leading to increased tumor aggressiveness and capacity for extensive genetic diversification [2]. Additional mechanistic information was provided through single-cell analyses that showed extensive heterogeneity of ecDNA copy number and CRISPR-based tracking, validating uneven and stochastic mitotic segregation. After mitosis, ecDNAs then became transcriptionally active hubs loaded with RNA polymerase II, enabling robust oncogene expression and sustained malignant progression [40].

Broadening these findings to additional cancer settings, comprehensive genomic studies in 3,212 cancer patients identified that ecDNA amplifications were associated with increased chromatin accessibility levels, greater frequencies of transcript fusions, and significantly poorer clinical outcomes, all collectively driving accelerated genetic diversification and aggressive phenotypic evolution [4]. Likewise, in vivo experiments with 4T1 triple-negative breast cancer (TNBC) cells implanted into mouse models showed that ecDNA changes caused early and persistent transcriptional reprogramming that selectively initiated immune evasion pathways, allowing highly heterogeneous and therapy-resistant tumor cell subpopulations to develop [41]. These findings were supported by a large-scale analysis of 14,778 patients with 39 tumor types in the 100,000 Genomes Project and showed that ecDNA was present in 17.1% of cancers and not only amplified oncogenes but also immunomodulatory and inflammatory genes. It was associated with promoting reduced infiltration of T-cells, increased immune evasion, metastatic capability, and further advanced stages of disease [42]. Furthermore, cooperative ecDNA species harboring other oncogenes were found to co-segregate together in an orderly manner during mitosis, allowing for common amplification and equal inheritance of multiple oncogenic and control components into daughter cells, further enhancing oncogene cooperation and additional intratumoral heterogeneity [43].

15.6 DIAGNOSTIC AND PROGNOSTIC POTENTIAL OF ECDNA

ecDNA has revolutionized molecular diagnosis as a strongly dynamic biomarker of real-time tumor evolution and systemic disease states. In a landmark pan-cancer analysis of 72,651 tumors, the ECHO algorithm identified focal high-level oncogene amplifications (FH-amp) on ecDNA in 4.2% of specimens with very high frequencies in GB (72%), gastroesophageal adenocarcinoma (57%), liposarcoma (43%), and breast cancer (23%). Notably, EGFR FH-amp on ecDNA were found in 90% of gastroesophageal adenocarcinoma and GB samples, validating the inclusion of ecDNA profiling in precision therapy and diagnostic procedures [44]. A study on Barrett's esophagus found ecDNA in 33% of the patients who later developed esophageal adenocarcinoma, identifiable even in prediagnostic biopsies, suggesting its usefulness as an early prognostic biomarker [45]. In addition to their diagnostic significance, ecDNA has also proved to be a valuable prognostic biomarker indicating correlations with tumor aggressiveness and clinical outcome. In ovarian cancer, the patients with reduced baseline plasma nuclear DNA showed better progression-free survival (hazard ratio [HR] = 0.35; 95% CI: 0.14–0.86) and overall survival (HR = 0.18; 95% CI: 0.04–0.77) independent of the tumor stage. Also, decreases in total ecDNA, nDNA, and mitochondrial DNA content following chemotherapy (54%, 63%, and 52%, respectively; $p < 0.05$) were also associated with successful outcomes [46]. In multiple myeloma, ecDNA was present in 6.8% of the patients at diagnosis and was highly correlated with reduced progression-free survival (median 22 vs. 41 months) and overall survival (median 41 vs. 105 months). Multivariate analysis also validated ecDNA as an independent predictor of poor survival (PFS: HR = 2.6; 95% CI: 1.26–5.6; $p = 0.0082$; OS: HR = 7.94; 95% CI: 3.5–17.9; $p < 0.0001$), with the patients having higher mutational burdens and greater chromosomal fragmentation [47]. Furthermore, in medulloblastoma, ecDNA was found in 18% of the tumors

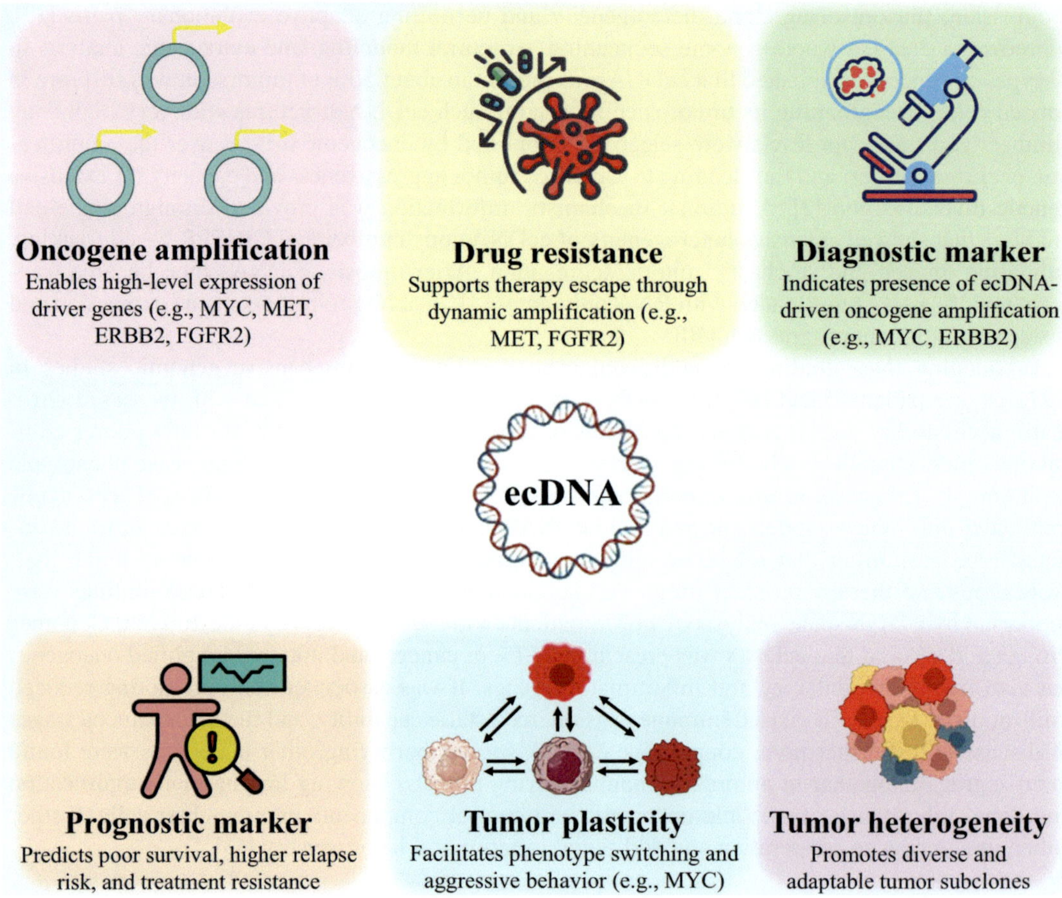

FIGURE 15.2 Multifaceted roles of extrachromosomal DNA in cancer biology and clinical implications.

and in patients with tumors that had ecDNA. The risk of relapse was over two times higher, and mortality threefold higher, within five years following diagnosis in comparison to ecDNA-negative patients. These effects were mechanistically associated with enhancer rewiring events and high ecDNA copy number heterogeneity, which together form aggressive tumor phenotypes [48]. These integrated biological functions and clinical implications of ecDNA in cancer are summarized in Figure 15.2.

15.7 CHALLENGES AND FUTURE PERSPECTIVES

15.7.1 Limitations in Current Detection and Quantification Technologies

The existing methods of detection and quantification of ecDNA have a number of limitations hampering their broad use in clinical practice or research. These issues are primarily caused by the nature of ecDNA, including small size, dynamic nature, and structural variability. Consequently, identification and precise measurement of ecDNA in cancerous tissues is a challenging process, despite the extensive progress in the fields of genomics and molecular biology [49]. The main challenge is that some forms of ecDNA are small in size. Most types of ecDNA are small (usually of less than 1 kb in size), and so are difficult to detect with traditional genomic methods that are generally optimized to detect larger, linear DNA fragments [50]. Such smaller DNA molecules are usually invisible to the light microscope and difficult to isolate and study. As an example, small polydispersed circular DNA (spcDNA), linked to chromosomal instability (CIN), and prevalent in cancer cells, is difficult to detect because of its size and its scattered distribution throughout the genome [51]. The use of fluorescence in situ hybridization (FISH) has been expanded to find specific ecDNA in tumor cells and especially in the case of double-minute chromosomes (DMs), a type of ecDNA that can contain oncogenes such as MYCN and EGFR. Despite this, FISH is limited, especially when used to detect small or low-abundance ecDNA [52]. The method needs special probes, not always accessible to all variants of ecDNA. Also, FISH-based detection is time-consuming and only semi-quantitative. As an example, the FISH-based MYCN amplification in NB has been successful, but the method fails to identify smaller concentrations of ecDNA or to diagnose its presence within heterogeneous tumors [53]. The NGS is more likely to show a more detailed view of ecDNA detection, particularly of larger ecDNA molecules, because the whole genome (extrachromosomal areas included) can be sequenced. Although NGS has enabled a major improvement in the ecDNA detection capability, accurately quantifying ecDNA remains a challenge since they are dynamic molecules [54]. The quantification is complicated by the ecDNA's propensity to alter depending on the environmental stress, like chemotherapy. Also, the original ecDNA signal may be lost due to the presence of ecDNA in micronuclei or as an integrated element in chromosomes, which complicates detection or quantification by conventional sequencing methods. The other significant weakness is the non-homogeneity of ecDNA in and between cancer cells [55]. ecDNA can be circular, linear, and fragmented, and can carry different oncogenes or regulatory sequences. This nonuniformity interprets findings gathered in high-throughput sequencing or other molecular methods [54]. As an illustration, the amplification of EGFR in GB is frequently undertaken on ecDNA, although the size and structure of the ecDNA in individual cancer cells may differ, which may result in variation in the amplification pattern and complicate the quantification [55]. Additionally, the fact that ecDNA is lowly abundant in most tumor samples may complicate detection and quantification further. ecDNA is generally a low percentage of the total genomic DNA, and this implies that its signals may be overwhelmed by the abundance of the chromosomal DNA in the analysis process. Consequently, there is a need to use sensitive and specific detection methods to distinguish the ecDNA from the background of chromosomal DNA [20]. Methods such as circulating tumor DNA (ctDNA) analysis have been suggested to detect ecDNA in blood samples, but such methods are experimental and still need optimization before they can be used in the clinic. Also, the absence of common guidelines to detect and quantify ecDNA in various forms of cancer further complicates the comparison of the results of different studies [52]. Due to the disparity in the methods used by different research groups, different techniques may employ different methods (including circle-sequencing (Circle-seq) or single-cell analyses), which introduce inconsistencies in data and a lack of consensus on the most effective methods to detect and quantify ecDNA with precision [22]. Another barrier is the measurement of ecDNA in clinical specimens. Although NGS-based and PCR-based techniques have the potential to give information about the presence of ecDNA, they also fail at giving reliable values of abundance, with low abundance or variable abundance of ecDNA. The poorly quantified samples may give misleading results concerning the functions of ecDNA in tumor development or resistance to therapy [17]. Moreover, the fact that ecDNA is dynamic with the potential to be lost through the formation of micronuclei or gained by incorporation into chromosomes presents additional complexity to quantification initiatives [19].

15.7.2 Standardization and Clinical Translation Hurdles

The translation of ecDNA as a biomarker or therapeutic target in the clinical setting has enormous challenges, especially standardization and regulatory approval. Such difficulties are mostly attributed to the complexities of the detection, quantification, and interpretation of the ecDNA role in cancer biology and the necessity of the development of reliable, reproducible methods that could be implemented in clinical practice [56]. The variation in the method of detection is one of the major issues in the clinical translation of ecDNA. Furthermore, ecDNA can also differ in size and organization according to the cancer and the tumor's evolutionary stage. As an example, MYCN amplification on ecDNA in NB is characteristic of the disease, but the amplification pattern can vary widely among individual patients or even varying cells within the same tumor [57]. The use of methods like FISH or NGS is widely employed to determine the presence of ecDNA currently, though these techniques lack sensitivity, specificity, and reproducibility. FISH, especially, has the risk of missing low-abundance or small ecDNA, whereas NGS uses costly machines and a lot of bioinformatics knowledge to interpret its results [58]. These issues complicate the standardization of ecDNA detection approaches in various laboratories and institutions, which prevents the spread of ecDNA-based diagnostic tests. In addition, there is no consensus regarding the most appropriate biomarkers or cutoff to identify ecDNA in clinical samples. Although in certain cancers (e.g., GB, NB, and breast cancer) amplification of certain oncogenes (e.g., EGFR, MYCN, and HER2) may occur on ecDNA, ecDNA amplification remains poorly understood clinically [59]. A case in point is that EGFR amplification on ecDNA is involved in drug resistance, though it is not yet clear whether the presence of the amplified EGFR on ecDNA should be taken as a standard biomarker of drug resistance or prognosis in GB. The ecDNA detection remains difficult to implement into the clinical practice routine with no rigorous recommendations on the markers of ecDNA to be targeted and the level of ecDNA to be interpreted [60]. In addition to the problems involved in detecting ecDNA, clinical translation of the ecDNA as a therapeutic target also faces regulatory obstacles. Therapy against or depletion of ecDNA, for example, by use of gene-editing tools or induction of micronuclei, has remained at an early stage. Treatment plans of such nature are largely experimental, and the clinical evidence on their mainstream use is minimal [61]. The level of safety and effectiveness of such treatments is not yet fully established, particularly when used in human beings [12]. To cite a particular example, even though micronucleus formation was shown to clear ecDNA in some cancer cell lines, its application in patients is complicated by the potential off-target effects as well as the apparent inability to target specifically the ecDNA and not the chromosomal DNA [53]. Moreover, a direct attack on ecDNA can also lead to undesirable genetic consequences, such as the deactivation of vital genes or the development of other mutations that can facilitate tumor development. Clinical translation of ecDNA-based therapies is another challenge of tumor heterogeneity [18]. Tumors are not always homogeneous, and the ecDNA composition of tumors of the same patient can vary considerably; the ecDNA quantity in each cancer cell can vary, and also the ecDNA oncogenes can vary [15]. This is because of the diversity, which hinders the development of a universal form of therapy. This is true in the case of GB, where ecDNA may harbor amplifications of EGFR and ABCG2, but not all the tumor cells possess the same amplifications, and it is not easy to develop a single therapeutic modality that can target all types of ecDNA [55]. Tumors are dynamic and will tend to change with time, and this can also change the composition of ecDNA and make treatment even more challenging. The dynamic nature of ecDNA that enables it to be eliminated or integrated into chromosomes presents additional complications in predicting the outcomes of treatment [20]. Moreover, the control of ecDNA-based diagnostics or interventions is being developed. Regulatory authorities such as the U.S. Food and Drug Administration (FDA) or the European Medicines Agency (EMA) have not yet devised regulatory measures for the use of ecDNA-based biomarker assays or treatment [57]. Such agencies require high levels of evidence of safety and efficacy before approving any new diagnostic tool or treatment, and since the role of ecDNA in cancer is a rather new finding, it will take some time to generate sufficient clinical evidence [50]. Clinical trials on a large scale that will demonstrate the effectiveness of the ecDNA-based diagnostics or therapy are time and resource-intensive and require collaboration between academic institutions, the pharmaceutical industry, and regulators. The lack of uniform methodology and the heterogeneity of ecDNA within and across tumors, and the unpredictable clinical utility of ecDNA-based biomarkers are the most important barriers to the clinical translation of ecDNA work [24]. The future then is to create standard methods of identifying and measuring ecDNA and to establish its predictive and treatment response in cancer. Optimization of more sensitive and specific diagnostic technologies and development of clinical guidelines will be the next step toward the introduction of the ecDNA-based techniques in clinical practice [16].

15.7.3 Emerging Tools

New ecDNA analysis tools are transforming our capability to analyze the ecDNA contribution to cancer biology, tumor heterogeneity, and drug resistance. Such new tools include single-cell ecDNA analysis and live-cell tracking, which are leading the way in bringing our insight into ecDNA dynamics to new levels of real time and resolution. The technologies enable researchers to gain an unprecedented level of detail into the complex dynamics of ecDNA behavior, and new information on the molecular mechanisms of cancer progression and resistance to therapy [62]. Single-cell ecDNA analysis is a sensitive tool that allows for to detection and measurement of ecDNA at the single-cell level, which is one of the key limitations of classical bulk DNA analysis. ecDNA is very heterogeneous, and various tumor cells have different amounts and variants of ecDNA, which commonly harbor amplified oncogenes that induce tumorigenesis [63]. Single-cell analysis of ecDNA will allow investigators to achieve this heterogeneity and understand how ecDNA amplification and its oncogenes develop over time, especially in response to therapies. To give an example, in GB, EGFR amplification on ecDNA is frequently present in special sub-populations of cells, and its distribution and dynamics may differ across individual cells [64]. Single-cell sequencing has given us insight into the role of EGFRvIII (mutant version of EGFR that is found on ecDNA) in drug resistance in GB, which helps cancer cells survive and grow in spite of targeted therapeutics. Also, single-cell technologies have enabled the detection of dynamic regulation of ecDNA, for example, reversible amplification of MDM2 in response to chemotherapy, which is challenging to observe with bulk sequencing methods [65]. The other important contribution to the ecDNA research is the development of live-cell tracking technologies. The real-time interaction of ecDNA in living cells is possible through the live-cell imaging technique [18]. The methodology has been important in gaining a better understanding of the mechanism of ecDNA segregation and the role of ecDNA in cancer development. As an example, the movement and dynamics of a form of ecDNA called DMs have been observed using fluorescence microscopy and live-cell imaging applications in cancer cells [50]. These DMs carrying amplified oncogenes such as MYCN in NB are very dynamic and may form a micronucleus during cell division or be integrated into the chromosomes. Monitoring of such processes in live cells would enable the scientists to better understand the role of ecDNA in tumor cell survival, genetic instability, and resistance to therapeutic approaches [51]. The live-cell tracking has also played a significant role in examining the impacts of chemotherapy on ecDNA dynamics. As an example, live-cell imaging of GB revealed the dynamics of EGFRvIII amplification on ecDNA in response to EGFR inhibitors in real time and how tumor cells respond and evolve with therapeutic pressure [23]. This dynamic monitoring enables a greater insight into the manner ecDNA amplification may be used to provide resistance to therapy and how it can either revert or change when exposed to a drug [21]. In addition, single-cell ecDNA profiling and in situ live-cell tracing allow exploring intratumoral heterogeneity with a degree of granularity that was previously inaccessible. Such methods enable the researchers to investigate the effect of ecDNA on the evolution of a tumor at the level of a single cell of a tumor and comprehend how resistance to drugs could develop in a group of particular cells [62]. The real-time monitoring of cancer cells' genetic and phenotypic diversity will enable scientists to acquire the critical information about the resistance mechanisms to chemotherapy and targeted treatments. A new possibility of personalized cancer treatments is also a result of the combination of these emerging tools [18]. Monitoring how the ecDNA behaves to treatment in single cells may provide clinicians with a better picture of how tumors develop and which treatments could be most effective in attacking ecDNA-dependent resistance pathways [50]. To give an example, in case live-cell tracking suggests that ecDNA is an essential component of a specific subpopulation of treatment-resistant tumor cells, personalized treatment regimens can be designed to attack these cells specifically [51]. More to it, single-cell ecDNA can be used to detect particular oncogenes or mutations that are contained in ecDNA, and lead to the implementation of more targetable and effective treatments to patients [23].

15.7.4 Need for ecDNA-Targeted Drug Development and Personalized Therapy

This is because the increasing awareness of ecDNA, as a key mechanism of tumor development and therapeutic resistance, puts a demand on the development of ecDNA-specific drugs and personalized treatment. ecDNA causes amplification of oncogenes and is central to encouraging tumor heterogeneity, genetic instability, and drug resistance [11]. Because ecDNA increases key oncogenes such as MYCN, EGFR, HER2, and MDM2, it forms an adaptive advantage to cancer cells, which are then able to survive the pressures of treatment and evolve at high speed and continue to develop tumors even under treatment pressure [66]. One of the major problems in treating most cancers, and in particular cancers that experience high frequencies of ecDNA amplifications, is drug resistance. The ecDNA-bearing cancer cells tend to be less sensitive to conventional therapies, such as chemotherapy and targeted therapy, due to a lack of drug efficacy caused by the amplification of oncogenes on ecDNA [67]. As an example, in GB, resistance to EGFR inhibitors like Erlotinib has been reported to be associated with amplification of EGFR on ecDNA, when the drug directly targets the EGFR pathway. Tracking of EGFRvIII, an EGFR mutant amplified on ecDNA in live cells, has shown that amplification of EGFRvIII enables GB cells to adapt and escape targeted therapy, leading to persistence of the tumor and a poor outcome [68]. MycN amplification is associated with poor prognosis and chemotherapy resistance in NB, and this underlies a requirement to develop therapeutic approaches that specifically target oncogene amplification, specifically involving ecDNA. Along with the role of ecDNA in therapeutic resistance and tumor evolution, a dire need to develop ecDNA-specific therapeutic solutions is becoming evident, as the approaches have the potential to target the specific factors through which ecDNA influences cancer [52]. The discovery of drugs that target the inhibition of ecDNA formation or the removal of ecDNA in tumor cells can be one of the strategies. Among the approaches under investigation is the selective destruction of the ecDNA containing amplified oncogenes, using gene-editing tools, like CRISPR-Cas9, to directly target these ecDNA and destroy them in cancer cells [24]. As an example, the combination and targeting of MYCN amplification in NB or EGFR amplification in GB would largely mitigate the tumor response to treatment and enhance the efficacy of the current regimes [16]. Nevertheless, the strategies remain under development and need improvement to make them specific and safe, as gene-editing therapies might also produce off-target effects that might complicate the application in clinical practice [63]. The other strategy under exploration is the utilization of small molecules capable of destabilizing ecDNA or inhibiting its amplification. As an example, the formation of micronuclei was found to eradicate ecDNA in some cancer cell lines, and thus induction of this process may be a possible therapeutic option [55]. An example of how the use of drug-induced micronuclei may be used to decrease the tumorigenicity of ecDNA is the use of hydroxyurea (HU) treatment, which has been demonstrated to enhance the clearance of MYCN-containing ecDNA in NB [22]. The difficulty, however, is how to specifically target ecDNA without affecting the chromosomal DNA or causing further mutations that would contribute to tumor evolution. The fact that ecDNA contributes to tumor heterogeneity also compels the importance of personalized cancer treatment, which can be customized to the individual genetic composition of a tumor in a patient [17]. The conventional methods of treatment are usually founded on blanket treatment, which might not specifically affect the mechanisms that cause cancer in each patient. Using single-cell sequencing, live-cell tracking, and other tools, clinicians can gain more insight into the ecDNA landscape within the tumor of the patient and can find out which oncogenes are specifically amplified on ecDNA [12]. This information may then be utilized to develop more specific, personalized treatments that can address the root of the problem in terms of drug resistance in the tumor. As an example, in GB, where amplification of EGFR by ecDNA is a frequent driver of therapy resistance, it is possible to make personalized therapies that specifically target EGFR amplification on ecDNA, improving the efficacy of existing EGFR inhibitors and improving patient outcome [59]. In addition, liquid biopsy technologies are presenting themselves as a potential means of identifying ecDNA in ctDNA as a non-invasive method of tracking the presence and level of ecDNA in real time [60,69]. This technology might be used to monitor the development of ecDNA amplification during treatment and assist clinicians in detecting drug resistance as it arises, and make adjustments to the treatment regimen before resistance develops to dangerous proportions [66]. The implementation of ecDNA monitoring as a part of daily clinical practice would allow clinicians to create an evolving, real-time, responsive approach to the treatment process that would adjust to the changing environment of the patient's tumor [62].

15.8 CONCLUSION

Recent advances in the study of ecDNA have radically changed our perception of cancer biology. As explained in various parts of this chapter, ecDNA is central in the amplification of oncogenes, spearheading tumor heterogeneity, therapeutic resistance, and tumor evolution. These circular DNA elements facilitate high-level, focal, oncogene amplification of MYC, EGFR, and ERBB2, among others, in many cases in a reversible and dynamic fashion, which contributes to aggressive phenotypes and poor clinical outcomes. More recently, it was found that the unusual non-Mendelian transmission of ecDNA to the daughter cells by cell division is another source of increased genetic diversity in tumors, contributing to adaptive evolution and immune escape.

In a translational perspective, ecDNA is an emerging category of cancer-related biomarkers in the diagnostic, prognostic, and therapeutic stratification processes. The finding of ecDNA abundance in pre-diagnostic tissue biopsies and circulating DNA specimens highlights its usefulness in the early detection and follow-up of cancers. Additionally, it has been found that the patterns of amplification of ecDNA are highly associated with clinical outcome in various cancers and therefore could be used as prognostic markers and future therapy responses. Notably, these revelations support the idea of including the profiling of ecDNA in precision oncology pipelines.

Although these developments have been made, a number of limitations limit the clinical translation of ecDNA research. Existing methods of detection, although being more advanced, are not standardized and sensitive enough to be easily used in routine clinical setups. Their heterogeneous and transient structure also proves a problem in quantifying and interpreting reliably. Also, ecDNA-based therapies are still mostly in the trial stages, and their progress is slowed by the poor knowledge of ecDNA biogenesis, maintenance, and susceptibility.

Moving forward, the top priority of future studies should be advanced methods of ecDNA detection at the single-cell and real-time magnification level to better describe ecDNA dynamics in vivo. Alternative therapeutic measures that may specifically target ecDNA, like selective inhibition of ecDNA replication machinery, use of synthetic lethal interactions, or the use of CRISPR/Cas-based editing methods, are also urgently required. Incorporation of the data provided by ecDNA into clinical decision-making systems will involve multi-omic correlation, clinical validation, and harmonization of regulations.

REFERENCES

[1] Wu S, Turner KM, Nguyen N, Raviram R, Erb M, Santini J, et al. Circular ecDNA promotes accessible chromatin and high oncogene expression. Nature. 2019;575:699–703.

[2] Turner KM, Deshpande V, Beyter D, Koga T, Rusert J, Lee C, et al. Extrachromosomal oncogene amplification drives tumour evolution and genetic heterogeneity. Nature. 2017;543:122–5.

[3] Nathanson DA, Gini B, Mottahedeh J, Visnyei K, Koga T, Gomez G, et al. Targeted therapy resistance mediated by dynamic regulation of extrachromosomal mutant EGFR DNA. Science (New York, NY). 2014;343:72–6.

[4] Kim H, Nguyen NP, Turner K, Wu S, Gujar AD, Luebeck J, et al. Extrachromosomal DNA is associated with oncogene amplification and poor outcome across multiple cancers. Nat Genet. 2020;52:891–7.

[5] Bao Y, Sui X, Wang X, Qu N, Xie Y, Cong Y, et al. Extrachromosomal circular DNA landscape of breast cancer with lymph node metastasis. Int J Cancer. 2024;155:756–65.
[6] Cox D, Yuncken C, Spriggs AI. Minute chromatin bodies in malignant tumours of childhood. Lancet (London, England). 1965;1:55–8.
[7] Koche RP, Rodriguez-Fos E, Helmsauer K, Burkert M, MacArthur IC, Maag J, et al. Extrachromosomal circular DNA drives oncogenic genome remodeling in neuroblastoma. Nat Genet. 2020;52:29–34.
[8] Møller HD, Mohiyuddin M, Prada-Luengo I, Sailani MR, Halling JF, Plomgaard P, et al. Circular DNA elements of chromosomal origin are common in healthy human somatic tissue. Nat Commun. 2018;9:1069.
[9] deCarvalho AC, Kim H, Poisson LM, Winn ME, Mueller C, Cherba D, et al. Discordant inheritance of chromosomal and extrachromosomal DNA elements contributes to dynamic disease evolution in glioblastoma. Nat Genet. 2018;50:708–17.
[10] Zhu Y, Gujar AD, Wong CH, Tjong H, Ngan CY, Gong L, et al. Oncogenic extrachromosomal DNA functions as mobile enhancers to globally amplify chromosomal transcription. Cancer Cell. 2021;39: 694–707.e7.
[11] Yan X, Mischel P, Chang H. Extrachromosomal DNA in cancer. Nat Rev Cancer. 2024;24:261–73.
[12] Shoshani O, Brunner SF, Yaeger R, Ly P, Nechemia-Arbely Y, Kim DH, et al. Chromothripsis drives the evolution of gene amplification in cancer. Nature. 2021;591:137–41.
[13] Zhang CZ, Spektor A, Cornils H, Francis JM, Jackson EK, Liu S, et al. Chromothripsis from DNA damage in micronuclei. Nature. 2015;522:179–84.
[14] Kadoch C, Copeland RA, Keilhack H. PRC2 and SWI/SNF chromatin remodeling complexes in health and disease. Biochemistry. 2016;55:1600–14.
[15] Wu P, Liu Y, Zhou R, Liu L, Zeng H, Xiong F, et al. Extrachromosomal circular DNA: A new target in cancer. Front Oncol. 2022;12:814504.
[16] Wu S, Bafna V, Mischel PS. Extrachromosomal DNA (ecDNA) in cancer pathogenesis. Curr Opin Genet Dev. 2021;66:78–82.
[17] Dong Y, He Q, Chen X, Yang F, He L, Zheng Y. Extrachromosomal DNA (ecDNA) in cancer: Mechanisms, functions, and clinical implications. Front Oncol. 2023;13:1194405.
[18] Li Z, Wang B, Liang H, Han L. Pioneering insights of extrachromosomal DNA (ecDNA) generation, action and its implications for cancer therapy. Int J Biol Sci. 2022;18:4006–25.
[19] Robert M, Crasta K. Breaking the vicious circle: Extrachromosomal circular DNA as an emerging player in tumour evolution. Semin Cell Dev Biol. 2022;123:140–50.
[20] Wu S, Tao T, Zhang L, Zhu X, Zhou X. Extrachromosomal DNA (ecDNA): Unveiling its role in cancer progression and implications for early detection. Heliyon. 2023;9:e21327.
[21] Xing J, Ning Q, Tang D, Mo Z, Lei X, Tang S. Progress on the role of extrachromosomal DNA in tumor pathogenesis and evolution. Clin Genet. 2021;99:503–12.
[22] Bafna V, Mischel PS. Extrachromosomal DNA in cancer. Ann Rev Genomics Hum genet. 2022;23:29–52.
[23] Gu X, Yu J, Chai P, Ge S, Fan X. Novel insights into extrachromosomal DNA: Redefining the oncodrivers of tumor progression. J Exp Clin Cancer Res. 2020;39:215.
[24] Ashique S, Upadhyay A, Garg A, Mishra N, Hussain A, Negi P, et al. Impact of ecDNA: A mechanism that directs tumorigenesis in cancer drug Resistance-A review. Chem Biol Interact. 2022;363:110000.
[25] Pongor LS, Schultz CW, Rinaldi L, Wangsa D, Redon CE, Takahashi N, et al. Extrachromosomal DNA amplification contributes to small cell lung cancer heterogeneity and is associated with worse outcomes. Cancer Discov. 2023;13:928–49.
[26] Zhang J, Jin Y, Lin H, Deng J, Ju Y, Hu X, et al. Amplification of extrachromosomal MYC paralogs shapes immunosuppressive tumor microenvironment in small cell lung cancer. Clin Cancer Res. 2025:OF1–OF14.
[27] Tyler LC, Le AT, Chen N, Nijmeh H, Bao L, Wilson TR, et al. MET gene amplification is a mechanism of resistance to entrectinib in ROS1+ NSCLC. Thoracic Cancer. 2022;13:3032–41.
[28] Tsai C, Suman S, Chu C, Chou JF, Krein PM, Capanu M, et al. Association of ERBB2 extrachromosomal DNA (ecDNA) with first-line HER2-targeted treatment outcomes in advanced esophagogastric adenocarcinoma (EGC). J Clin Oncol. 2024;42(16):4041.
[29] Kanayama K, Imai H, Hashizume R, Matsuda C, Usugi E, Hirokawa YS, et al. Extrachromosomal DNA dynamics contribute to intratumoral receptor tyrosine kinase genetic heterogeneity and drug resistance in gastric cancer. Mol Cancer Res. 2025;23:503–14.
[30] Christiansen JH, Nguyen N-p, Verhaak R, Kim H, Bafna V, Mischel P, et al. Extrachromosomal DNA (ecDNA) carrying amplified oncogenes as a biomarker for insensitivity to pembrolizumab treatment in gastric cancer patients. J Clin Oncol. 2020;38(15):3123.

[31] Starble RM, Sun EG, Jensen TB, Gbyli R, Sun N, Xiao AZ. Abstract B016: A novel epigenetic factor is required for the acquisition of TKI resistance in lung adenocarcinoma via the regulation of oncogene amplicons. Cancer Res. 2022;82:B016.

[32] Turner K, Hemmati H, DuPai C, Bibay J, Klebs C, Chen J, et al. Abstract B092: Intrinsic genomic plasticity of extrachromosomal DNA (ecDNA) enables oncogene amplified tumor cells to develop rapid acquired resistance to targeted therapy. Mol Cancer Ther. 2023;22:B092.

[33] Yong J-X, Zeng Z-L. IDDF2024-ABS-0202 Multi-omics profiling reveals characteristics and associations of gastric cancer with extrachromosomal DNA (ecDNA) and drug resistance. Gut. 2024;73:A142–A143.

[34] Meng X-N, Ma J-F, Liu Y-H, Li S-Q, Wang X, Zhu J, et al. Dynamic genomic changes in methotrexate-resistant human cancer cell lines beyond DHFR amplification suggest potential new targets for preventing drug resistance. Br J Cancer. 2024;130:1819–27.

[35] Wang X, Qu Y, Xing R, Zhou J, Liu Y, Zhang H, et al. Novel insights into the ecDNA formation mechanism involving MSH3 in methotrexate-resistant human colorectal cancer cells. Int J Oncol. 2023;63:134.

[36] Wang Y, Li H, Li Q, Li Y, Wu H, Ge Y, et al. Characterization of extrachromosomal circular DNA in primary and cisplatin-resistant high-grade serous ovarian cancer. Genes. 2025;16:517.

[37] Lin C, Chen Y, Zhang F, Liu B, Xie C, Song Y. Encoding gene RAB3B exists in linear chromosomal and circular extrachromosomal DNA and contributes to cisplatin resistance of hypopharyngeal squamous cell carcinoma via inducing autophagy. Cell Death & Disease. 2022;13:171.

[38] Holmes K, Perusse D, DuPai C, Hemmati H, Weiss J, Kuelbs C, et al. Taxane-based chemotherapy leads to acquired resistance through ecDNA-based amplification of multi-drug resistance genes. Cancer Res. 2024;84:5870.

[39] deCarvalho AC, Kim H, Poisson LM, Winn ME, Mueller C, Cherba D, et al. Discordant inheritance of chromosomal and extrachromosomal DNA elements contributes to dynamic disease evolution in glioblastoma. Nat Genet. 2018;50:708–17.

[40] Yi E, Gujar AD, Guthrie M, Kim H, Zhao D, Johnson KC, et al. Live-cell imaging shows uneven segregation of extrachromosomal DNA elements and transcriptionally active extrachromosomal DNA hubs in cancer. Cancer Discov. 2022;12:468–83.

[41] Iftehimul M, Muganda PM, Newman RH, Rorie CJ, Harrison SH, Hossain MT, et al. Molecular phenotypic plasticity informs possible adaptive change of triple-negative breast cancer cells in vivo. bioRxiv: The Preprint Server for Biology. 2025:2025.05. 31.657204.

[42] Bailey C, Pich O, Thol K, Watkins TBK, Luebeck J, Rowan A, et al. Origins and impact of extrachromosomal DNA. Nature. 2024;635:193–200.

[43] Hung KL, Jones MG, Wong IT-L, Curtis EJ, Lange JT, He BJ, et al. Coordinated inheritance of extrachromosomal DNAs in cancer cells. Nature. 2024;635:201–9.

[44] Repetto M, Wagner K, Suman S, Krein P, Rao P, Arcila M, et al. Reliable detection of potentially therapeutically actionable ecDNA using clinical-grade NGS in a large pan-cancer cohort. Cancer Res. 2024;84:3365.

[45] Luebeck J, Ng AWT, Galipeau PC, Li X, Sanchez CA, Katz-Summercorn AC, et al. Extrachromosomal DNA in the cancerous transformation of Barrett's oesophagus. Nature. 2023;616:798–805.

[46] Kalavska K, Minarik T, Vlkova B, Manasova D, Kubickova M, Jurik A, et al. Prognostic value of various subtypes of extracellular DNA in ovarian cancer patients. J Ovarian Res. 2018;11:85.

[47] Shah P, Aktas-Samur A, Fulciniti M, Szalat R, Shammas MA, Richardson PG, et al. Presence of extrachromosomal DNA (ecDNA) impacts both progression free and overall survival and is an independent poor prognostic marker in multiple myeloma. Blood. 2021;138:461.

[48] Chapman OS, Luebeck J, Sridhar S, Wong IT, Dixit D, Wang S, et al. Circular extrachromosomal DNA promotes tumor heterogeneity in high-risk medulloblastoma. Nat Genet. 2023;55:2189–99.

[49] Zhu H, Huangfu L, Chen J, Ji J, Xing X. Exploring the potential of extrachromosomal DNA as a novel oncogenic driver. Sci China Life Sci. 2025;68:144–57.

[50] Liao Z, Jiang W, Ye L, Li T, Yu X, Liu L. Classification of extrachromosomal circular DNA with a focus on the role of extrachromosomal DNA (ecDNA) in tumor heterogeneity and progression. Biochim Biophys Acta Rev Cancer. 2020;1874:188392.

[51] Karami Fath M, Akbari Oryani M, Ramezani A, Barjoie Mojarad F, Khalesi B, Delazar S, et al. Extra chromosomal DNA in different cancers: Individual genome with important biological functions. Crit Rev Oncol Hematol. 2021;166:103477.

[52] Karami Fath M, Karimfar N, Fazlollahpour Naghibi A, Shafa S, Ghasemi Shiran M, Ataei M, et al. Revisiting characteristics of oncogenic extrachromosomal DNA as mobile enhancers on neuroblastoma and glioma cancers. Cancer Cell Int. 2022;22:200.

[53] Chen Z, Qi Y, He J, Xu C, Ge Q, Zhuo W, et al. Distribution and characterization of extrachromosomal circular DNA in colorectal cancer. Mol Biomed. 2022;3:38.

[54] Zhao Y, Yu L, Zhang S, Su X, Zhou X. Extrachromosomal circular DNA: Current status and future prospects. eLife. 2022;11.

[55] Guo J, Zhang Z, Li Q, Chang X, Liu X. TeCD: The eccDNA collection database for extrachromosomal circular DNA. BMC Gen. 2023;24:47.

[56] Li F, Yang L, Han J, Han X, Peng L, Du Y, et al. Characterization of extrachromosomal circular DNA in cattle using 676 whole genome sequencing datasets. Anim Genet. 2022;53:761–8.

[57] Mann L, Seibt KM, Weber B, Heitkam T. ECCsplorer: A pipeline to detect extrachromosomal circular DNA (eccDNA) from next-generation sequencing data. BMC Bioinformatics. 2022;23:40.

[58] Ling X, Han Y, Meng J, Zhong B, Chen J, Zhang H, et al. Small extrachromosomal circular DNA (eccDNA): Major functions in evolution and cancer. Mol Cancer. 2021;20:113.

[59] Wang M, Chen X, Yu F, Ding H, Zhang Y, Wang K. Extrachromosomal circular DNAs: origin, formation and emerging function in cancer. Int J Biol Sci. 2021;17:1010–25.

[60] Yüksel A, Altungöz O. Gene amplifications and extrachromosomal circular DNAs: Function and biogenesis. Mol Biol Rep. 2023;50:7693–703.

[61] Pourrajab F, Zare-Khormizi MR. Extrachromosomal circular DNAs, amplified oncogenes, and CRISPR-Cas9 system. Mol Pharmacol. 2022;102:209–15.

[62] Chen JP, Diekmann C, Wu H, Chen C, Della Chiara G, Berrino E, et al. scCircle-seq unveils the diversity and complexity of extrachromosomal circular DNAs in single cells. Nat Commun. 2024;15:1768.

[63] Lv W, Pan X, Han P, Wang Z, Feng W, Xing X, et al. Circle-Seq reveals genomic and disease-specific hallmarks in urinary cell-free extrachromosomal circular DNAs. Clin Trans Med. 2022;12:e817.

[64] Li F, Ming W, Lu W, Wang Y, Dong X, Bai Y. Bioinformatics advances in eccDNA identification and analysis. Oncogene. 2024;43:3021–36.

[65] Gao X, Liu K, Luo S, Tang M, Liu N, Jiang C, et al. Comparative analysis of methodologies for detecting extrachromosomal circular DNA. Nat Commun. 2024;15:9208.

[66] Wu S, Bafna V, Chang HY, Mischel PS. Extrachromosomal DNA: An emerging hallmark in human cancer. Annu Rev Pathol. 2022;17:367–86.

[67] Lange JT, Rose JC, Chen CY, Pichugin Y, Xie L, Tang J, et al. The evolutionary dynamics of extrachromosomal DNA in human cancers. Nat Genet. 2022;54:1527–33.

[68] Padfield E, Ellis HP, Kurian KM. Current therapeutic advances targeting EGFR and EGFRvIII in glioblastoma. Front Oncol. 2015;5:5.

[69] Hussain MS, Rejili M, Khan A, Alshammari SO, Tan CS, Haouala F, Ashique S, Alshammari QA. AI-powered liquid biopsy for early detection of gastrointestinal cancers. Clin Chim Acta. 2025;120484.

16 Targeting ecDNA
Novel Therapeutic Approach

Oyindamola John Samson,[1] *John Olusola Olayiwola,*[2]
Ayush Madan,[3] *Titilola Fausat Salisu,*[4]
and Ismail Babatunde Onajobi[5]

1 Department of Microbiology and Biotechnology
Faculty of Natural Sciences
Ajayi Crowther University, Oyo, Nigeria
Department of Microbiology
Faculty of Science, Olabisi Onabanjo University
Ago-Iwoye, Nigeria
2 Department of Microbiology and Biotechnology
Faculty of Natural Sciences, Ajayi Crowther University
Oyo, Nigeria
3 Department of Biotechnology
School of Research & Technology, People's University
Bhanpur, Bhopal, India
4 Department of Zoology and Environmental Biology
Faculty of Science, Olabisi Onabanjo University
Ago-Iwoye, Nigeria
5 Department of Microbiology
Faculty of Science
Olabisi Onabanjo University
Ago-Iwoye, Nigeria

16.1 INTRODUCTION

Cancer remains one of the most formidable challenges in modern medicine due to its complexity, heterogeneity, and adaptability. Traditional paradigms of cancer genetics have long focused on chromosomal mutations and structural rearrangements as the primary drivers of tumorigenesis (Imodoye et al., 2024). However, recent discoveries have unveiled a distinct and dynamic player in cancer biology: extrachromosomal DNA (ecDNA). These circular, non-chromosomal DNA elements are increasingly recognized for their ability to harbor oncogenes, amplify gene expression, and influence cellular behavior in ways that exceed the capabilities of chromosomal DNA alterations (Yang et al., 2025).

Unlike the linear and tightly regulated chromosomal genome, ecDNA exists independently in the nucleus, escaping many of the epigenetic and topological constraints imposed on chromosomal DNA. This unique positioning allows ecDNA to replicate more freely and express oncogenes at significantly higher levels. Such properties give cancer cells harboring ecDNA a potent evolutionary advantage, contributing to aggressive growth, therapeutic resistance, and enhanced adaptability under selective pressures, including chemotherapy and radiation (Dong et al., 2023).

Recent advances in high-throughput sequencing, fluorescence imaging, and single-cell analysis have significantly improved our understanding of the formation, structure, and function of ecDNA (Mao et al., 2025). These tools have revealed that ecDNA is not a rare anomaly but a prevalent feature

DOI: 10.1201/9781003655299-16

in many cancer types, particularly in high-grade glioblastomas, neuroblastomas, and some epithelial cancers. Moreover, ecDNA can dynamically re-integrate into chromosomes or exist as persistent extrachromosomal entities, further contributing to genomic instability (Noorani et al., 2022).

The oncogenic potential of ecDNA has sparked considerable interest in developing targeted therapies that exploit its vulnerabilities. Unlike chromosomal mutations, which are often irreversible and difficult to target, ecDNA's structural uniqueness and reliance on specific pathways for replication and maintenance present multiple therapeutic entry points (Noorani et al., 2022). Early studies suggest that disrupting ecDNA biogenesis, impairing its replication, or enhancing its degradation can selectively suppress tumor growth without affecting normal cells (Vízkeleti and Spisák, 2023; Yan et al., 2024).

This review provides a comprehensive overview of ecDNA in the context of cancer biology and highlights emerging therapeutic strategies aimed at targeting this powerful genetic element. By integrating insights from molecular biology, genomics, and pharmacology, we aim to shed light on the promise and challenges of ecDNA-based cancer therapies. As the field rapidly evolves, targeting ecDNA may represent a paradigm shift in precision oncology, especially for cancers that are currently difficult to treat with conventional approaches.

16.2 STRUCTURE AND BIOGENESIS OF ECDNA

Extrachromosomal DNA (ecDNA) is a circular DNA element that exists independently of the chromosomal genome within the nucleus of cancer cells. Unlike chromosomal DNA, ecDNA lacks centromeres and telomeres, rendering it acentric and prone to uneven segregation during mitosis. This structural uniqueness contributes to intratumoral heterogeneity, as daughter cells may inherit differing numbers of ecDNA copies. Moreover, ecDNAs range in size from a few kilobases to several megabases and can harbor entire oncogenes or regulatory regions, allowing for substantial transcriptional output (Hung et al., 2022).

The biogenesis of ecDNA is a complex and not yet fully understood process, but several mechanisms have been proposed. One widely accepted model is the *chromothripsis-like shattering* of chromosomes followed by random reassembly, leading to circular DNA fragments (de Groot et al., 2023). Another mechanism involves *breakage–fusion–bridge (BFB) cycles*, where telomere erosion or chromosomal breaks trigger cycles of fusion and re-breakage, resulting in the formation of ecDNA. Illegitimate recombination events and replication stress have also been implicated in the generation of these elements, particularly in rapidly dividing tumor cells (Kolb and Ernst, 2021).

Once formed, ecDNA can replicate autonomously during S-phase using the host cell's replication machinery. Lacking centromeres, ecDNAs do not segregate evenly during cell division, causing unequal distribution among progeny cells. This random inheritance promotes intratumoral heterogeneity and accelerates tumor evolution. Furthermore, ecDNA can reintegrate into chromosomal loci or persist independently, which allows cancer cells to flexibly modulate oncogene dosage and rapidly adapt to environmental stressors such as drug treatment (Buss et al., 2024).

One of the defining features of ecDNA is its *epigenetic accessibility*. Compared to chromosomal DNA, ecDNA is generally more open and transcriptionally active, which enhances oncogene expression. This is partly because ecDNA lacks the repressive histone modifications typically associated with tightly packed chromatin. In addition, ecDNA often contains enhancers and super-enhancers that further amplify gene expression (Wu et al., 2021). These features make ecDNA a potent oncogenic driver and a key target in therapeutic development. Mechanisms of ecDNA formation in tumor cells. Multiple genomic instability processes contribute to ecDNA biogenesis, including chromothripsis, breakage–fusion–bridge (BFB) cycles, and aberrant recombination, as shown in Figure 16.1.

16.3 FUNCTIONAL ROLES OF ECDNA IN CANCER

Extrachromosomal DNA (ecDNA) plays a central role in the aggressive behavior of many cancers by amplifying oncogenes beyond levels typically observed in chromosomal DNA. These circular

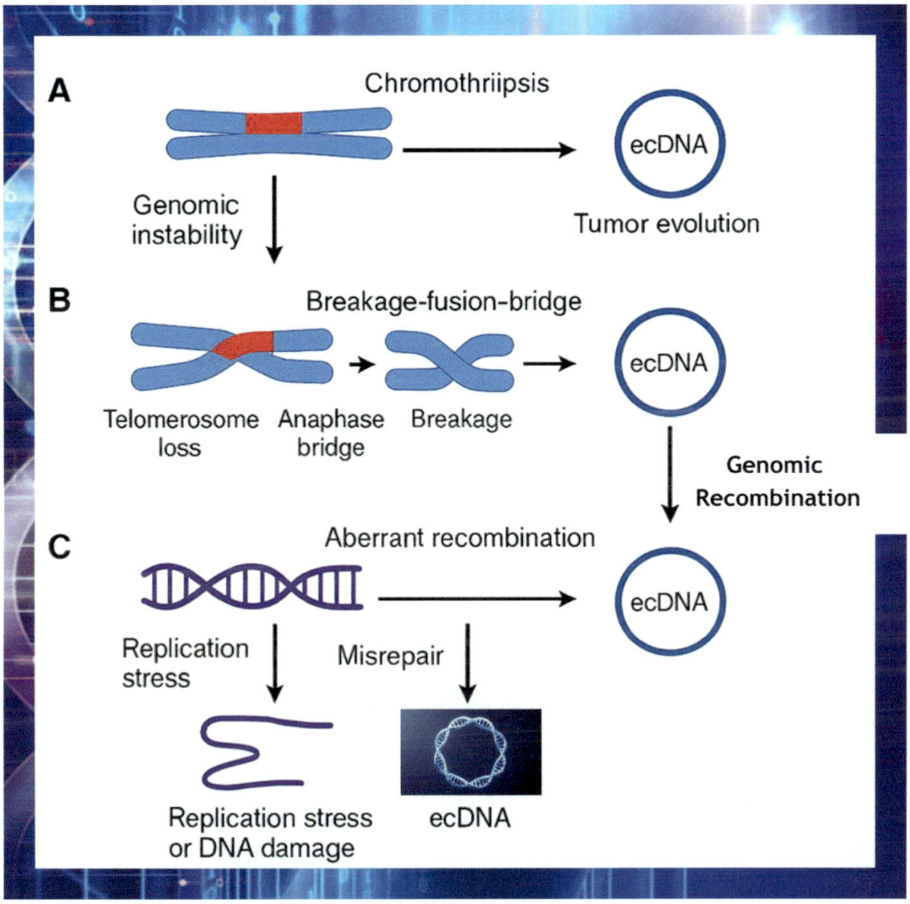

FIGURE 16.1 Biogenesis pathways of ecDNA in cancer cells.

DNA elements can harbor multiple copies of oncogenes such as *MYC*, *EGFR*, or *CDK4*, leading to their overexpression without the need for chromosomal mutations or translocations (Robert and Crasta, 2022). Because ecDNA lacks the repressive chromatin structures found in chromosomes, the oncogenes it carries are more transcriptionally active. This unregulated overexpression contributes significantly to uncontrolled proliferation, angiogenesis, and evasion of growth suppressors in tumor cells (Bafna and Mischel, 2022).

The random segregation of ecDNA during cell division introduces variability in oncogene dosage among daughter cells, thereby promoting *intratumoral heterogeneity*. This phenomenon contributes to the dynamic nature of tumors, where subclones with different growth rates, metabolic demands, and drug responses coexist. ecDNA-driven heterogeneity is a key reason why tumors can rapidly adapt to environmental stresses, including immune responses or nutrient limitations, enhancing their survival and expansion (Liao et al., 2020).

Therapy resistance is another critical consequence of ecDNA-mediated oncogene amplification. Cancer cells with high ecDNA content often demonstrate reduced sensitivity to chemotherapy, targeted inhibitors, and radiation. For example, increased *EGFR* or *MET* copy number on ecDNA has been associated with resistance to tyrosine kinase inhibitors. In many cases, when therapy pressure is applied, ecDNA content increases, allowing tumor cells to survive and repopulate the tumor mass with more aggressive clones (Ashique et al., 2022).

Additionally, ecDNA facilitates *tumor evolution* by acting as a mobile genetic reservoir. These circular DNA elements can reintegrate into chromosomal locations or form complex ecDNA clusters

TABLE 16.1
Implicative activity by Functional Roles of ecDNA in Cancer

Functional Role	Strategies	Implication for Cancer Progression	References
Oncogene Amplification	ecDNA harbors multiple copies of oncogenes (e.g., *MYC, EGFR*)	Drives overexpression, increased proliferation	Bafna and Mischel (2022)
Transcriptional Hyperactivity	Lacks repressive chromatin, allowing high gene expression	Enhances tumor cell fitness and aggressiveness	Robert and Crasta (2022)
Intratumoral Heterogeneity	Uneven segregation during mitosis leads to variable ecDNA copy numbers	Promotes clonal diversity, adaptability, and evolution	Liao et al. (2020)
Therapy Resistance	Elevated oncogene expression from ecDNA undermines drug efficacy	Leads to relapse, survival of resistant clones	Ashique et al. (2022)
Tumor Evolution	Dynamic replication and reintegration fuel genetic remodeling	Supports metastasis and long-term adaptation	Yan et al. (2024)

with additional genetic material. This plasticity enables rapid genome remodeling and adaptation, accelerating malignancy progression and metastasis. The non-Mendelian inheritance pattern of ecDNA ensures selective amplification of beneficial traits without needing to fix mutations in the entire tumor population (Yan et al., 2024).

Given these roles, ecDNA is now viewed not just as a byproduct of genomic instability but as a powerful driver of cancer aggressiveness (Chiu et al., 2020). Understanding how ecDNA shapes tumor behavior is critical for designing more effective therapeutic strategies (Table 16.1). Targeting the mechanisms of ecDNA maintenance and function may offer a new approach to controlling tumor growth, minimizing heterogeneity, and overcoming drug resistance.

16.4 TECHNOLOGICAL ADVANCES IN ECDNA RESEARCH

The study of extrachromosomal DNA (ecDNA) has witnessed significant breakthroughs due to recent advances in genomic and imaging technologies. Traditional cytogenetics and fluorescence in situ hybridization (FISH) were among the first methods to detect ecDNA as double minutes in cancer cells (Purshouse et al., 2024). However, high-resolution whole-genome sequencing (WGS) has now become the cornerstone for identifying ecDNA with greater accuracy. Techniques like Circle-Seq, AmpliconArchitect, and Optical Mapping allow researchers to distinguish circular DNA from chromosomal counterparts based on breakpoint junctions and structural signatures. These tools have been essential in confirming the circular architecture and oncogene enrichment of ecDNA (Kayode-Edwards et al., 2024).

Advanced imaging techniques have also revolutionized ecDNA visualization. Multiplexed FISH, immuno-DNA FISH, and confocal microscopy enable real-time localization of ecDNA in the nucleus. More recently, super-resolution microscopy and CRISPR-based live imaging methods have allowed researchers to track ecDNA movement, replication dynamics, and distribution during mitosis (Lacen and Lee, 2024). These insights have confirmed that ecDNA clusters into transcriptional hubs, termed "ecDNA hubs," which facilitate high transcriptional activity and may serve as functional nuclear substructures unique to cancer cells (Thuma et al., 2023).

Single-cell technologies have emerged as powerful tools for dissecting ecDNA heterogeneity across individual tumor cells. Single-cell whole-genome sequencing (scWGS) allows for the detection of ecDNA copy number variations at the cellular level, revealing profound intra-tumoral diversity. Furthermore, single-cell ATAC-seq and RNA-seq have shown that ecDNA not only varies in number but also contributes to differential chromatin accessibility and gene expression across cancer cell populations. This single-cell resolution is essential for understanding how ecDNA contributes to drug resistance and tumor evolution (Pongor et al., 2023).

Spatial omics approaches further expand the understanding of ecDNA by integrating the physical and functional context of ecDNA-bearing cells within the tumor microenvironment. Techniques such as spatial transcriptomics, multiplexed RNA imaging, and 3D genome architecture mapping have revealed how ecDNA-driven gene expression correlates with tumor niches and cell-cell interactions. These approaches are critical for identifying ecDNA-associated microenvironments that may foster therapeutic resistance or immune evasion (Behrouzi et al., 2024).

In parallel, computational and bioinformatics tools have become indispensable in analyzing ecDNA from large datasets. Algorithms like AmpliconArchitect and AmpliconClassifier reconstruct ecDNA architecture from sequencing data, while tools such as CircleMap identify ecDNA breakpoints. Machine learning models are being developed to predict ecDNA formation, oncogene localization, and chromatin accessibility patterns. Integrating multi-omic datasets—genomic, transcriptomic, and epigenomic—has enhanced the predictive power of computational models in ecDNA research (Zhao et al., 2023).

Together, these technological advances are reshaping our understanding of ecDNA biology. They provide unprecedented insight into how ecDNA arises, propagates, and exerts its effects within tumor ecosystems. These tools also hold promise for translational applications, including early cancer detection, ecDNA-based biomarkers, and the development of targeted therapeutics aimed at disabling ecDNA-related oncogenic pathways (Arshadi et al., 2023).

16.5 THERAPEUTIC STRATEGIES TARGETING ECDNA

Extrachromosomal DNA (ecDNA) is a major driver of cancer aggressiveness, therapy resistance, and tumor evolution, making it a compelling target for novel therapeutic strategies. Recent research has focused on disrupting the fundamental processes that enable ecDNA formation, replication, stability, and oncogenic expression. These targeted interventions aim not only to eliminate ecDNA but also to re-sensitize ecDNA-harboring tumors to conventional treatments, such as chemotherapy, radiotherapy, and targeted therapies (Curtis et al., 2024).

One promising approach is the *disruption of ecDNA biogenesis*, which involves targeting the mechanisms that generate ecDNA in the first place. ecDNA formation is often associated with chromothripsis (massive chromosome fragmentation) and breakage–fusion–bridge (BFB) cycles. Inhibitors of DNA damage response (DDR) pathways, such as ATR or ATM inhibitors, may reduce the frequency of chromosomal rearrangements and ecDNA formation. Additionally, targeting proteins involved in DNA repair (e.g., RAD51, PARP) may limit the misrepair events that lead to ecDNA circularization (Huang et al., 2024).

Another therapeutic avenue focuses on the *inhibition of ecDNA replication*. Because ecDNA replicates autonomously using the host cell's replication machinery, it is susceptible to agents that target DNA replication enzymes. Inhibitors of DNA polymerase α, replication protein A (RPA), and the minichromosome maintenance (MCM) complex have shown promise in reducing ecDNA abundance. Disrupting replication can prevent ecDNA amplification and diminish oncogene expression, ultimately reducing tumor growth and progression (Bailey et al., 2020).

A third strategy involves *inducing the degradation of ecDNA*. Studies have shown that ecDNA is vulnerable to nucleases and other DNA-damaging agents due to its acentric and unprotected structure (Dong et al., 2023). Therapeutic agents that induce oxidative DNA damage, such as certain chemotherapy drugs (e.g., cisplatin) or targeted endonucleases, can selectively degrade ecDNA without affecting the genome's stable chromosomal regions. Additionally, engineered nucleases like CRISPR/Cas9 can be programmed to target ecDNA-specific sequences, enabling precise cleavage and elimination of oncogenic circles (Mao et al., 2025).

Re-sensitization of ecDNA-positive tumors to therapy is another critical strategy. Tumors with high ecDNA content often develop resistance to targeted therapies due to oncogene amplification. However, by combining ecDNA-targeted therapies with conventional drugs, it may be possible to overcome resistance. For example, targeting ecDNA hubs or enhancer elements can suppress the transcriptional hyperactivity of ecDNA, rendering tumors more responsive to EGFR or MYC

TABLE 16.2
Therapeutic Strategies Targeting ecDNA

Strategy	Mechanism of Action	Potential Agents or Tools	Therapeutic Goal	References
Disruption of ecDNA Biogenesis	Blocks chromothripsis or BFB cycles involved in ecDNA formation	ATR inhibitors, PARP inhibitors, RAD51 blockers	Prevent formation of new ecDNA	Huang et al. (2024)
Inhibition of ecDNA Replication	Targets DNA replication machinery required for ecDNA duplication	DNA pol α inhibitors, RPA inhibitors, MCM blockers	Suppress ecDNA amplification and oncogene dosage	Bailey et al. (2020)
Induction of ecDNA Degradation	Promotes breakdown of ecDNA via DNA damage or nucleases	Cisplatin, CRISPR-Cas9, oxidative stress inducers	Eliminate existing ecDNA	Dong et al. (2023)
Re-sensitization to Therapy	Reduces ecDNA-driven gene expression to restore drug sensitivity	HDAC inhibitors, BET inhibitors, MYC/EGFR inhibitors	Enhance response to standard therapies	Zeng et al. (2020)
Immunotherapeutic Interventions	Targets neoantigens arising from ecDNA-mediated expression	Checkpoint inhibitors, CAR-T cells, tumor vaccines	Boost immune recognition of ecDNA-rich tumors	Pecorino et al. (2022)

inhibitors. Epigenetic drugs, such as histone deacetylase inhibitors (HDACi), may also downregulate ecDNA-driven gene expression and restore sensitivity to chemotherapy (Zeng et al., 2020).

Emerging strategies also include immunotherapeutic interventions. Because ecDNA can lead to aberrant protein expression and neoantigen formation, immune checkpoint inhibitors (e.g., anti-PD-1, anti-CTLA-4) might be more effective in ecDNA-rich tumors. Additionally, tumor vaccines or CAR-T cells could potentially be engineered to recognize ecDNA-driven epitopes, opening a novel avenue for immuno-oncology (Pecorino et al., 2022).

Despite promising preclinical data, several challenges remain. These include the identification of ecDNA-specific biomarkers, delivery mechanisms for ecDNA-targeted agents, and the avoidance of toxicity to normal cells (Table 16.2). Nonetheless, a better understanding of ecDNA biology continues to inform innovative drug development and personalized therapeutic approaches for cancer treatment.

16.6 EMERGING TOOLS AND MOLECULES IN ECDNA TARGETING

The therapeutic landscape for cancer is rapidly evolving, and targeting extrachromosomal DNA (ecDNA) has emerged as a promising frontier. Recent discoveries have revealed the critical role of ecDNA in amplifying oncogenes, promoting tumor heterogeneity, and driving resistance to treatment. Consequently, the development of specialized molecular tools and small molecules that can selectively disrupt ecDNA offers a novel therapeutic approach. Among these, CRISPR-Cas systems, DNA-damaging agents, and inhibitors of DNA repair and stress response pathways have shown particular promise.

CRISPR-Cas systems represent one of the most revolutionary gene-editing tools and have been successfully adapted to target ecDNA. By designing guide RNAs specific to sequences found uniquely or abundantly on ecDNA, researchers can direct the Cas nucleases to introduce double-strand breaks within the circular DNA. This targeted cleavage can lead to ecDNA degradation or loss of oncogene expression. Unlike chromosomal DNA, ecDNA lacks telomeric protection and centromeres, making it especially vulnerable to double-strand breaks. Furthermore, advanced CRISPR-based platforms such as

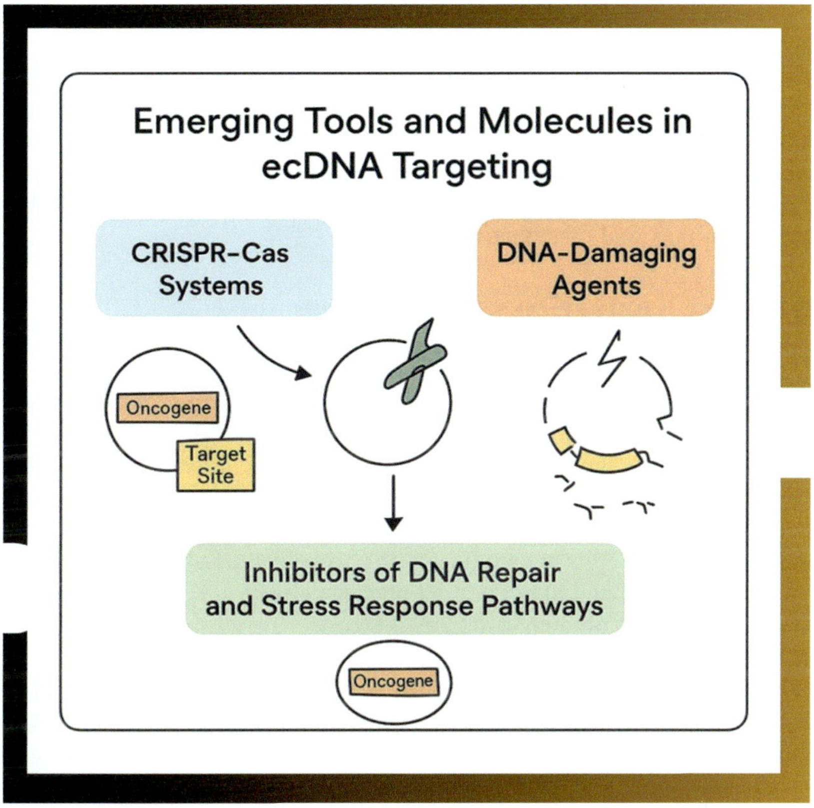

FIGURE 16.2 Technological advances in ecDNA research.

CRISPR interference (CRISPRi) and CRISPR-associated transposases expand the potential to silence or excise ecDNA without altering chromosomal integrity (Song et al., 2025).

DNA-damaging agents have long been a staple in oncology, and their relevance in targeting ecDNA is gaining renewed interest. Agents like cisplatin, doxorubicin, and etoposide introduce lesions that ecDNA is less equipped to repair, leading to fragmentation and loss of oncogenic activity. Because ecDNA lacks the structured chromatin and repair scaffolds present in chromosomes, it is more prone to degradation when exposed to genotoxic stress. Moreover, certain drugs can be optimized or encapsulated in delivery systems (e.g., nanoparticles) to enhance specificity and reduce off-target toxicity. Combining DNA-damaging agents with inhibitors of DNA repair could further enhance ecDNA vulnerability (Dong et al., 2023).

Inhibitors of DNA repair and stress response pathways target the cellular mechanisms that stabilize ecDNA and promote its replication. These include molecules that inhibit poly (ADP-ribose) polymerase (PARP), ATM, ATR, and DNA-PK—all of which are crucial to repairing double-strand breaks and managing replication stress. Inhibiting these pathways creates a synthetic lethal environment for ecDNA-containing cells, as they rely more heavily on these repair mechanisms due to their unstable and repetitive DNA structures. When combined with genotoxic agents or CRISPR-based approaches, these inhibitors can synergistically promote the collapse of ecDNA maintenance (Drew et al., 2025).

Interestingly, these molecular tools also offer opportunities for combinatorial therapies. For instance, CRISPR-mediated cleavage followed by inhibition of repair proteins like PARP may amplify ecDNA fragmentation. Alternatively, pre-treating tumors with DNA-damaging agents may increase ecDNA accessibility to CRISPR machinery, enhancing editing efficiency. This strategy underscores the importance of designing sequential or co-delivery systems to target ecDNA at multiple regulatory nodes (Thompson et al., 2021).

While these emerging tools are promising, challenges such as delivery specificity, potential off-target effects, and tumor heterogeneity remain. Advancements in delivery vectors (e.g., lipid nanoparticles, viral vectors), next-generation sequencing, and bioinformatics will be essential for identifying ecDNA-rich regions and designing precision-targeted interventions (Figure 16.2). Additionally, patient stratification based on ecDNA content or signature could help personalize treatment regimens for maximal therapeutic efficacy.

16.7 CHALLENGES, LIMITATIONS, AND FUTURE PERSPECTIVES IN ECDNA TARGETING

Despite the rapid advancements in ecDNA research and the promise of novel therapeutic strategies, several challenges and limitations still hinder the full clinical application of ecDNA-targeted interventions. One of the primary concerns is *specificity and off-target effects*, especially with tools like CRISPR-Cas systems and DNA-damaging agents (Karn et al., 2022). While these tools can effectively target ecDNA, they may also interact with chromosomal DNA if sequence homology exists, leading to unintended genome edits or cytotoxicity in normal cells. Enhancing sequence specificity, developing safer Cas variants, and incorporating precise delivery mechanisms are ongoing strategies to mitigate these risks (Balaraman et al., 2024).

Another major obstacle is *tumor heterogeneity and resistance mechanisms*. Tumors containing ecDNA often display high levels of genetic variability within and between individual cancer cells. This heterogeneity is partly due to the uneven segregation and amplification of ecDNA during cell division. As a result, some tumor cells may lose or gain different ecDNA elements in response to therapy, leading to adaptive resistance. Moreover, ecDNA can re-integrate into chromosomes or re-circularize after damage, further complicating treatment. Addressing this requires longitudinal monitoring of ecDNA dynamics and multi-targeted strategies that anticipate clonal evolution (El-Sayes et al., 2021).

The *delivery systems* used to administer ecDNA-targeting molecules present another challenge. Efficient and selective delivery to tumor tissues—especially those with blood-tumor barriers or complex microenvironments—remains a significant hurdle. Viral vectors, lipid nanoparticles, and engineered exosomes are being explored to improve delivery efficiency and tissue targeting. However, immune responses, limited payload capacity, and scale-up challenges must be resolved before these systems can be routinely employed in clinical settings. An ideal delivery system would ensure cell-type specificity, minimal immunogenicity, and controlled release of therapeutic agents (Ukidve et al., 2021).

From a translational standpoint, *integrating ecDNA targeting into precision oncology* requires the development of robust diagnostic tools to detect and quantify ecDNA in real time. Liquid biopsies, ecDNA-specific probes, and high-resolution imaging could allow for non-invasive detection and monitoring of ecDNA burden during treatment. Stratifying patients based on ecDNA profiles would facilitate personalized treatment regimens, ensuring that those most likely to benefit from ecDNA-targeting receive tailored therapies. This integration will require interdisciplinary collaboration among molecular biologists, bioinformaticians, and clinicians (Santamaria et al., 2025).

The *potential for combination therapies* also represents a promising yet complex frontier. Given that ecDNA contributes to therapy resistance, combining ecDNA-targeting agents with existing chemotherapies, targeted inhibitors, or immunotherapies may yield synergistic effects. For example, DNA repair inhibitors can enhance the efficacy of CRISPR-mediated ecDNA degradation, while checkpoint inhibitors might boost immune responses against ecDNA-driven neoantigens. Careful optimization of dosing, scheduling, and sequencing is necessary to avoid additive toxicities and ensure therapeutic synergy (Noorani et al., 2022).

Looking ahead, *clinical trial prospects* for ecDNA-targeted therapies are on the horizon but remain in early phases (Wu et al., 2021). Most current research is preclinical, involving in vitro and animal models. Moving into human trials will require comprehensive safety and efficacy assessments, regulatory approval, and large-scale manufacturing of therapeutic constructs. Adaptive trial designs and real-time biomarker monitoring may accelerate the evaluation of ecDNA-targeted interventions and improve response assessment.

16.8 CONCLUSION

The development of tools and molecules to selectively target ecDNA is a major leap forward in precision oncology. CRISPR systems, DNA-damaging agents, and DNA repair inhibitors, individually or in combination, can exploit the vulnerabilities of ecDNA to suppress tumor growth and overcome therapy resistance. Continued research and clinical translation of these tools hold promise for reshaping the future of cancer treatment by turning ecDNA from a driver of malignancy into a therapeutic liability. While the concept of targeting ecDNA is revolutionary, its clinical realization faces several biological, technical, and regulatory challenges. Addressing these issues will require continued research into ecDNA biology, improved delivery systems, and the integration of emerging technologies in genomics and nanomedicine. With strategic investment and collaborative efforts, ecDNA-targeting could become a cornerstone of next-generation cancer therapy—transforming one of cancer's greatest weapons into its Achilles' heel.

REFERENCES

Arshadi, A., Tolomeo, D., Venuto, S., & Storlazzi, C. T. (2023). Advancements in focal amplification detection in tumor/liquid biopsies and emerging clinical applications. *Genes*, *14*(6), 1304. https://doi.org/10.3390/genes14061304

Ashique, S., Upadhyay, A., Garg, A., Mishra, N., Hussain, A., Negi, P., Hing, G. B., Bhatt, S., Ali, M. K., Gowthamarajan, K., & Singh, S. K. (2022). Impact of ecDNA: A mechanism that directs tumorigenesis in cancer drug Resistance-A review. *Chemico-Biological Interactions*, *363*, 110000. https://doi.org/10.1016/j.cbi.2022.110000

Bafna, V., & Mischel, P. S. (2022). Extrachromosomal DNA in cancer. *Annual Review of Genomics and Human Genetics*, *23*(1), 29–52. https://doi.org/10.1146/annurev-genom-120821-100535

Bailey, C., Shoura, M. J., Mischel, P. S., & Swanton, C. (2020). Extrachromosomal DNA—Relieving heredity constraints, accelerating tumour evolution. *Annals of Oncology*, *31*(7), 884–893. https://doi.org/10.1016/j.annonc.2020.03.303

Balaraman, A. K., Babu, M. A., Moglad, E., Mandaliya, V., Rekha, M. M., Gupta, S., Prasad, G. S., Kumari, M., Chauhan, A. S., Ali, H., & Goyal, K. (2024). Exosome-mediated delivery of crispr-Cas9: A revolutionary approach to cancer gene editing. *Pathology-Research and Practice*, 155785. https://doi.org/10.1016/j.prp.2024.155785

Behrouzi, R., Clipson, A., Simpson, K. L., Blackhall, F., Rothwell, D. G., Dive, C., & Mouliere, F. (2024). Cell-free and extrachromosomal DNA profiling of small cell lung cancer. *Trends in Molecular Medicine*. https://doi.org/10.1016/j.molmed.2024.08.004

Buss, J. H., Begnini, K. R., & Lenz, G. (2024). The contribution of asymmetric cell division to phenotypic heterogeneity in cancer. *Journal of Cell Science*, *137*(5), jcs261400. https://doi.org/10.1242/jcs.261400

Chiu, R. W., Dutta, A., Henssen, A. G., Lo, Y. D., Mischel, P., & Regenberg, B. (2020). What is extrachromosomal circular DNA and what does it do? *Clinical Chemistry*, *66*(6), 754–759. https://doi.org/10.1093/clinchem/hvaa096

Curtis, E. J., Rose, J. C., Mischel, P. S., & Chang, H. Y. (2024). Extrachromosomal DNA: Biogenesis and functions in cancer. *Annual Review of Cancer Biology*, *8*. https://doi.org/10.1146/annurev-cancerbio-070620-092730

de Groot, D., Spanjaard, A., Hogenbirk, M. A., & Jacobs, H. (2023). Chromosomal rearrangements and chromothripsis: The alternative end generation model. *International Journal of Molecular Sciences*, *24*(1), 794. https://doi.org/10.3390/ijms24010794

Dong, Y., He, Q., Chen, X., Yang, F., He, L., & Zheng, Y. (2023). Extrachromosomal DNA (ecDNA) in cancer: Mechanisms, functions, and clinical implications. *Frontiers in Oncology*, *13*, 1194405. https://doi.org/10.3389/fonc.2023.1194405

Drew, Y., Zenke, F. T., & Curtin, N. J. (2025). DNA damage response inhibitors in cancer therapy: Lessons from the past, current status and future implications. *Nature Reviews Drug Discovery*, *24*(1), 19–39. https://doi.org/10.1038/s41573-024-01060-w

El-Sayes, N., Vito, A., & Mossman, K. (2021). Tumor heterogeneity: A great barrier in the age of cancer immunotherapy. *Cancers*, *13*(4), 806. https://doi.org/10.3390/cancers13040806

Huang, Q., Zhang, S., Wang, G., & Han, J. (2024). Insight on ecDNA-mediated tumorigenesis and drug resistance. *Heliyon*. https://doi.org/10.1016/j.heliyon.2024.e27733

Hung, K. L., Mischel, P. S., & Chang, H. Y. (2022). Gene regulation on extrachromosomal DNA. *Nature Structural & Molecular Biology*, *29*(8), 736–744. https://doi.org/10.1038/s41594-022-00806-7

Imodoye, S. O., Adedokun, K. A., & Bello, I. O. (2024). From complexity to clarity: Unravelling tumor heterogeneity through the lens of tumor microenvironment for innovative cancer therapy. *Histochemistry and Cell Biology*, *161*(4), 299–323. https://doi.org/10.1007/s00418-023-02258-6

Karn, V., Sandhya, S., Hsu, W., Parashar, D., Singh, H. N., Jha, N. K., Gupta, S., Dubey, N. K., & Kumar, S. (2022). CRISPR/Cas9 system in breast cancer therapy: Advancement, limitations and future scope. *Cancer Cell International*, *22*(1), 234. https://doi.org/10.1186/s12935-022-02654-3

Kayode-Edwards, I. I., Isibor, P. O., & Akinduti, P. A. (2024). Omics technologies in aquatic and marine toxicology. In: Isibor, P. O. (eds) *Arctic Marine Ecotoxicology*. Springer, Cham. https://doi.org/10.1007/978-3-031-73584-4_24

Kolb, T., & Ernst, A. (2021). Cell-based model systems for genome instability: Dissecting the mechanistic basis of chromothripsis in cancer. *International Journal of Cancer*, *149*(4), 754–759. https://doi.org/10.1002/ijc.33618

Lacen, A. N., & Lee, H. T. (2024). Tracing the chromatin: From 3C to live-cell imaging. *Chemical & Biomedical Imaging*, *2*(10), 659–682. https://doi.org/10.1021/cbmi.4c00033

Liao, Z., Jiang, W., Ye, L., Li, T., Yu, X., & Liu, L. (2020). Classification of extrachromosomal circular DNA with a focus on the role of extrachromosomal DNA (ecDNA) in tumor heterogeneity and progression. *Biochimica et Biophysica Acta (BBA)-Reviews on Cancer*, *1874*(1), 188392. https://doi.org/10.1016/j.bbcan.2020.188392

Mao, X., Rao, G., Li, G., & Chen, S. (2025). Insights into extrachromosomal DNA in cancer: Biogenesis, methodologies, functions, and therapeutic potential. *Advanced Biology*, *9*(3), 2400433. https://doi.org/10.1002/adbi.202400433

Noorani, I., Mischel, P. S., & Swanton, C. (2022). Leveraging extrachromosomal DNA to fine-tune trials of targeted therapy for glioblastoma: Opportunities and challenges. *Nature Reviews Clinical Oncology*, *19*(11), 733–743. https://doi.org/10.1038/s41571-022-00679-1

Pecorino, L. T., Verhaak, R. G., Henssen, A., & Mischel, P. S. (2022). Extrachromosomal DNA (ecDNA): An origin of tumor heterogeneity, genomic remodeling, and drug resistance. *Biochemical Society Transactions*, *50*(6), 1911–1920. https://doi.org/10.1042/BST20221045

Pongor, L. S., Schultz, C. W., Rinaldi, L., Wangsa, D., Redon, C. E., Takahashi, N., Fialkoff, G., Desai, P., Zhang, Y., Burkett, S., & Hermoni, N. (2023). Extrachromosomal DNA amplification contributes to small cell lung cancer heterogeneity and is associated with worse outcomes. *Cancer Discovery*, *13*(4), 928–949. https://doi.org/10.1158/2159-8290.CD-22-0796

Purshouse, K., Pollard, S. M., & Bickmore, W. A. (2024). Imaging extrachromosomal DNA (ecDNA) in cancer. *Histochemistry and Cell Biology*, *162*(1), 53–64. https://doi.org/10.1007/s00418-024-02280-2

Robert, M., & Crasta, K. (2022, March). Breaking the vicious circle: Extrachromosomal circular DNA as an emerging player in tumour evolution. *Seminars in Cell & Developmental Biology*, *123*, 140–150. Academic Press. https://doi.org/10.1016/j.semcdb.2021.11.015

Santamaria, S., Cardinali, B., Rovere, M., Marconi, S., Nardin, S., Sacco, G., Barcellini, L., Del Mastro, L., Genova, C., & Coco, S. (2025). New insight in early detection and precision medicine in small cell lung cancer: Liquid biopsy as innovative clinical tool. *Critical Reviews in Clinical Laboratory Sciences*, 1–25. https://doi.org/10.1080/10408363.2025.2493121

Song, Y., Guan, C., Zhang, Y., Xu, Y., Li, P., Luo, L., Feng, C., & Chen, G., 2025. A novel CRISPR-Cas9 nickase-mediated rolling circle amplification (CRIRCA) technique for gene identification and quantitative analysis of extrachromosomal DNA. *Journal of Advanced Research*. https://doi.org/10.1016/j.jare.2025.04.031

Thompson, M. K., Sobol, R. W., & Prakash, A. (2021). Exploiting DNA endonucleases to advance mechanisms of DNA repair. *Biology*, *10*(6), 530. https://doi.org/10.3390/biology10060530

Thuma, J., Chung, Y. C., & Tu, L. C. (2023). Advances and challenges in CRISPR-based real-time imaging of dynamic genome organization. *Frontiers in Molecular Biosciences*, *10*, 1173545. https://doi.org/10.3389/fmolb.2023.1173545

Ukidve, A., Cu, K., Kumbhojkar, N., Lahann, J., & Mitragotri, S. (2021). Overcoming biological barriers to improve solid tumor immunotherapy. *Drug Delivery and Translational Research*, 1–26. https://doi.org/10.1007/s13346-021-00923-8

Vízkeleti, L., & Spisák, S. (2023). Rewired metabolism caused by the oncogenic deregulation of MYC as an attractive therapeutic target in cancers. *Cells*, *12*(13), 1745. https://doi.org/10.3390/cells12131745

Wu, S., Bafna, V., & Mischel, P. S. (2021). Extrachromosomal DNA (ecDNA) in cancer pathogenesis. *Current Opinion in Genetics & Development*, *66*, 78–82. https://doi.org/10.1016/j.gde.2021.01.001

Yan, X., Mischel, P., & Chang, H. (2024). Extrachromosomal DNA in cancer. *Nature Reviews Cancer*, *24*(4), 261–273. https://doi.org/10.1038/s41568-024-00669-8

Yang, Q. L., Xie, Y., Qiao, K., Lim, J. Y. S., & Wu, S. (2025). Modern biology of extrachromosomal DNA: A decade-long voyage of discovery. *Cell Research*, 1–12. https://doi.org/10.1038/s41422-024-01054-8

Zeng, X., Wan, M., & Wu, J. (2020). ecDNA within tumors: A new mechanism that drives tumor heterogeneity and drug resistance. *Signal Transduction and Targeted Therapy*, *5*(1), 277. https://doi.org/10.1038/s41392-020-00403-4

Zhao, X., Zhao, H., Liu, Y., & Guo, Z. (2023). Methods, bioinformatics tools and databases in ecDNA research: An overview. *Computers in Biology and Medicine*, *167*, 107680. https://doi.org/10.1016/j.compbiomed.2023.107680

Index

Note: Page numbers in **bold** indicate a table and page umbers in *italics* indicate a figure on the corresponding page.

D

E

F

U

V

W

For Product Safety Concerns and Information please contact our
EU representative GPSR@taylorandfrancis.com Taylor & Francis
Verlag GmbH, Kaufingerstraße 24, 80331 München, Germany